202 Current Topics in Microbiology and Immunology

AF785822

Editors

Springer
Berlin
Heidelberg
New York
Barcelona
Budapest
Hong Kong
London
Milan
Paris
Tokyo

HIV and Dementia

Proceedings of the NIMH-Sponsored Conference "Pathogenesis of HIV Infection of the Brain: Impact on Function and Behavior"

Edited by M.B.A. Oldstone and L. Vitkovic

With 40 Figures

Springer

Michael B.A. Oldstone
Department of Neuropharmacology
Division of Virology
The Scripps Research Institute
10666 North Torrey Pines Road
La Jolla, CA 92037, USA

Ljubiša Vitković
Chief, Neuroimmunology/Neurovirology Program
Molecular and Cellular Neuroscience Research Branch
Division of Neuroscience and Behavioral Science
National Institutes of Health, National Institute of Mental Health
5600 Fishers Lane, Rm 11C-06
Rockville, MD 20857, USA

Cover illustration: Cover depicts a prototypic immunodeficiency virus and its "host range": humans, monkeys, and cats. Human immunodeficiency virus (HIV) causes acquired immune deficiency syndrome (AIDS), a disease ultimately resulting in the death of a host, whereas simian (SIV) and feline (FIV) viruses cause a mild, if any, disease. However, scientists have discovered that some isolates of SIV and FIV can cause a disease in monkeys and cats, respectively, that recapitulate aspects of AIDS. These animal models of AIDS are yielding important information on potential therapies of the human disease. Courtesy of M.B. Gardner.

Cover design: Künkel+Lopka, Ilvesheim

ISSN 0070-217X
ISBN-13: 978-3-642-79659-3 e-ISBN-13: 978-3-642-79657-9
DOI: 10.1007/978-3-642-79657-9

Softcover reprint of the hardcover 1st edition 1995
Library of Congress Catalog Card Number 15-12910

Typesetting: Thomson Press (India) Ltd, Madras
SPIN: 10490029 27/3020/SPS – 5 4 3 2 1 0 – Printed on acid-free paper

Foreword

The National Institute of Mental Health (NIMH) AIDS Program is the fourth largest acquired immune deficiency syndrome (AIDS) program within the National Institutes of Health (NIH). Since 1983, our program's contributions have concentrated on two major areas. The first has been to develop effective strategies to prevent or reduce behaviors that place individuals at risk for human immunodeficiency virus type 1 (HIV-1) infection. The second has been to support and foster research to enhance our understanding of the profound impact of HIV-1 infection on the central nervous system (CNS). The brain appears to be a prime target of the virus and may serve as a reservoir for the virus. Post mortem examination of brain tissue has provided evidence of CNS cell damage in 80%–90% of people who die with AIDS. For about 10%–20% of people with AIDS, mild neuropsychological symptoms are the first signs of the disease. Approximately 20% of individuals infected with HIV-1 develop AIDS dementia complex. The symptoms include apathy, difficulty concentrating, irritability, depression, and personality changes. In the later stages of the disease, people may experience psychiatric disorders. Death usually occurs within 6 months of the appearance of those more severe symptoms. Up to 90% of children infected with HIV-1 experience attention and concentration difficulties and often experience neurodevelopmental delay or regression over time.

Investigators supported by NIMH have provided evidence that HIV-1 reaches the CNS within days of penetrating the body, and that there are high levels of HIV-1 expression and replication in the cerebral spinal fluid throughout the time course of the disease. Investigations comparing infected brain tissue from adults and fetuses have found that the developing CNS clears the virus more rapidly than the adult CNS; however, the mechanism by which this occurs remains unknown. The exact mechanism(s) by which HIV-1 crosses the blood-brain barrier and ultimately produces CNS dysfunction remain unknown. It is essential that the mode of entry and the direct as well as indirect mechanisms leading to the CNS damage are identified so that the infection of the CNS can be prevented and/or the consequences of the infection be therapeutically ameliorated.

The objectives of the conference *Pathogenesis of HIV Infection of the Brain: Impact on Function and Behavior* were to examine current knowledge, to attract more neuroscientists to AIDS research, and to generate new ideas and collaborations. To further these objectives, the organizers asked the plenary speakers to contribute papers to these proceedings. The papers in this volume address the central questions in the neuroscience of AIDS and are likely to lead to research that will discover ways to treat HIV/CNS disease.

Acknowledgements. These proceedings and the conference "Pathogenesis of HIV infection of the brain: impact on function and behavior", held from April 4–7, 1994 in Chantilly, VA, USA, could not have been successfully completed without the diligent contributions from many individuals. Special thanks are due to the editors of these proceedings who with the participation of Walter Goldschmidts, Ph.D., organized the scientific program of the conference.

The organizers are extremely indebted to Ms. Sherry Roberts and Ms. Susan Kennell, Office on AIDS, National Institute of Mental Health, National Institutes of Health, USA, for their scrupulous attention to organizational details before and after the meeting, respectively. Special thanks are due to Ms. Debra Woods, Division of Neuroscience and Behavioral Science, National Institute of Mental Health, National Institutes of Health, USA, for coordinating voluminous correspondence with the invited participants in the conference and contributors to this volume.

Rockville

E. Stover
Director, Office on AIDS, NIMH, NIH
S.H. Koslow
Director, Division of Neuroscience and Behavioral Science, NIMH, NIH

Preface

Almost 13 million people worldwide have been infected by HIV-1, and nearly 6 million of these individuals have disorders of the nervous system caused by the virus. The nervous system illness ranges from the AIDS dementia complex, in which the clinical disease is excessive for the observed end stage neuropathologic changes and occurs often in the absence of a perivascular infiltration, to the occurrence of peripheral neuropathology, presenting a Guillian-Barre state-like picture that is frequently associated with perivascular cuffing and inflammation.

A current challenge is to understand how such disparate neuropsychiatric disorders occur, with the hope of obtaining rational therapies. In addition, it is clear that by preventing HIV infection or amplification, the nervous system disorders can be modulated. Hence, those interested in dementia need to be interested in the host's immune response to combat the virus and in the virus' countermove to cause immunosuppression. Understanding the mechanism(s) by which the virus causes or is associated with dementia may shed light on the mechanism(s) underlying dementia in general.

For the above reasons, a conference was sponsored by the National Institute of Mental Health, NIH. The agenda was to bring together three groups of scientists that often do not meet jointly. One group, as represented by Drs. Richard Johnson, Richard Price, Robert Heaton, Clayton Wiley, Diane Griffin, Ashley Haase, Anthony Fauci, Bruce Chesebro, and Bruce Walker, has worked with and studied HIV-1-infected patients with associated mental and neurologic illnesses and immunosuppression and/or investigated markers in nervous tissues from such patients. The second group, as represented by Drs. Murray Gardner, John Elder, Bill Narayan, Lennart Mucke, and David Pauza, has developed or employed lentivirus animal models, i.e., SIV in macaques, FIV in cats, and HIV gene expression in brain using transgenic technology, or studied HIV replication in vitro. The last group, as represented by

Drs. Bernard Fields, Milton Brightman, William Hickey, Richard Compans, and Michael Oldstone, has been involved in understanding general principles of viral pathogenesis, such as viral entry and/or exit into the brain; blood-brain barrier organization or disorganization; and penetration of immunocompetent cells across the blood-brain barrier, including the state of their recognition of CNS cells and the ability or inability of CNS cells to present viral antigens to such immunocompetent cells.

Such cross-talk among these groups and other participants attending this conference and participating in workshop discussions and poster presentations promoted lively discussions and debates frequently as reflected in the accompanying cartoon. Bringing these disparate groups together is likely to lead to rethinking and redefining several aspects of the problem. By integrating the many approaches addressed at this meeting, collaborations previously not considered are likely to be forged. This will result in a better understanding of and novel clinical and experimental approaches to AIDS CNS disease. Hopefully, the end result will eventually lead to the control and prevention of AIDS and its CNS manifestations.

Lastly, a personal comment. AIDS is a devastating disease. Enormous research efforts have correctly been directed towards understanding the lifestyle of the virus that causes it—HIV—and towards its control by public health measures and drug therapy, and its prevention by vaccine. Yet, much of what

is being learned about HIV-1 and AIDS and other lentiviral infections concerning persistence, immunosuppression, accompanying disease manifestation and immune therapies to prevent or treat opportunistic infections has already been revealed in data obtained from studies with other animal models (murine leukemia viruses, lactic dehydrogenase virus, lymphocytic choriomeningitis virus) and human disease (measles virus). Thus, it seemed quite sensible to have included in the program and volume a number of these insights.

La Jolla/Rockville

M.B.A. OLDSTONE
L. VITKOVIĆ

List of Contents

B Lentivirus Animal Models of HIV Central Nervous System Disease

C Tissue Culture and Animal Models to Generate Basic Concepts in Viral Immune Cell: Central Nervous System Interactions

List of Contributors

(Their addresses can be found at the beginning of their respective chapters.)

A The Human Problem: HIV, Neuropsychiatric Disorders, and Dementia

The Pathogenesis of HIV Infections of the Brain

R.T. Johnson

1 Introduction

Last summer, the World Health Organization estimated that over 13 million persons worldwide were infected with the human immunodeficiency virus (HIV) (Merson 1993). From prospective studies, it is reasonable to assume that the majority of these persons already have infection of the central nervous system; therefore, HIV is the most prevalent viral infection of the nervous system. It also can be assumed that during the next decade over 50% of these patients will develop disabling neurological diseases (dementia with HIV encephalopathy, paraparesis with vacuolar myelopathy, and disabling pain with predominantly sensory neuropathy); therefore, neurological complications of the acquired immunodeficiency syndrome (AIDS) will be among the commonest neurological diseases of the future.

Such a scenario seemed unimaginable in the summer of 1981, when reports appeared in the *Morbidity and Mortality Weekly Reports* of clusters of Karposi's sarcoma and pneumocystis pneumonia in otherwise healthy young gay men (Centers for Disease Control 1981a,b). Indeed, even with further case recognition and the spread of the epidemic, neurological disease was of little concern. When HIV was isolated, the virus was shown to replicate in the CD4-positive lymphocytes, and the destruction of these cells appeared to cause immunodeficiency which, in turn, was complicated by opportunistic infections that led to

The Departments of Neurology, Molecular Biology and Genetics, and Neuroscience, The Johns Hopkins University School of Medicine, 600 North Wolfe Street, Meyer 6–113, Baltimore, MD 21287–7613, USA death. Neurological interest was limited to unusual opportunistic infections of the nervous system, such as the strange presentations of toxoplasmosis, unusual forms of cryptoccal meningitis, and remarkable numbers of cases of progressive multifocal leukoencephalopathy, and increased rates of cerebral lymphoma (SNIDER et al. 1983). During the early 1980s, patients were described with withdrawal, apathy, and apparent dementia who had marked atrophy of the brain and the presence of microglial nodules, but it was assumed these represented opportunistic infections with cytomegalovirus or unidentified agents (SNIDER et al. 1983; NIELSEN et al. 1984).

2 Nervous System Infection

The fixation on the CD4-positive lymphocyte and the hematological manifestation of the disease changed in 1985. First, HIV was recovered from brain tissue, spinal fluid, spinal cord, and peripheral nerves of patients with neurological complications of AIDS (HO et al. 1985; LEVY et al. 1985); HIV-RNA was demonstrated in the microglial nodules in the brain, and HIV-DNA was found by Southern blot to be more abundant in the brains of some patients than in lymph node or spleen (SHAW et al. 1985); and intrathecal synthesis of antibody against HIV was demonstrated (RESNICK et al. 1985). These findings suggested a primary infection of the nervous system with HIV. Second, in 1985, the sequences of HIV and other known retroviruses were compared, and sequence similarities showed that HIV was not an oncovirus related to HTLV-1, as previously suspected, but was a member of the lentivirus family (GONDA et al. 1985), a group of viruses that are highly species specific, that infect macrophages which play a central role in pathogenesis, and that cause chronic neurological infection and encephalitis. As Gardner pointed out in this meeting, animal studies of the pathogenesis of lentiviruses were limited prior to this time, but it is of note that studies by Narayan in our laboratories and those by Haase and Nathanson were not being pursued because any of those investigators foresaw the evolution of HIV, but because visna and caprine arthritis-encephalitis viruses permitted in vivo investigations of virus persistence, relapses and remissions, the evolution of new mutants, and virus-induced demyelination in the brain. These studies were supported primarily by the National Multiple Sclerosis Society, the Kroc Foundation, and the National Institute of Neurologic Diseases and Stroke as animal models of multiple sclerosis.

Once HIV was associated with primary neurological disease, two questions became pivotal:

1. Which neurological diseases are due to opportunistic infections and which diseases are the result of HIV infection?

2. How early is the central nervous system infected? Initially, some thought that the virus might simply enter the nervous system with inflammatory cells at the time of opportunistic infection late in the course of AIDS. Under these circumstances nervous system infection would be a preterminal event of little concern.

Defining the spectrum of diseases associated with HIV infection of the nervous system continues, but a variety of clinical syndromes and pathological findings affecting both the central and peripheral nervous systems have been ascribed to HIV infection. Some are seen at the time of initial seroconversion, such as acute encephalopathy (CARNE et al. 1985) and acute demyelinating poly-neuritis resembling the Guillian-Barré syndrome (CORNBLATH et al. 1987). Others are seen at time of lymphadenopathy in the AIDS-related complex, such as a chronic meningitis and a vasculitis with multiple nerve infarcts leading to mononeuritis multiplex, and these are thought to be immune-related diseases occurring prior to full blown immunosuppression (MCARTHUR 1987). In immunodeficient patients dementia occurs in 20%–50%, a vacuolar myelopathy in about 20%, and painful sensory neuropathy in 30%, although the axonal degeneration underlying this neuropathy is found in all autopsies of AIDS patients (GRIFFIN et al. 1995). Since these diseases occur at disparate times in the course of infection and show different pathological changes, it must be assumed that the mode of pathogenesis is different. Never before in the history of virology have so many different neurological illnesses been associated with a single virus.

The multicenter AIDS Cohort Study is a prospective study of gay men which was initiated even before isolation of the virus. After recognition of nervous system involvement, a subgroup of these men was selected for periodic neuro-cognitive tests, neurological exams and lumbar punctures at 6 month intervals. Some of these men were seropositive, some were seronegative; but sero status was determined after the initial recruitment limiting selection bias. Studies of cerebrospinal fluid in healthy men infected for less than a year or two showed that two thirds had abnormalities of spinal fluid with pleocytoses, abnormal spinal fluid immunoglobulins, and/or the ability to recover virus (MCARTHUR et al. 1988). This and similar findings by others indicate that the nervous system is infected early in the majority of asymptomatic infected persons. This surprising finding immediately led to the question whether a decline in cognitive function begins at the time of infection and progresses over the 8–10 years of the incubation period. Although the studies presented from Grant's group show a higher rate of minor neuropsychiatric abnormalities in HIV-infected persons, most of the large cross-sectional studies and the longitudinal studies have not found major or discernable differences (MCARTHUR et al. 1989; GOETHE et al. 1989; SELNES et al. 1990). All agree that the development of severe, rapidly progressive dementia is limited to patients with marked reductions of CD4 counts. Dementia occurs as the AIDS-defining illness in only 3% of AIDS patients; thereafter, approximately 7% of AIDS patients develop dementia each year of survival (MCARTHUR et al. 1993).

3 Clinical and Pathological Studies

A number of clinical questions persist. For example, what are the determinants of the different neurological diseases? Age of infection appears important, since infections in the perinatal period lead, in some cases, to a more rapid course with retardation and developmental microcephaly with mineralization within the brain (EPSTEIN et al. 1985; BELMAN et al. 1988). Also, comparisons of intravenous drug users with gay men have shown striking differences in the rates of Karposi's sarcoma, but similar comparisons have not shown significant differences in the rates or forms of neurological illness (ROYAL et al. 1991). Further comparisons of neurological diseases in different risk groups may clarify host determinants. Why do some patients develop dementia and others not? Are persons who develop dementia more prone to develop myelopathy and sensory neuropathy? This is not totally clarified, but these illnesses are not tightly linked, since each can and do appear independent of the others.

Clinical-pathological correlations of cognitive deficits are not fully defined. Wiley discussed the relative rarity of multinucleated giant cells in patients who have HIV encephalitis as defined by the presence of HIV antigen by immunoperoxidase staining. The Hopkins prospective study showed that, although almost all patients with multinucleated giant cells or diffuse myelin pallor were demented, fully half of the patients who were clinically demented failed to show either of these pathological abnormalities (GLASS et al. 1993).

The question remains as to the role of antiviral agents in modifying the course of these diseases. Data are available only for the dementing illness. The original clinical trial of AZT covered only a 16 week period but nonetheless showed a clear improvement in cognitive function (SCHMITT et al. 1988). A subsequent study in children documented improvement in IQ with AZT treatment (PIZZO et al. 1988). The data presented by Lloente and others suggest this initial improvement is transient and that no improvement is detectable 18 months later; this is not surprising since the evolution of drug resistant strains of HIV and renewed progression of disease occur within this time frame. In closing the AZT study, those of us on the Data and Safety Monitoring Board were concerned that information on the possible efficacy of AZT to improve the neuropsychiatric abnormalities would be lost. Instead, a rather robust improvement was evident from the limited data (SCHMITT et al. 1988), but those limited data gave us false hope since long-term open studies now show this improvement to be short-lived.

4 Virological Studies

There are several important virological questions that need to be answered. First is the query as to whether or not macrophages and cells of macrophage origin are the only cells infected in the nervous system. Increasing evidence suggests that astrocytic infection does occur, particularly in children (BLUMBERG et al. 1992; SAITO et al. 1994; TORNATORE et al. 1994), and possibly endothelial cell infection as well (MOSES et al. 1993), but the rarity of these infected cells raises the question of biological relevance. With exception to NUOVO's work (NUOVO et al. 1994), cited by HAASE, there is no evidence of infection of neurons. Yet clearly morphological changes in neurons are found with attenuation of their dendritic processes and cell loss (MASLIAH et al. 1992).

The second virological question is whether or not neurological disease is directly related to amount of virus. Our preliminary data using reverse transcription-PCR show HIV/gag and the tat/rev transcripts more frequently in demented than nondemented patients; using semiquantitation with laser scanning densitometry, the quantities of RNA appeared to be greater in those patients who were demented than in seropositive nondemented patients (WESSELINGH and colleagues, unpublished data). Similar correlations were not found, however, with spinal cords with and without vacuolar myelopathy or with nerves from painful sensory neuropathy, again suggesting us that the mechanisms of the pathogenesis may be quite different in different neurological diseases.

The third virological question was raised by Chesebro (see Power et al., this volume) as to whether or not strains show different neurovirulence. It is important to differentiate between neuroinvasiveness, neurotropism, and neurovirulence. Neuroinvasiveness is the ability of virus to enter the nervous system; neurotropism simply means the ability to infect neural cells whereas neuronotropic means the ability to infect neurons; and neurovirulence is defined as the ability to cause neurological disease. Thus mumps virus is one of the most neuroinvasive viruses in humans, since it enters the nervous system in a majority of persons but usually causes mild disease with infection limited to ependymal and meningeal cells; thus it is neurotropic but not neuronotropic. It seldom causes serious disease; therefore it is not highly neurovirulent. In contrast, herpes simplex virus in humans is an ubiquitous agent that only rarely invades the nervous system; therefore it is not highly neuroinvasive. It infects all cells including neurons so it is neurotropic and neuronotropic, and in untreated cases kills 70% of those infected, so it is highly neurovirulent. In this context HIV causes confusion; as documented by studies referred to above, early in the course of infection HIV shows great neruroinvasiveness, but since it rarely causes serious neurological disease it lacks neurovirulence. There is no evidence at any stage that it is neuronotropic. Late in the infection, however, severe neurological disease develops in many patients, so HIV acquires a high degree of neurovirulence. This change may result from crippling of the host with the onset of severe immunodeficiency so that the viral burden increases leading to neurological disease. Alternatively, this development of neurovirulence may result from a change in the virus. Over a period of time, virus in brain that is already neurotropic may evolve distinct sequences that confer neurovirulence (POWER et al. 1994). The latter alternative would be supported by the observation that there is little correlation between severity of immunodeficiency and the development of dementia. Indeed, in a large percentage of the patients followed prospectively to death, dementia does not evolve even though virus can be recovered from the brain.

5 Pathogenetic Mechanisms

Many questions regarding pathogenesis persist. How does the virus invade the nervous system? Since HIV is predominantly an infection of blood cells, spread to the nervous system is assumed to be via the blood, but little experimental data substantiate this assumption. Furthermore, it is not known whether invasion occurs from a plasma viremia or a cell-associated viremia. Central nervous system invasion may occur during the early plasma viremia or it may occur when infected monocytes migrate into the nervous system maturing into macrophages. This is not a meaningless question, since if virus enters the nervous system from a plasma viremia, neutralizing antibody may play a critical role in aborting this invasion, whereas if virus enters in infected mononuclear cells, adhesion molecules may be of greater importance. What the signal is for this entry, whether there is selective entry of infected monocytes or whether macrophages are infected in greater numbers once they are in the nervous system is unknown, but greater knowledge of the traffic of these cells early in the course of infection, similar to studies described by Hickey, would give important information.

Finally, since the neuron is affected but not infected, what are the mediators of the disease? The toxicity of viral proteins (Brenneman et al. 1988; Lipton 1991; Sabatier et al. 1991), an excitotoxin (Heyes et al. 1991), and cytokines (Tyor et al. 1992; Wesselingh et al. 1993) have all been incriminated. Neuronal damage may be the direct result of mediators released from macrophages and microglia or may be mediated through an intermediate reaction with astrocytes or other cells, as Price suggested.

To return to a point made at the beginning of this litany of questions, the spectrum of diseases due primarily to HIV may not be completely defined. Since different diseases appear to have different mechanisms of pathogenesis, the role of opportunistic infections needs evaluation in each neurological complication. This program has focused on infection of the brain and its impact on function and behavior. We must not assume that the lessons learned here are necessarily applicable to the spinal cord, nerve roots, or peripheral nerves where the virus also has profound effects by mechanisms yet unknown.

Acknowledgments. This work was supported in part by grant PO1 NS26643 from the National Institute of Neurological Diseases and Stroke of the National Institutes of Health.

References

Belman AL, Diamond G, Dickson D, Horoupian D, Llena J, Lantos G, Rubinstein A (1988) Pediatric acquired immunodeficiency syndrome: neurologic syndromes. Am J Dis Child 142: 29–35

Blumberg BM, Epstein LG, Saito Y, Chen D, Sharer LR, Anand R (1992) Human immunodeficiency virus type 1 nef quasispecies in pathological tissue. J Virol 66: 5256–5264

Brenneman DE, Weatbrook GL, Fitzgerald SP, Ennist DL, Elkins KL, Ruff MR, Pert CB (1988) Neuronal cell killing by the envelope protein of HIV and its prevention by vasoactive intestinal peptide. Nature 335: 639–642

Carne CA, Tedder RS, Smith A, Sutherland S, Elkington SG, Daly HM, Preston FE, Craske J (1985) Acute encephalopathy coincident with seroconversion for anti-HTLV III. Lancet 2: 1206–1208

Centers for Disease Control (1981a) Kaposi's sarcoma and pneumocystis pneumonia among homosexual men—New York City and California. Morb Mort Wkl Rep 30: 305–308

Centers for Disease Control (1981b) Pneumocystis pneumonia—Los Angeles. Morb Mort Wkl Rep 30: 250–252

Cornblath DR, McArthur JC, Kennedy PGE, Witte AS, Griffin JW (1987) Inflammatory demyelinating peripheral neuropathies associated with human T-cell lymphotropic virus type III infection. Ann Neurol 21: 32–40

Epstein LG, Sharer LR, Joshi VV, Fojas MM, Koenigsberger MR, Oleske JM (1985) Progressive encephalopathy in children with acquired immunodeficiency syndrome. Ann Neurol 17: 488–496

Glass JD, Wesselingh SL, Selnes OA, McArthur JC (1993) Clinical-neuropathologic correlation in HIV-associated dementia. Neurology 43: 2230–2237

Goethe KE, Mitchell JE, Marshall DW, Brey RL, Cahill WT, Leger GD, Hoy LJ, Boswell RN (1989) Neuropsychological and neurological function of human immunodeficiency virus seropositive asymptomatic individuals. Arch Neurol 46: 129–133

Gonda MA, Wong-Staal F, Gallo RC, Clements JE, Narayan O, Gilden RV (1985) Sequence homology and morphologic similarity of HTLV-III and visna virus, a pathogenic lentivirus. Science 227: 173–177

Griffin JW, Crawford TO, Tyor WR, Glass JD, Price DL, Cornblath DR, McArthur JC (1995) Sensory neuropathy in AIDS. I. Neuropathology. Brain (in press)

Heyes MP, Brew BJ, Martin A, Price RW, Salazar AM, Sidtis JJ, Yergey JA, Mouradian MM, Sadler AE, Keilp J, Rubinow D, Markey SP (1991) Quinolinic acid in cerebrospinal fluid and serum in HIV-1 infection: relationship to clinical and neurologic status. Ann Neurol 29: 202–209

Ho DD, Rota TR, Schooley RT, Kaplan JC, Allan JD, Groopman JE, Resnick L, Felnsenstein D, Andrews CA, Hirsch MS (1985) Isolation of HTLV-III from cerebrospinal fluid and neural tissues of patients with neurologic syndromes related to the acquired immunodeficiency syndrome. N Engl J Med 313: 1493–1497

Levy JA, Shimabukuro J, Hollander H, Mills J, Kaminsky L (1985) Isolation of AIDS-associated retroviruses from cerebrospinal fluid and brain of patients with neurological symptoms. Lancet 2: 586–588

Lipton SA (1991) HIV-related neurotoxicity. Brian Pathol 1: 193–199

Masliah E, Ge N, Morey M, Deteresa R, Terry RD, Wiley CA (1992) Cortical dendritic pathology in human immunodeficiency virus encephalitis. Lab Invest 66: 285–291

McArthur JC (1987) Neurologic manifestations of AIDS. Medicine 66: 407–437

McArthur JC, Cohen BA, Farzedegan H, Cornblath DR, Selnes OA, Ostrow D, Johnson RT, Phair J, Polk BF (1988) Cerebrospinal fluid abnormalities in homosexual men with and without neuropsychiatric findings. Ann Neurol 23 [Suppl]: S34–S37

McArthur JC, Cohen BA, Selnes OA, Kumar AJ, Cooper K, McArthur JH, Soucy G, Cornblath DC, Chmiel JS, Wang M-C, Starkey DL, Ginzburg H, Ostrow DG, Johnson RT, Phair JP, Polk BF (1989) Low prevalence of neurological and neuropsychological abnormalities in otherwise healthy HIV-1-infected individuals: results from the multicenter AIDS cohort study. Ann Neurol 26: 601–611

McArthur JC, Hoover DR, Bacellar H, Miller EN, Cohen BA, Becker JT, Graham NMH, McArthur JH, Selnes OA, Jacobson LP, Visscher BR, Concha M, Saah A (1993) Dementia in AIDS patients: incidence and risk factors. Neurology 43: 2245–2252

Merson MH (1993) Slowing the spread of HIV: agenda for the 1990s. Science 260: 1266–1268

Moses AV, Bloom FE, Pauza CD, Nelson JA (1993) Human immunodeficiency virus infection of human brain capillary endothelial cells occurs via a CD4/galactosylceramide-independent mechanism. Proc Natl Acad Sci USA 90: 10474–10478

Nielsen SL, Petito CK, Urmacher CD, Posner JB (1984) Subacute encephalitis in acquired immune deficiency syndrome: a postmortem study. Am J Clin Pathol 82: 678–682

Nuovo GJ, Gallery F, MacConnell P, Braun A (1994) In situ detection of polymerase chain reaction-amplified HIV-1 nucleic acids and tumor necrosis factor-α RNA in the central nervous system. Am J Pathol 144: 659–666

Pizzo PA, Eddy J, Falloon J, Balis FM, Murphy RF, Moss H, Wolter P, Brouwers P, Jarosinski P, Rubin M, Broder S, Yarchoan R, Brunetti A, Maha M, Nusinoff-Lehrman S, Poplack DG (1988) Effect of continuous intravenous infusion of zidovudine (AZT) in children with symptomatic HIV infection. N Engl J Med 319: 889–896

Power C, McArthur JC, Johnson RT, Griffin DE, Glass JD, Perryman S, Chesebro B (1994) Distinct brain-derived HIV-1 envelope sequences associated with HIV dementia. J Virol 68: 4643–4649
Resnick L, DiMarzo-Veronese F, Schüpbach J, Tourtellotte WW, Ho DD, Müller F, Shapshak P, Vogt M, Groopman JE, Markham PD, Gallo RC (1985) Intra-blood-brain-barrier synthesis of HTLV-III-specific IgG in patients with neurologic symptoms associated with AIDS or AIDS-related complex. N Engl J Med 313: 1498–1504
Royal W III, Updike M, Selnes OA, Procter TV, Nance-Sproson L, Solomon L, Vlahov D, McArthur JC (1991) HIV-1 infection and nervous system abnormalities among a cohort of intravenous drug users. Neurology 41: 1905–1910
Sabatier J-M, Vives E, Mabrouk K, Benjouad A, Rochat H, Duval A, Hue B, Bahraoui E (1991) Evidence for neurotoxic activity of tat from human immunodeficiency virus type I. J Virol 65: 961–967
Saito Y, Sharer LR, Epstein LG, Michaels J, Mintz M, Louder M, Golding K, Cvetkovich TA, Blumberg BM (1994) Overexpression of nef as a marker for restricted HIV-1 infection of astrocytes in postmortem pediatric central nervous tissues Neurology 44: 474–481
Schmitt FA, Bigley JW, McKinnis R, Logue PE, Evans RW, Drucker JL, AZT Collaborative Working Group (1988) Neuropsychological outcome of zidovudine (AZT) treatment of patients with AIDS and AIDS-related complex. N Engl J Med 319: 1573–1578
Selnes OA, Miller E, McArthur JC, Gordon B, Munoz A, Sheridan K, Fox R, Saah AJ (1990) HIV-1 infection: no evidence of cognitive decline during the asymptomatic stages. Neurology 40: 204–208
Shaw GM, Harper ME, Hahn BH, Epstein LG, Gajdusek DC, Price RW, Navia BA, Petito CK, O'Hara CJ, Cho E-S, Oleske JM, Wong-Staal F, Gallo RC (1985) HTLV-III infection in brains of children and adults with AIDS encephalopathy. Science 227: 177–182
Snider WD, Simpson DM, Nielsen S, Gold JWM, Metroka CE, Posner JB (1983) Neurological complications of acquired immune deficiency syndrome: analysis of 50 patients. Ann Neurol 14: 403–418
Tornatore C, Chandra R, Berger JR, Major EO (1994) HIV-1 infection of subcortical astrocytes in the pediatric central nervous system. Neurology 44: 481–487
Tyor WR, Glass JD, Griffin JW, Becker PS, McArthur JC, Bezman L, Griffin DE (1992) Cytokine expression in the brain during the acquired immunodeficiency syndrome. Ann Neurol 31: 349–360
Wesselingh SL, Power C, Glass JD, Tyor WR, McArthur JC, Farber JM, Griffin JW, Griffin DE (1993) Intracerebral cytokine mRNA expression in AIDS dementia. Ann Neurol 33: 576–582

Neurocognitive Disorders in HIV-1 Infection

I. GRANT[1,2], R.K. HEATON[1], J.H. ATKINSON[1,2], and the HNRC GROUP*

1 Introduction

Neurocognitive disorder was noted to be associated with acquired immune deficiency syndrome (AIDS) even before HIV-1 was suspected of being the agent responsible for immunosuppression. HOROWITZ et al. (1982) observed that seven of nine patients suffering from gay-related immunodeficiency disorder, and who had CMV retinopathy as a complication, had neurological disorders, including disturbance in higher mental functioning. At the same time BRITTON et al. (1982) observed that three of their four patients with "gay immunosuppressed syndrome" with neurological complications suffered from psychosis and

[1] Department of Psychiatry, University of California, San Diego, 9500 Gilman Drive, La Jolla, CA 92093-0680, USA

[2] Psychiatry Service, Department of Veterans Affairs Medical Center, 3350 La Jolla Village Drive, San Diego, CA 92161, USA

*The San Diego HIV Neurobehavioral Research Center (HNRC) Group is affiliated with the University of California, San Diego, the Naval Hospital, San Diego, and the San Diego VA Medical Center, and includes: I. GRANT, M.D., Director, J.H. ATKINSON, M.D., Co-Director; M. Kelly, Ph.D., Center Manager; J.A. MCCUTCHAN, M.D., P.I. Medical Core; S.A. SPECTOR, M.D., P.I. Virology Core; L. THAL, M.D., P.I. Neurology Core; R.K. HEATON, Ph.D., P.I. Neuropsychology Core; J. HESSELINK, M.D. and T. JERNIGAN, Ph.D., Co-P.I.s Imaging Core; J.H. ATKINSON, M.D., P.I. Psychiatry Core; E. MASLIAH, M.D. and C.A. WILEY, M.D., Ph.D., Co-P.I.s Neuropathology Core; I. ABRAMSON, Ph.D., P.I. Biostatistics Core; N. BUTTERS, Ph.D., P.I. Memory Project; R. DUPONT, M.D., P.I. SPECT Project; T. PATTERSON, Ph.D., P.I. Life Events Project; S. ZISOOK, M.D., P.I. Mood Project; D. JESTE, M.D., P.I. Psychosis Project. The principal support for the HNRC is provided by NIMH Center Grant 5 P50 MH45294 (HIV Neurobehavioral Research Center)

dementia. In the following year SNIDER et al. (1983) reported that 18 of a series of 50 brains examined neuropathologically had evidence of subacute encephalopathy; furthermore, many of these patients exhibited, prior to death, clinical evidence of disturbed cognitive functioning and impaired ability to look after themselves.

In the mid-1980s the Memorial Sloan-Kettering group of investigators made the first systematic effort to characterize the cognitive disorder associated with AIDS. Proposing the term "AIDS dementia complex" (ADC), these investigators emphasized that certain cognitive changes (such as mental slowing) were to be found in the context of motor changes (weakness, incoordination) as well as emotional/motivational disturbances. The rates of ADC in persons with AIDS were estimated variously, but a figure of 33% of at least "mild" ADC was commonly accepted (NAVIA et al. 1986; PRICE and BREW 1988).

At about the same time several research groups began the process of characterizing the neurocognitive disorder associated with HIV infection in more precise terms using neuropsychological testing. While the methodologies varied, there was general consensus that the most marked impairment among those who had symptomatic HIV infection was on tests measuring attention, speeded information processing, learning of new information, and tests of executive functioning (AYERS et al. 1986; JOFFE et al. 1986; GRANT et al. 1987; TROSS et al. 1988).

Since publication of this early work, research on HIV-related neurocognitive disorders has burgeoned. Readers interested in a summary of neuropsychiatric studies prior to 1990 might wish to consult previous reviews by our group (GRANT 1990; GRANT and ATKINSON 1990a, b; GRANT and HEATON 1990). A more extensive treatment of neuropsychological research through mid-1993 may be found in GRANT and MARTIN (1994). The purpose of the present article will be to focus on emerging insights on neurocognitive disorders associated with HIV infection, with some reference to recent findings from the San Diego HIV Neurobehavioral Research Center (HNRC).

2 Neurocognitive Disorders: Classification

Neurocognitive disorder refers to a disturbance in function resulting from presence of a neurocognitive impairment. A neurocognitive (neuropsychological) *impairment* exists when there is deficient performance in some area of cognitive functioning, which can include attention/speed of information processing, verbal/language skills, abstracting (executive abilities), complex perceptual/motor abilities, psychomotor skills, sensory/perceptual ability, and memory functions (including learning and recall of information). To classify performance on a cognitive task as being impaired, it is necessary to establish a criterion beyond which performance is considered to be abnormal.

While this notion might seem straightforward, putting it into operation has turned out to be remarkably complex. For instance, both age and education have significant effects on neuropsychological performance in people without any evidence of cerebral disturbance, and these effects differ in accordance with the specific demands of a particular test (HEATON et al. 1986). Until recently, large normative studies had only been accomplished for intelligence tests such as the Wechsler Adult Intelligence Scale. However, only some components of traditional intelligence tests are sensitive to many types of acquired brain dysfunction. For this reason, neuropsychologists typically use other tests thought to be more specific to the measurement of cognitive abilities of interest—such as acquisition of new information, rates of forgetting, and divided attention. While such specific tests may be highly suitable to assessing cognitive impairment in HIV-infected persons, many of them do not have norms developed across sufficiently diversified samples. An exception is the ensemble of tests grouped into the "extended Halstead-Reitan Battery" (HEATON et al. 1991), for which there exist age/education/gender-specific norms based on a sampling of over 500 nonneurologically impaired individuals.

An additional problem is that even where age/education-adjusted norms exist, they have usually been developed on Caucasian samples. Systematic normative studies of African-Americans, Hispanic/Latino groups, and other minority groups are still not readily available. Therefore, neuropsychologists are constantly struggling to infer what might or might not be impaired performance based on a still imperfectly developed normative database.

According to the World Health Organization (WHO), a *disability* exists when certain impairments interfere with a person's functioning. We believe that the term *disorder* (as in neurocognitive disorder) should be reserved for those individuals who have neurocognitive impairments that are significant enough to produce disability, i.e., to interfere to some extent at least with the day-to-day functioning. Unfortunately, as we shall discuss below, there is still insufficient systematic study on the implications of reliably determined neuropsychological impairment on day-to-day functioning.

Accepting for a moment the definition that a person suffers from a neurocognitive disorder if he/she has neuropsychological impairment sufficient to interfere to some degree with everyday functioning, how should these disorders be classified? There are two basic strategies, one based on considerations of severity of impairment and its duration, and the other on presumed etiology.

Based on severity/time course considerations, three forms of neurocognitive disorder can be found in HIV-infected persons. These are: (1) mild neurocognitive disorder (MND), (2) HIV-associated dementia (HAD), and (3) delirium associated with HIV disease. The criteria utilized by the San Diego HNRC for each of these disorders are pesented in Tables 1–3. Tables 1 and 2 also provide a summary of criteria proposed by the WORKING GROUP OF THE AMERICAN ACADEMY OF NEUROLOGY AIDS TASK FORCE (1991) for dementia (HIV-1 associated dementia complex) and mild neurocognitive disorder (HIV-1 associated minor cognitive/motor disorder).

Table 1. HIV-1 associated cognitive disorders

As defined by Grant and Atkinson (1995)	As proposed by the Working Group of the American Academy of Neurology AIDS Task Force (1991)
HIV-1 Associated Neurocognitive Disorders **HIV-1 Associated Dementia (HAD)**	*HIV-1 Associated Cognitive/Motor Complex* **Probable[a] HIV-1 Associated Dementia Complex**
1. *Marked acquired impairment in cognitive functioning,* involving at least two ability domains (e.g., memory, attention); typically the impairment is in multiple domains, especially in learning of new information, slowed information processing, and defective attention/concentration. The cognitive impairment can be ascertained by history, mental status examination, or neuropsychological testing.	1. *Acquired abnormality in at least two of the following cognitive abilities* (present for at least 1 month): attention/concentration; speed of information processing; abstraction/reasoning; visuospatial skills; memory/learning; speech/language. *Cognitive dysfunction causes impairment in work or activities of daily living.*
2. The cognitive impairment produces *marked interference with day to day functioning* (work, home life, social activities).	2. *At least one of the following*: (a) Acquired abnormality in motor functioning; (b) Decline in motivation or emotional control or change in social behavior.
3. The marked cognitive impairment has been present for at least 1 month.	3. Absence of clouding of consciousness during a period long enough to establish presence of #1.
4. The pattern of cognitive impairment *does not meet criteria for delirium* (e.g., clouding of consciousness is not a prominent feature); or, if delirium is present, criteria for dementia need to have been met on a prior examination when delirium was not present.	4. Absence of another cause of the above cognitive, motor, or behavioral symptoms or signs (e.g., active CNS opportunistic infection or malignancy, psychiatric disorders, substance abuse).
5. There is *no evidence of another, preexisting etiology* that could explain the dementia, e.g., other CNS infection, CNS neoplasm, cerebrovascular disease, preexisting neurological disease, or severe substance abuse compatible with CNS disorder.	

[a]The designation *probable* is used when criteria are met, there is no other likely cause, and data are complete. The designation *possible* is used if another potential etiology is present whose contribution is unclear, or where a dual diagnosis is possible, or when the evaluation is not complete.
From Grant and Martin (1994).

The main etiologies of the neurocognitive disorders are displayed in Table 4. It is convenient to divide the etiologies into primary (i.e., those thought to be linked directly to HIV infection of the brain) and those that are secondary (i.e., infections, tumors, etc., linked to immunodeficiency, or other adverse events associated with HIV disease or its treatment). For the sake of completeness, we have included a listing of key other HIV-related neurobiological complications; these may or may not coexist with the neurocognitive disorders.

The use of criteria presented in Tables 2–4 to ascertain the presence of MND, dementia, or delirium in HIV-1-infected persons will greatly facilitate both basic descriptive/epidemiological work and theoretical research into the

Table 2. HIV-1 associated cognitive disorders

As defined by Grant and Atkinson (1995)	As proposed by the Working Group of the American Academy of Neurology Aids Task Force (1991)
HIV-1 Associated Neurocognitive Disorders **HIV-1 Associated Mild Neurocognitive Disorder (MND)**	*HIV-1 Associated Cognitive/Motor Complex* **Probable[a] HIV-1 Associated Minor Cognitive/Motor Disorder**
1. Acquired impairment in cognitive functioning, involving at least two ability domains, documented by performance of at least 0.5 standard deviations below age-education-appropriate norms on standardized neuropsychological tests. The neuropsychological assessment must survey at least the following abilities: verbal/language; attention/speeded processing; abstraction; memory (learning; recall); complex perceptual-motor performance; motor skills	1. Acquired cognitive/motor/behavior abnormalities (must have both 1 and 2): a. At least two of the following symptoms present for at least one month verified by a reliable history: (i) impaired attention or concentration; (ii) mental slowing (iii) impaired memory (iv) slowedmovements; (v) incoordination b. Acquired cognitive/motor abnormality verified by clinical neurologic examination or neuropsychological testing.
2. The cognitive impairment produces at least mild interference in daily functioning (at least one of the following): (a) Self-report of reduced mental acuity, inefficiency in work, home-making, or social functioning; (b) Observation by knowledgeable others that the individual has undergone at least mild decline in mental acuity with resultant inefficiency in work, homemaking, or social functioning.	2. Cognitive/motor/behavioral abnormality causes mild impairment of work or activities of daily living (objectively verifiable or by report of key informant)
3. The cognitive impairment has been present at least 1 month.	3. Does not meet criteria for HIV-1 associated dementia complex or HIV-1-associated myelopathy.
4. Does not meet criteria for delirium or dementia.	4. Absence of another cause of the above cognitive/motor/ behavioral abnormality (e.g., active CNS opportunistic infection or malignancy, psychiatric disorders, substance abuse).
5. There is no evidence of another preexisting cause for the MND.[b]	

[a]The designation *probable* is used when criteria are met, there is no other likely cause, and data are complete. The designation *possible* is used if another potential etiology is present whose contribution is unclear, or where a dual diagnosis is possible, or when the evaluation is not complete.

[b]If the individual with suspected MND also satisfies criteria for a *major depressive episode* or *substance dependence*, the diagnosis of MND should be deferred to a subsequent examination conducted at a time when the major depression has remitted or at least 1 month has elapsed following termination of dependent-substance use.

From Grant and Martin (1994).

Table 3. Delirium associated with HIV-1 disease

Clouding of consciousness (examples: reduced alertness, disturbed awareness, disorientation).
Marked acquired impairment in cognitive functioning, involving at least two ability domains. Typically the impairment is in multiple domains, with defective attention/concentration and difficulty in learning and retaining information prominent. The cognitive impairment can be ascertained by history, mental status examination, or neuropsychological testing.
The marked impairment develops rapidly (i.e., over hours to days).
If dementia or mild neurocognitive disorder was previously diagnosed, there must be significant rapid cognitive worsening to support the additional diagnosis of delirium.
There is evidence from the history, physical exam, or laboratory findings of a neurological or medical condition to account for the delirium.

Table 4. Neurobiological complications of HIV-1 infection

- Primary neurobiological complications
  - HIV-1 neurocognitive disorders
    - HIV-1 mild neurocognitive disorder (MND)
    - HIV-1 associated dementia (HAD)
  - Other HIV-1 neurobiological complications
    - HIV-1 meningitis
    - HIV-1 vacuolar myelopathy
    - HIV-1 neuropathies: acute demyelinating (Guillain-Barré); relapsing or progressive demyelinating (e.g., mononeuritis multiplex); predominantly sensory polyneuropathy
    - HIV-1 myopathy
- Secondary neurobiological complications (generally causing delirium)
  - Infections
    - *Toxoplasma* encephalitis/abscess
    - *Cryptococcus* meningitis
    - Cytomegalovirus (CMV) encephalitis[a]
    - Progressive multifocal leukoencephalopathy (PML)[a]
    - Other infections of the CNS
  - Neoplasia
    - Primary or secondary CNS lymphoma
    - Kaposi's sarcoma of the CNS
    - Other neoplasia
  - Cerebrovascular disease related to HIV infection
  - Other delirium
    - Adverse effects of drugs
    - Hypoxemia, hypercapnia (e.g., with PCP pneumonia)
    - Other metabolic and nutritional disorders

From GRANT and ATKINSON (1995).
[a]Can produce a picture akin to MND, dementia, and/or delirium.

neurocognitive complications of HIV disease. In the past, failure to make these distinctions has led to wildly differing estimates of dementia in HIV-infected persons. For example, a World Health Organization report indicated that 8%–66% of persons with AIDS might develop AIDS dementia complex (WHO 1988). PRICE and BREW (1988) suggested that prevalence in AIDS patients might be in the order of 33%; while data from the U.S. Centers for Disease Control suggested a rate of 6.5% of all cases of AIDS reported to the CDC between September 1, 1987 and December 31, 1988 (JANSSEN et al. 1989a).

Since in these earlier studies the possible influence of zidovudine treatment would not yet have been a factor, the most likely sources of disagreement must stem from sampling (e.g., series based on neurological referrals would be expected to have higher rates of neurocognitive disorder) or ascertainment difficulties. Based on our literature review and discussions with many of the investigators who have been working in this area, we are convinced that at least some of the disagreements are attributable to overinclusive use of the term dementia by some vs more restricted use by other. In other words, some cases which would now be termed delirium or mild neurocognitive disorder were almost certainly classed as dementia in some earlier series.

The more recent practice of using structured diagnostic criteria, either based on the American Psychiatric Association's Diagnostic and Statistical Manual (DSM) or the International Classification of Diseases (ICD) or the AAN Working Group Criteria (all of which have essential features in common), has led to much more comparable estimates for dementia among recent reports. For example, a small series by DAY et al. (1992) indicated an annual dementia incidence among symptomatic HIV-infected patients of 14% using DSM-III criteria; MCARTHUR et al. (1993), reporting on data from the Multi-Center AIDS Cohort Study (MACS), noted an annual incidence of 7.1% using DSM-III-like criteria; and MAJ et al. (1994), reporting on data from Munich, Sao Paolo, Kinshasa, and Nairobi, found a prevalence among symptomatic HIV-infected persons ranging from 4.4% to 6.5% (DSM-III-R criteria) or 5.4% to 6.9% (ICD-10 criteria). Clearly, the use of well-defined criteria for dementia appears to have led to more similar, as well as lower, estimates for the prevalence of dementia both in the USA and internationally.

Unfortunately, the concept of mild neurocognitive disorder (minor cognitive motor disorder) has been articulated only recently. Therefore, comparable data on its prevalence are not available at this time. Similarly, while the concept of delirium was articulated in more or less its present form in 1980 (AMERICAN PSYCHIATRIC ASSOCIATION 1980), it appears not to have been regularly utilized in work with AIDS patients. Hopefully, this practice will change in the near future so that the epidemiology of this serious neurocognitive complication can also be defined.

3 Neuropsychological Impairment in HIV-1 Disease

Although the relative newness of the concept of mild neurocognitive disorder (minor cognitive motor disorder) and, in particular, lack of information on real life consequences of neuropsychological impairment preclude a detailed analysis of the epidemiology and associated features of MND, it is possible to make some plausible inferences through a consideration of findings regarding neuropsychological *impairment* associated with HIV-1 infection. Impairment might not prove that a neurocognitive *disorder* exists, but it is a necessary first step in establishing such a diagnosis.

3.1 Prevalence

Most observers agree that symptomatic HIV-1 infection is accompanied by increased rate of neuropsychological impairment (GRANT and MARTIN 1994). Exactly how much of an increased risk AIDS patients experience is not yet agreed. Some studies have indicated that at least a third of persons with symptomatic HIV infection have at least mild neuropsychological impairment (GRANT et al. 1987; TROSS et al. 1988; JANSSEN et al. 1989b; LUNN et al. 1991; REINVANG et al. 1991; HEATON et al. 1994). Other studies, while documenting neuro-psychological impairment in AIDS, have estimated somewhat lower rates, i.e., in the order of 20% (MILLER et al. 1990; BORNSTEIN et al. 1993). A few studies have failed to detect any differences in rate of impairment between AIDS patients and controls (RICCIO et al. 1993). A review of 11 neuropsychological studies of AIDS patients found that the rate of neurocognitive impairment ranged from 12% to 87% with a median of 53% (HEATON et al. 1995a). Parametric and nonparametric comparisons of test scores derived from persons with AIDS have consistently indicated reduced performance among such patients (GRANT et al. 1987; JANSSEN et al. 1989b; MILLER et al. 1990; STERN et al. 1991; HANDELSMAN et al. 1992; BORNSTEIN et al. 1993; MARTIN 1994).

With respect to the asymptomatic phase of HIV disease, neuropsychological results have been mixed. Initially, there were reports of increased prevalence of impairment among asymptomatic persons (GRANT et al. 1987). This was followed by a number of reports, including some from very large studies, indicating no differences in neuropsychological status between asymptomatic HIV+ persons and controls (MCARTHUR et al. 1989; MILLER et al. 1990; COLLIER et al. 1992; MCALLISTER et al. 1992; RICCIO et al. 1993). More recently, newer studies have again detected performance differences between asymptomatics and controls (STERN et al. 1991; HANDELSMAN et al. 1992; BORNSTEIN et al. 1993; MARTIN et al. 1992, 1993; HEATON et al. 1994; MAJ et al. 1994). For those recent, large-scale studies reporting elevated rate of impairment among asymptomatic persons, the rates have been estimated to range from a low of 9.1% in the WHO Neuropsychiatric AIDS Study (MAJ et al. 1994) to 30.5% in the San Diego HNRC Cohort (Fig. 1). A review of 30 studies with data on rates of neuropsychological impairment in asymptomatic carriers and controls revealed a median rate of 12% impaired among samples of HIV– controls (range 0%–42%) and 35% for asymptomatic HIV+ (range 0%–50%) (WHITE et al. 1995).

3.2 Qualitative Features

The notion that there are subcortical features to the brain disease associated with HIV-1 infection was proposed almost a decade ago (NAVIA et al. 1986). Again, some definitional issues arise. As noted by A. MARTIN (1994) the original notion behind subcortical was that a particular disease was neuropathologically characterized by white matter or subcortical gray matter involvement primarily.

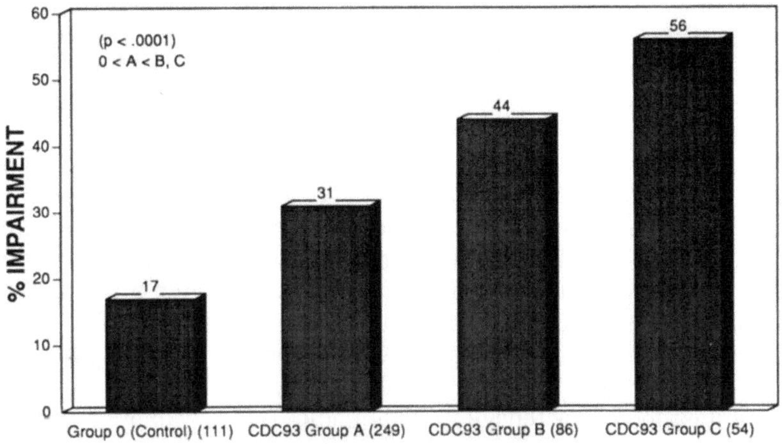

CDC93 STATUS (N = 389)

Fig. 1. Results from HNRC 500 series: global neuropsychological impairment at entry into study in relation to stage of disease. (From HEATON et al. 1995b)

The clear examples of such subcortical dementias are Huntington's disease and Parkinson's disease.

There is also a neuropsychological concept of subcortical dementia whose principal features include reduced attention, slowness of information processing, reduced fluency, impaired acquisition of motor skills, and difficulty on tests involving "egocentric spatial processing" (MARTIN 1994). Another feature is impaired free recall of recently learned information in the context of good recognition recall (BUTTERS et al. 1987). In contrast, persons with a cortical neuropsychological pattern of deficits (e.g., Alzheimer's disease) tend to have aphasia, apraxia, agnosia, alexia, and memory defects characterized both by severe difficulties in learning and also rapid forgetting with a tendency toward intrusions (BUTTERS et al. 1987; MARTIN 1994).

While neurologists, neuropathologists, and neuropsychologists use apparently the same terms (subcortical, cortical), their meanings are actually somewhat different. From a functional standpoint, a subcortical pattern of neuropsychological deficit appears to reflect primarily disturbances in cognitive functions attributed to integrity of the striatum and fronto-striatal connections. Thus, diseases that involve the frontal cortex may still present as subcortical from a neuropsychological perspective. However, white matter diseases such as multiple sclerosis can sometimes present with problems in naming or praxis, reflecting interference with pathways connecting parietal, temporal, and other cortical areas.

In the case of HIV-1 disease of the brain, there is evidence from multiple sources that the striatum is involved early and frequently. A. MARTIN (1994) cites convergent information from studies on regional distribution of neuropathology at autopsy (DE LA MONTE et al. 1987), regional distribution of HIV antigen

(PUMAROLA-SUNE et al. 1987; KURE et al. 1990), regional abnormalities in glucose utilization (ROTTENBERG et al. 1987; VAN GORP et al. 1992), increased sensitivity to the extrapyramidal side effects of antipsychotics (BREITBART et al. 1988; MAJ 1990), and, of course, the neuropsychological pattern of impairment manifested by persons with AIDS.

Figure 2, from the HNRC 500 series, shows the most common cognitive abilities that are impaired in HIV-infected persons. It will be seen that defects in attention/speeded information processing and learning of new information are most common followed by difficulties in the verbal area (often fluency) and motor speed. These types of impairment are generally consistent with a subcortical pattern, though it must be said that some HIV-infected persons, especially those with frank AIDS, exhibit cortical features as well—for example, problems in tasks requiring naming, rapid forgetting, or impaired abstraction.

In order to test the consistency of memory impairment with a subcortical paradigm, HNRC investigators attempted to classify a subset of HIV+ and HIV– participants on the basis of their performance on the California Verbal Learning Test (CVLT). The CVLT is a standardized list learning procedure that involves administration of the names of common objects over five trials, followed by immediate and delayed recall of these lists, interspersed with an interference trial. Both free recall and cued (i.e., semantic cueing) are tested, as is recognition recall (DELIS et al. 1987). Previous research with the CVLT has shown that patients with Huntington's disease (HD) differ from those with Alzheimer's disease in their capacity to store and retrieve information, and that failures in memory occur

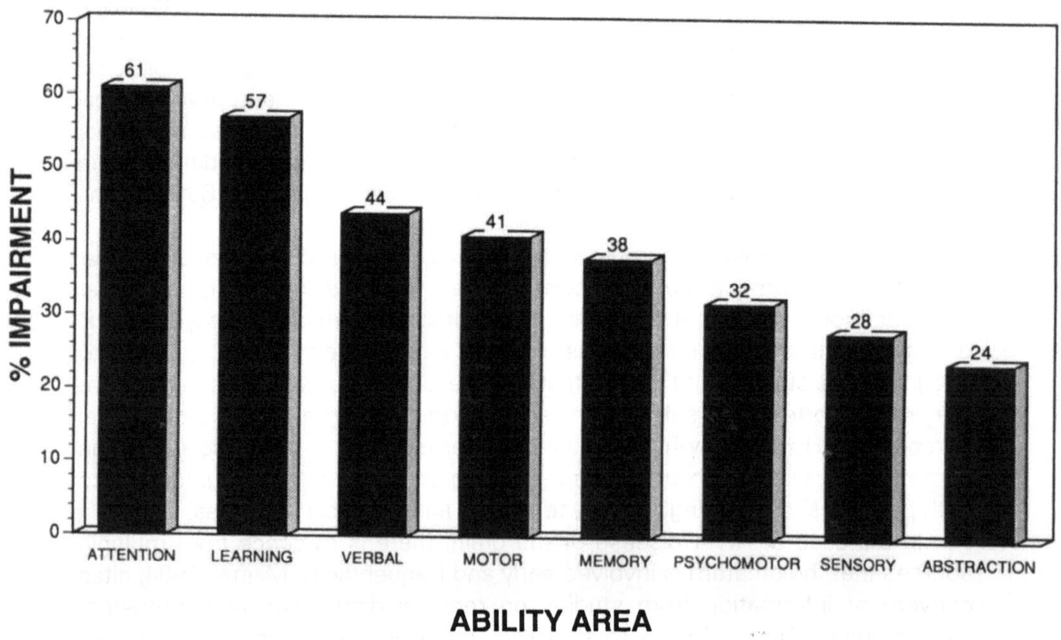

Fig. 2. Results from HNRC 500 series: percent impairment by ability area (globally impaired HIV+ only). (From HEATON et al. 1995b)

for different reasons in the two groups of dementia patients. Indeed, MASSMAN et al. (1992) were able to develop a discriminant function approach to classification of results from the CVLT that successfully discriminated most patients with HD from Alzheimer's disease.

Utilizing this discriminant function approach, HNRC investigators determined that 39 of 40 seronegative controls were classified as "normal" and one as "HD." In contrast, 20 of 31 HIV+ symptomatic participants were classified as normal vs 11 of 31 as impaired in memory, with 10 of those 11 being classified as HD. A similar pattern was seen for the asymptomatic participants. Of 94 participants, 16 were classified as impaired, and of those 15 had an HD pattern. These results are summarized in Table 5, taken from PEAVY et al. (1994).

In one of the most detailed examinations of possible subcortical patterns of cognitive impairment associated with HIV-1 infection, A. Martin and colleagues administered tests of reaction time, visual attention, extrapersonal space, egocentric space, visual recognition, verbal learning, and cognitive flexibility to a group of 52 HIV+ and 33 HIV– persons. The results (Fig. 3, taken from A. MARTIN 1994) indicate that both simple and choice reaction time are slowed in HIV+ individuals as is visual attention. Interestingly, a test of egocentric space (which emphasizes analysis of objects in relation to the frame of reference defined by the observer—which seems more dependent on frontal lobe functioning than tests of extrapersonal space—such as map reading and block design—which depend largely on parietal-cortical integrity) discriminated HIV+ from controls. Recent work with the pursuit rotor task, another test sensitive to integrity of the striatum, has also been promising in discriminating HIV+ from HIV– persons (see review by A. MARTIN 1994).

In sum, while neuropsychological data do not imply that HIV-1-associated brain disease produces a "pure" subcortical profile, there is certainly good evidence for the inference that early cognitive changes closely resemble disorders seen in fronto-striatal pathology. As the brain disease evolves, there tend to appear impairments in more cognitive areas, and in some cases a mixed cortical/subcortical picture may emerge.

3.3 Progression

The progression of neurocognitive complications is poorly understood. Investigators from the Multicenter AIDS Cohort Study (MACS) found no evidence for

Table 5. Discriminant function classification results (from PEAVY et al. 1994)

Actual Group	Predicted Group		
	Normal	HD	AD
HIV– NC	39	1	0
HIV+ Symptomatics	20	10	1
HIV+ Asymptomatics	78	15	1

NC, non-infected controls; HD, Huntington's disease; AD, Alzheimer's disease.

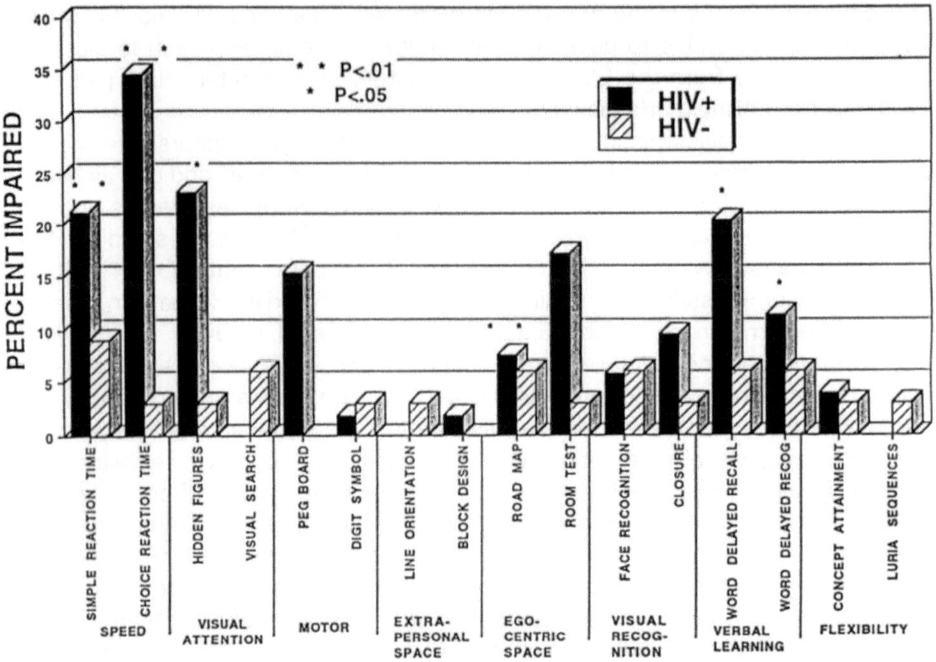

Fig. 3. Percentage of subjects in the HIV+ ($n = 52$) and HIV– ($n = 33$) groups who performed two or more SDs below the control group mean on select neuropsychological measures. Also indicated are significant group differences. (From GRANT and MARTIN 1994)

cognitive decline in a large group of asymptomatic patients followed for approximately 2 years (SELNES et al. 1990). By contrast, DUNBAR et al. (1992), following up a group of ARC and AIDS patients for about a year, concluded that the seropositive subjects declined significantly on neuropsychological tests compared to seronegative controls.

In the HNRC series we have completed at least 2 years of follow-up in over 150 participants. The annual rate of neuropsychological decline among seropositive individuals is in the order of 20%, and this decline does not seem to be particularly associated with stage of disease. DUNBAR et al. (1992) made a similar observation: neither initial classification as AIDS-related complex (ARC) or AIDS, nor whether they progressed from one diagnosis to the other, were associated with likelihood of neurocognitive decline.

There is even less information on likelihood of neuropsychological change as a function of estimated date of seroconversion. In the HNRC series (HEATON et al. 1995b), we have preliminary data to suggest that the likelihood of manifesting neuropsychological impairment increases starting about 3 years from the estimated date of seroconversion: a 28% impairment rate for subjects tested within 3 years of seroconversion ($n = 131$) vs a 44% rate for those with a longer duration of infection ($n = 45$).

3.4 Associated Factors

There has been considerable interest in ascertaining variables that may increase or attenuate risk of developing neurocognitive disorders in the context of HIV-1 infection.

3.4.1 Age

Advancing age is associated with declining performance on some neuropsychological tests, particularly those stressing psychomotor speed (HEATON et al. 1986). It is theoretically possible that advancing age might decrease "cerebral reserve" (SATZ 1993) allowing subtle impairments to be detected more easily.

To date, little systematic research has been done on this issue. VAN GORP et al. examined 1004 seronegative and 1066 seropositive, primarily asymptomatic, HIV+ persons on a set of brief neuropsychological tests and assessments of simple and complex reaction time. They found the predicted age-associated declines, but no interaction with serostatus. In the same article, a second study of 76 seropositive men of whom 29 were over age 55 and a comparison of 47 seronegative controls also failed to reveal any consistent suggestion of a serostatus by age interaction.

Preliminary analyses of HNRC data agree with VAN GORP et al. (VELIN et al. 1992). Our older participants tended to do worse on some neuropsychological tests, but the rate of impairment between clinical groups was similar so long as age-adjusted norms were used in classifying neuropsychological status.

3.4.2 Education

Here is another factor that powerfully influences neuropsychological performance (HEATON et al. 1986). The meaning of the education factor is complex. In part, educational level may reflect economic/cultural disadvantages which can manifest themselves in inferior performance on certain verbal tests and those requiring accumulation of knowledge. For some persons, however, low education may reflect subtle CNS injury or inherently poor intelligence and in this sense may be another indicator of lessened cerebral reserve.

SATZ et al. (1993) addressed this issue with the MACS cohort by dividing the sample into 172 men who had grade 12 education or less and comparing their performance to those 1601 who had greater than 12 years of schooling. The less educated group consisted of 60 seronegative and 112 HIV+ subjects, whereas the higher educated group contained 825 HIV– and 776 HIV+ subjects.

Among the higher educated persons, 15.9% of HIV– and 15.8% of HIV+ were judged to be neuropsychologically impaired on the MACS battery. Among the less educated group, 16.9% of HIV– and 38.1% of HIV+ were impaired. These investigators suggested that low education might be a risk factor for HIV-associated cognitive impairment in the sense that those with less cerebral reserve might reach threshold or impairment earlier in the course of their disease. While this might be so, another likely explanation is that the brief MACS battery was simply too easy to reveal HIV-associated impairment in the very well educated majority of the MACS participants (see discussion of this issue in GRANT and HEATON 1990 and GRANT et al. 1992).

3.4.3 Drug Abuse

Most neurocognitive investigations of HIV-1 effects have been confined to those with presumed sexual mode of transmission. Since both alcoholism (see review by GRANT 1987) and multiple substance use (GRANT et al. 1978; REED and GRANT 1990) are associated with neuropsychological impairment, it may be speculated that heavy concurrent or previous substance use might facilitate the emergence of HIV-associated neurocognitive disorder.

At this point there is no convincing evidence to suggest that this is the case. For example, WHITE et al. (1995) compared findings from seven neuropsychological studies of HIV+ intravenous drug users and 35 studies involving sexual transmission. They concluded that the rates of impairment were comparable from both sets of studies. Indeed, White et al. concluded that the likelihood of detecting impairment in seropositive asymptomatic individuals depended largely on comprehensiveness of the neuropsychological assessment, and not on issues such as risk factor, sample size, and method of data analysis.

The HNRC sample was not selected specifically to address the issue of heavy substance use. However, there was a considerable variation in recent and past experience with various substances. In Table 6, a subset of our participants is classified as impaired or unimpaired. Participants are compared on summary estimates of degree of recent, 5-year cumulative, and 11-year cumulative immersion aggregated for five classes of drugs between impaired and unimpaired participants. It will be seen that cumulative estimates of experience with various classes of drugs did not differ between neuropsychologically impaired and unimpaired persons (HEATON et al. 1995b). The remaining data in Table 6 indicate no difference in rate of abuse of any of these substances related to impairment.

3.4.4 Race and Ethnicity

Since at least some neuropsychological tests are culture-specific, i.e., require familiarity with the dominant or "Anglo" culture, it may be expected that unacculturated persons may have more difficulties with such procedures. An obvious example is unfamiliarity with language, compromising a person's ability to respond to verbally demanding tasks. Since a growing number of new seroconversions appear to be occurring among Hispanic/Latino minority groups, this is an area of pressing need for future research. Unfortunately, we are not aware of systematic published data on the relationship of HIV-associated impairment to ethnicity. Indirectly, the recently reported work by MAJ et al. (1994) from the WHO study suggests that if culturally relevant instruments are used, rates of impairment are fairly comparable from centers in Africa, Asia and Europe.

ɔle 6. Relationship between neuropsychological (NP) impairment and drug and alcohol histories for five drug classes. ɔm HEATON et al. 1995b)

	11 yr Average				1 yr Average				2 wk Average			
	HIV+ *n* % Impaired		HIV– *n* % Impaired		HIV+ *n* % Impaired		HIV– *n* % Imparied		HIV+ *n* % Impaired		HIV– *n* % Impaired	
ɱulants (mg/wk)												
Ɲone	171	40.9	53	24.5	264	38.6	81	19.8	279	38.4	83	19.3
≤ 10	88	34.1	26	11.5	16	31.3	3	0.0	2	100.0	0	0.0
>10	26	38.5	5	0.0	4	50.0	0	0.0	4	25.0	1	0.0
ırijuana (joints/wk)												
Ɲone	110	43.6	34	23.5	232	37.9	66	18.2	269	38.3	80	18.8
≤ 1	116	32.8	37	16.2	47	40.4	16	25.0	3	0.0	1	0.0
>1	59	40.7	13	15.4	5	40.0	2	0.0	13	53.9	3	33.3
dative/Hypnotics (mg/wk)												
None	214	39.7	61	21.3	258	36.7	79	20.3	271	38.4	81	18.5
≤ 5	61	32.8	20	15.0	20	40.0	5	0.0	9	44.4	2	0.0
>5	10	50.0	3	0.0	6	66.7	0	0.0	5	40.0	0	0.0
ıcaine (mg/wk)												
None	170	40.0	47	25.5	273	39.2	83	19.3	283	38.9	84	19.1
≤10	90	36.7	28	7.1	8	12.5	1	0.0	0	0.0	0	0.0
>10	25	36.0	9	22.2	3	33.3	0	0.0	2	0.0	0	0.0
ɔohol (g/wk)												
None	8	62.5	5	60.0	45	53.3	18	27.8	99	50.5	27	25.9
≤ 210	253	36.8	72	18.1	231	35.1	64	17.2	181	31.5	54	16.7
>210	24	50.0	7	0.0	8	50.0	2	0.0	5	60.0	3	0.0
ıtal substance use[a]												
NP impaired [*m* (SD)]	8.04	(2.23)	7.19	(1.87)	6.28	(1.11)	5.94	(0.77)	5.81	(0.89)	5.67	(0.82)
NP unimpaired [*m* (SD)]	8.21	(2.06)	8.26	(2.15)	6.34	(0.96)	6.21	(0.94)	5.94	(0.89)	5.88	(0.91)

ımulants have been converted to dextroamphetamine equivalents; sedative hypnotics have been converted to azepam equivalents (GRANT et al. 1978).
ʌmount of substance use was coded as none=1,middle range=2, and high range=3. Total substance use was calculated or each subject as the sum of substance use scores across all substances.

An even more complex issue concerns race, which in the United States may reflect altered educational and social opportunities and which might therefore cause African-Americans to appear more impaired based on established norms. In the only study that we found that provided comparisons between seropositive African-Americans and Caucasians, KLUSMAN et al. (1991) noted that 50% of African-American asymptomatic HIV+ persons were impaired vs 20% of Caucasians. Clearly, this is another area that is important for future systematic investigation.

3.4.5 Antiretroviral Drug Treatment

The early enthusiasm that zidovudine and other antiretrovirals may slow the progress of HIV disease brought with it the hope that CNS complications might also be ameliorated. Despite almost a decade of experience, the protectiveness of antiretrovirals in regard to CNS effects remains unclear.

Schmitt et al. (1988) reported that patients in the phase II licensing trial of zidovudine performed better on Trailmaking Part B if they were on active treatment than on placebo. At the same time a prospective 2 year follow-up of licensing trial participants from the San Diego Center revealed no difference in incidence of dementia between arms of the study (Day et al. 1992).

Portegies et al. (1989) published an influential article suggesting dramatic decline in incidence of HIV dementia related to availability of zidovudine treatment in Holland. These observations were echoed by the Nordic Medical Research Council's HIV Therapy Group (1992). However, a recent analaysis of MACS data reveals no association between dementia incidence and zidovudine exposure; and analysis of HNRC data among largely asymptomatic individuals or those with "early AIDS" also failed to establish an association between current or cumulative AZT treatment and neuropsychological impairment.

There is some lack of clarity in regard to benefits in pediatric AIDS, also. For example, early favorable experience reported by Pizzo et al. (1988) and Brouwers et al. (1990) was not confirmed by McKinney et al. (1991), the latter reporting for the protocol 043 study group.

The reasons for these disagreements are unclear. In part, it may be that benefits ascribed to AZT in fact reflect the operations of more general factors—better medical management, earlier aggressive treatment of opportunistic infection, or perhaps even genetic shifts in HIV itself away from "neuropathic" strains. However, it is also possible that suboptimal doses of AZT are being administered, at least insofar as neuroprotection is concerned. For example, McArthur et al. (1993) note that doses of 1000 mg of zidovudine daily may be required to achieve CNS benefit. It may also be that the neurological benefit is time-limited. For example, HNRC investigations with single photon computed emitted tomography (SPECT) suggest that HIV-associated reductions in tracer uptake are reversed in the first year of AZT treatment; after that, tracer uptake is again reduced, suggesting either perfusion or CNS metabolic abnormality (Dupont et al., in press).

3.4.6 Other Risk Factors

There has been considerable effort to identify other variables which might predict neuropsychological decline or actual incidence of dementia. Data from the MACS study indicate that increasing age, lower hemoglobin and body mass index, and higher numbers of constitutional symptoms all predicted greater likelihood of dementia (McArthur et al. 1993). Prediction of neuropsychological impairment or change has proved difficult. Boccellari et al. (1993) found that rise in serum β_2 microglobulin was associated with greater likelihood of neuropsychological impairment, while Bornstein et al. (1991) noted that rate of decline of CD4 cell count was related to neuropsychological abnormality.

For some time there was considerable speculation that depressed mood might contribute to or in some way explain neuropsychological impairment in

HIV+ persons, especially those who were asymptomatic. However, several studies have now failed to detect a clinically significant association between mood disturbance and neuropsychological impairment in seropositive individuals (see GRANT et al. 1993 for review). Therefore, although mood disorder is certainly increased both in seropositive individuals and those coming from risk groups for HIV infection (ATKINSON et al. 1988), the presence of mood disturbance does not appear to confound interpretation of neuropsychological test results in these populations.

4 Conclusions

There has been gradual progress in our understanding of the quantitative and qualitative features and epidemiology of neurocognitive disorders associated with HIV infection. Part of the progress has been permitted by the development and application of more rigorous and reliable techniques for diagnosis. Here, we recommend that a distinction be made between three commonly occurring disorders: mild neurocognitive disorder, HIV-associated dementia, and delirium associated with HIV infection. By demarcating these conditions, we can avoid mistakes of the past which have led, among other things, to widely ranging estimates of dementia prevalence in HIV-infected persons.

It is evident that the rate of AIDS dementia is significantly lower than was originally estimated. Whether this reflects actual decline in dementia, associated perhaps with zidovudine treatment or other medical improvements, or whether what has actually disappeared is the delirium associated with HIV, due largely to other causes which are now better treated, is hard to say. Be that as it may, it appears that about 5%–7% of persons with AIDS will develop dementia on an annual basis. The rate of mild neurocognitive disorders among those with AIDS is probably in the order of 50%, based on review of all available neuropsychological literature. The rate of impairment among asymptomatic HIV carriers remains a matter of controversy, but the median rate based on review of 30 studies is 35% as compared to 12% for seronegative controls.

The features of neurocognitive impairment strongly suggest a subcortical pattern, but it is not exclusively so. Future research needs to consider how currently available techniques of MRI morphometry or functional brain imaging techniques might shed light on progressive regional anatomic changes in relation to specific neuropsychological test data. By emphasizing greater specificity both in cognitive assessment and in imaging, it may be possible to determine whether currently available or newly developed treatments actually benefit these CNS complications.

Future research on neurocognitive impairment must address itself to the changing face of the AIDS epidemic. Increasingly, underserved minorities, women, and persons of color are at risk for infection. These are often exactly the groups for whom adequate norms for neuropsychological tests do not exist. Such normative studies are urgently needed and should be combined with neuropsychological assessments of seropositive minorities, and related to measures of socioeconomic status and acculturation.

The real life implications of mild neurocognitive impairment are also in urgent need of definition. The diagnosis of mild neurocognitive *disorder* (minor cognitive motor disorder) requires some effect of neuropsychological impairment on day-to-day functioning. However, there is almost no research on occupational or other social functioning in relationship to neuropsychological impairment. In one exception to this rule, HEATON et al. (1994) have demonstrated that the rate of unemployment is twice as high among neuropsychologically impaired HIV+ persons as those who are unimpaired, despite the fact that the cases studied were not medically advanced. More data are needed linking neuropsychological status to more specific indicators of work efficiency as well as health-related functioning.

The availability of such improved outcomes measures can then strengthen future clinical trials aimed at neuroprotection or neuroremediation. Such future trials should ideally deemphasize end stage HIV dementia, since the likelihood of gaining significant benefit here is not strong. Rather, future trials should focus on early and middle phases of disease and on those suffering from mild neurocognitive impairment, since it is in these instances that we can hope to achieve real benefits in terms of improved life quality for a prolonged period of time. Trials such as the recently completely NIMH Peptide T trial serve as a good model in this regard (the results remained blinded at the time of this writing). In the future, it is hoped that emphasis for neuroremediation trials will be given to those antiretrovirals that cross the blood-brain barrier best (e.g., d4T), or other agents that may be suggested by current theories on the pathogenesis of mild cognitive disorder. For example, if we are to follow the argument that release of certain cytokines by activated macrophages may impair CNS function (GENIS et al. 1992), then it may be timely to consider trials of agents such as pentoxifylline (a putative anti-tumor necrosis factor α compound) to determine its possible benefits. If we postulate that release of excitotoxins, such as quinolinic acid, may be important, then agents which interfere with such excitotoxic injury (e.g., NMDA antagonists, calcium channel blockers) deserve further scrutiny (LIPTON 1992).

Fundamental to continued progress in all of these issues is the requirement that neurocognitive assessment be performed in a standardized manner that balances comprehensiveness and specificity against realistic time and effort requirements. In this regard, an ensemble of assessments recommended by the National Institute of Mental Health group (BUTTERS et al. 1990) has much to commend it. Just as in other parts of medicine, in which standardization of laboratory procedures has been a prerequisite to proper definition of disease indicators, so those working in neurology and neuropsychology will have to come to an agreement on common methodologies of neurocognitive assessment in order to gain further progress.

References

American Psychiatric Association (1980) Diagnostic and statistical manual of mental disorders, 3rd edn. American Psychiatric Press, Washington DC

Atkinson JH, Grant I, Kennedy CJ, Richman DD, Spector SA, McCutchan JA (1988) Prevalence of psychiatric disorders among men infected with human immunodeficiency virus: a controlled study. Arch Gen Psychiatry 45: 859–864

Ayers M, Newell T, Beardslee D, Mills CM, Abrams DI (1986) Neuropsychological deficits associated with AIDS retrovirus infection. Poster presentation at the International Conference on AIDS, Paris, June 23, 1986, p 161 (Poster 219)

Boccellari AA, Dilley JW, Chambers DB, Yingling CD, Tauber MA, Moss AR, Osmond DH (1993) Immune function and neuropsychological performance in HIV-1-infected homosexual men. J Acquir Immune Defic Syndr 6: 592–601

Bornstein RA, Nasrallah HA, Para M, Fass R, Whitacre C, Rice R (1991) Rate of CD4 decline and neuropsychological performance in HIV infection. Arch Neurol 48: 704–707

Bornstein RA, Nasrallah HA, Para MF, Whitacre CC, Rosenberger P, Fass RJ (1993) Neuropsychological performance in symptomatic and asymptomatic HIV infection. AIDS 7: 519–524

Breitbart W, Marotta RF, Call P (1988) AIDS and neuroleptic malignant syndrome. Lancet 2: 1488–1489

Britton CB, Marquardt MD, Koppel B, Garvey G, Miller JR (1982) Neurological complications of the gay immunosuppressed syndrome: clinical and pathological features (abstract). Ann Neurol 12: 80

Brouwers P, Moss H, Wolters P, Eddy J, Balis F, Poplack DG, Pizzo PA (1990) Effect of continuous-infusion zidovudine therapy on neuropsychologic functioning in children with symptomatic human immunodeficiency virus infection. J Pediatr 117: 980–985

Butters N, Granholm E, Salmon DP, Grant I, Wolfe J (1987) Episodic, semantic, and procedural memory: some comparisons of Alzheimer and Huntington disease patients. In: Terry RD (ed) Aging and the brain. Raven, New York, pp 63–87

Butters N, Grant I, Haxby J, Judd LL, Martin A, McClelland J, Pequegnat W, Schacter D, Stover E (1990) Assessment of AIDS-related cognitive changes: recommendations of the NIMH workshop on neuropsychological assessment approaches. J Clin Exp Neuropsychol 12: 963–978

Collier AC, Marra C, Coombs RW, Claypoole K, Cohen W, Longstreth WT, Townes BD, Maravilla KR, Critchlow C, Murphy VL, Handsfield HH (1992) Central nervous system manifestations of human immunodeficiency virus infection without AIDS. J Acquir Immune Defic Syndr 5: 229–241

Day JJ, Grant I, Atkinson JH, Brysk LT, McCutchan JA, Hesselink JR, Heaton RK, Weinrich JD, Spector SA, Richman DD (1992) Incidence of AIDS dementia in a two year follow-up of AIDS and ARC patients on an initial phase II AZT placebo-controlled study: San Diego cohort. J Neuropsychiatry Clin Neurosci 4: 15–20

de la Monte SM, Schooley RT, Hirsch MS, Richardson EP (1987) Subacute encephalomyelitis of AIDS and its relation to HTLV-III infection. Neurology 37: 562–569

Delis DC, Kramer JH, Kaplan E, Ober BA (1987) California verbal learning test: research edition. Psychological Corporation, New York

Dunbar N, Perdices M, Grunseit A, Cooper DA (1992) Changes in neuropsychological performance of AIDS-related complex patients who progress to AIDS. AIDS 6: 691–700

Dupont RM, Lamoureux G, Grant I, Lehr P, Dixon L, Golshan S, Chandler JL, Halpern S, the HNRC Group (to be published) SPECT imaging in HIV infection. Psychiatr Res

Genis P, Jett M, Bernton EW, Boyle T, Gelbard HA, Dzenko K, Keane RW, Resnick L, Mizrachi Y, Volsky DJ, Epstein LG, Gendelman HE (1992) Cytokines and arachidonic metabolites produced during human immunodeficiency virus (HIV)-infected macrophage-astroglia interactions: Implications for the neuropathogenesis of HIV disease. J Exp Med 176: 1703–1718

Grant I (1987) Alcohol and the brain: neuropsychological correlates. J Consult Clin Psychol 55: 310–324

Grant I (1990) The neuropsychiatry of human immunodeficiency virus. Semin Neurol 3: 267–275

Grant I, Atkinson JH (1990a) Neurogenic and psychogenic behavioral correlates of HIV infection. In: Waksman BH (ed) Immunologic mechanisms in neurologic and psychiatric disease. vol 68. Raven, New York, pp 291–304

Grant I, Atkinson JH (1990b) The evolution of neurobehavioral complications of HIV infection. Psychol Med 20: 747–754

Grant I, Atkinson JH (1995) Psychobiology of HIV infection. In: Kaplan HI, Sadock BJ (eds) Comprehensive textbook of psychiatry, vol VI. Williams and Wilkins, Baltimore (in press)

Grant I, Heaton RK (1990) Human immunodeficiency virus-1 (HIV-1) and the brain. J Consult Clin Psychol 58: 22–30

Grant I, Martin A (1994) Neuropsychology of HIV infection. Oxford University Press, New York

Grant I, Adams KM, Carlin AS, Rennick P, Judd LL, Schooff K (1978) The collaborative neuropsychological study of polydrug users. Arch Gen Psychiatry 35: 1063 –1074

Grant I, Atkinson JH, Hesselink JR, Kennedy CJ, Richman DD, Spector SA, McCutchan JA (1987) Evidence for early central nervous system involvement in the acquired immunodeficiency syndrome (AIDS) and other human immunodeficiency virus (HIV) infections: studies with neuropsychologic testing and magnetic resonance imaging. Ann Intern Med 107: 828–836

Grant I, Caun K, Kingsley DPE, Winer J, Trimble MR, Pinching AJ (1992) Neuropsychological and NMR abnormalities in HIV infection. The St. Mary's-Queen Square study. Neuropsychiatry Neuropsychol Behav Neurol 5: 185–193

Grant I, Olshen RA, Atkinson JH, Heaton RK, Nelson J, McCutchan JA, Weinrich JD (1993) Depressed mood does not explain neuropsychological deficits in HIV-infected persons. Neuropsychology 7: 53–61

Handelsman L, Aronson M, Maurer G, Wiener J, Jacobson J, Bernstein D, Ness R, Herman S, Losonczy M, Song IS, Holloway K, Horvath T, Donnelly N, Hirschowitz J, Rowan AJ (1992) Neuropsychological and neurological manifestations of HIV-1 dementia in drug users. J Neuropsychiatry Clin Neurosci 4: 21–28

Heaton RK, Grant I, Mathews CG (1986) Differences in neuropsychological test performance associated with age, education and sex. In: Grant I, Adams KM (eds) Neuropsychological assessment of neuropsychiatric disorders. Oxford University Press, New York, pp 100–120

Heaton RK, Grant I, Matthews CG (1991) Comprehensive norms for an expanded Halstead-Reitan battery: demographic corrections, research findings, and clinical applications. Psychological Assessment Resources, Odessa

Heaton RK, Velin RA, McCutchan JA, Gulevich SJ, Atkinson JH, Wallace MR, Godfrey HPD, Kirson DA, Grant I, NHRC Group (1994) Neuropsychological impairment in human immunodeficiency virus-infection: Implications for employment. Psychosom Med 56: 8–17

Heaton RK, Velin RA, Atkinson JH, Gulevich SJ, McCutchan JA, Hesselink JR, Chandler JL, Grant I, HNRC Group (1995a) Neuropsychological impairment in an HIV-positive male cohort. In: Stein M, Baum A (eds) Perspectives on behavioral medicine. East Sussex, England (in press)

Heaton RK, Grant I, Butters N, White DA, Kirson D, Atkinson JH, McCutchan JA, Taylor MJ, Kelly MD, Ellis RJ, Wolfson T, Velin R, Marcotte TD, Hesselink JR, Jernigan TL, Chandler J, Wallace M, Abramson I, HNRC Group (1995b). The HNRC 500—Neuropsychology of HIV infection at different disease stages J Int Neuropsychol Soc 1: 231–251

Horowitz SL, Benson DF, Gottleib MS, Davos I, Bentson JR (1982) Neurological complications of gay-related immunodeficiency disorder (abstract). Ann Neurol 12: 80

Janssen RS, Stehr-Green J, Starcher T (1989a) Epidemiology of HIV encephalopathy in the United States. Abstracts, 5th International Conference on AIDS, p 50

Janssen RS, Saykin AJ, Cannon L, Campbell J, Pinsky PF, Hessol NA, O'Malley PM, Lifson AR, Doll LS, Rutherford GW, Kaplan JE (1986b) Neurological and neuropsychological manifestations of HIV-1 infection: association with AIDS-related complex but not asymptomatic HIV-1 infection. Ann Neurol 26: 592–600

Joffe RT, Rubinow DR, Squillace K, Lane CH, Duncan CC, Fauci AS (1986) Neuropsychiatric manifestations of the acquired immune deficiency syndrome (AIDS). Psychopharmacol Bull 22: 684–688

Klusman LE, Moulton JM, Hornbostel LK, Picano JJ, Beattie MT (1991) Neuropsychological abnormalities in asymptomatic HIV seropositive military personnel. J Neuropsychiatry Clin Neurosci 3: 422–428

Kure K, Weidenheim KM, Lyman WD, Dickson DW (1990) Morphology and distribution of HIV-1 gp41-positive microglia in subacute AIDS encephalitis. Neuropathologica 80: 393–400

Lipton SA (1992) Models of neuronal injury in AIDS: another role for the NMDA receptor? Trends Neurosci 15: 75–79

Lunn S, Skydsbjerg M, Schulsinger H, Parnas J, Pedersen C, Mathiesen L (1991) A preliminary report on the neuropsychologic sequelae of human immunodeficiency virus. Arch Gen Psychiatry 48: 139–142

Maj M (1990) Psychiatric aspects of HIV-1 infection and AIDS. Psychol Med 20: 547–563

Maj M, Satz P, Janssen R, Zaudig M, Starace F, D'Elia L, Sughondhabirom B, Mussa M, Naber D, Ndetei D, Schulte G, Sartorius N (1994) WHO neuropsychiatric AIDS study, cross-sectional phase II. Arch Gen Psychiatry 51: 51–61

Martin A (1994) HIV, cognition, and the basal ganglia. In: Grant I, Martin AM (eds) Neuropsychology of HIV infection. Oxford University Press, New York, pp 234–259

Martin A, Heyes MP, Salazar AM, Law WA, Williams J (1993) Impaired motor-skill learning, slowed reaction time, and elevated cerebrospinal fluid quinolinic acid in a subgroup of HIV-infected individuals. Neuropsychology 7: 149–157

Martin EM, Robertson LC, Edelstein HE, Jagust WJ, Sorenson DJ, San Giovanni D, Chirurgi VA (1992) Performance of patients with early HIV-1 infection on the Stroop task. J Clin Exp Neuropsychol 14: 857–868

Massman PJ, Delis DC, Butters N, Dupont RM, Gillin JC (1992) The subcortical dysfunction hypothesis of memory deficits in depression: neuropsychological validation in a subgroup of patients. J Clin Exp Neuropsychol 14: 687–706

McAllister RH, Herns MV, Harrison MJG, Newman SP, Connolly S, Fowler CJ, Fell M, Durrance P, Manji H, Kendall BE, Valentine AR, Weller IVD, Adler M (1992) Neurological and neuropsychological performance in HIV seropositive men without symptoms. J Neurol Neurosurg Psychiatry 55: 143–148

McArthur JC, Cohen BA, Selnes OA, Kumar AJ, Cooper K, McArthur JH, Soucy G, Cornblath DR, Chmiel JS, Wang MC, Starkey DL, Ginzburg H, Ostrow DG, Johnson RT, Phair JP, Polk BF (1989) Low prevalence of neurological and neuropsychological abnormalities in otherwise healthy HIV-1-infected individuals: results from the Multicenter AIDS Cohort Study. Ann Neurol 26: 601–611

McArthur JC, Hoover DR, Bacellar H, Miller EN, Cohen BA, Becker JT, Graham NMH, McArthur JH, Selnes OA, Jacobson LP, Visscher BR, Concha M, Saah A (1993) Dementia in AIDS patients: incidence and risk factors. Neurology 43: 2245–2252

McKinney RE, Maha MA, Connor EM, Feinberg J, Scott GB, Wulfsohn M, McIntosh K, Borkowsky W, Modlin JF, Weintrub P, O'Donnell K, Gelber RD, Rogers GK, Lehrman SN, Wilfert CM, Protocol 043 Study Group (1991) A multicenter trial of oral zidovudine in children with advanced human immunodeficiency virus disease. N Engl J Med 324: 1018–1025

Miller EN, Selnes OA, McArthur JC, Satz P, Becker JT, Cohen BA, Sheridan K, Machado AM, van Gorp WG, Visscher B (1990) Neuropsychological performance in HIV-1-infected homosexual men: The Multicenter AIDS Cohort Study (MACS). Neurology 40: 197–203

Navia BA, Jordan BD, Price RW (1986) The AIDS dementia complex: I. Clinical features. Ann Neurol 19: 517–524

Nordic Medical Research Council's HIV Therapy Group (1992) Double blind, dose-response study of zidovudine in AIDS and advanced HIV infection. BMJ 304: 13–17

Peavy G, Jacobs D, Salmon DP, Butters N, Delis DC, Taylor M, Massman P, Stout JC, Heindel WC, Kirson D, Atkinson JH, Chandler JL, Grant I, HNRC Group (1994) Verbal memory performance of patients with human immunodeficiency virus infection: evidence of subcortical dysfunction. J Clin Exp Neuropsychol 16: 508–523

Pizzo PA, Eddy J, Falloon J, Balis F, Murphy R, Moss H, Wolters P, Brouwers P, Jarosinski P, Rubin M, Broder S, Yarchoan R, Burnetti A, Maha M, Nusinoff-Lehrman S, Poplack D (1988) Effect of continuous infusion of zidovudine (AZT) in children with symptomatic HIV infection. N Engl J Med 319: 889–896

Portegies P, de Gans J, Lange JM, Derix MMA, Speelman H, Bakker M, Danner SA, Goudsmit J (1989) Declining incidence of AIDS dementia complex after introduction of zidovudine treatment (published erratum appears in BMJ 299: 1141, 1989). BMJ 299: 819–821

Price RW, Brew BJ (1988) The AIDS dementia complex. J Infect Dis 158: 1079–1083

Pumarola-Sune T, Navia BA, Cordon-Cardo C, Cho ES, Price RW (1987) HIV antigen in the brains of patients with the AIDS dementia complex. Ann Neurol 21: 490–496

Reed RJ, Grant I (1990) The long-term neurobehavioral consequences of substance abuse: conceptual and methodological challenges for future research. NIDA Res Monogr 101: 10–56

Reinvang I, Froland SS, Skripeland V (1991) Prevalence of neuropsychological deficit in HIV infection. Incipient signs of AIDS dementia complex in patients with AIDS. Acta Neurol Scand 83: 289–293

Riccio M, Pugh K, Jadresic D, Burgess A, Thompson C, Wilson B, Lovett E, Baldeweg T, Hawkins DA, Catalan J (1993) Neuropsychiatric aspects of HIV-infection in gay men: controlled investigation of psychiatric, neuropsychological and neurological status. J Psychosom Res 37: 819–830

Rottenberg DA, Moeller JR, Strother SC, Sidtis JJ, Navia BA, Dhawan V, Ginos JZ, Price RW (1987) The metabolic pathology of the AIDS dementia complex. Ann Neurol 22: 700–706

Satz P (1993) Brain reserve capacity on symptom onset after brain injury: a formulation and review of evidence for threshold theory. Neuropsychology 7: 273–295

Satz P, Morgenstern HJ, Miller EN, Selnes OA, McArthur JC, Cohen BA, Wesch J, Becker JT, Jacobson L, D'Elia LF, van Gorp W, Visscher B (1993) Low education as a possible risk factor for cognitive abnormalities in HIV-1: findings from the Multicenter AIDS Cohort Study (MACS). J Acquir Immune Defic Syndr 6: 503–511

Schmitt FA, Bigley JW, McKinnis R, Logue PE, Evans RW, Deucker JL, AZT Collaborative Working Group (1988) Neuropsychological outcome of zidovudine (AZT) treatment of patients with AIDS and AIDS-related complex. N Engl J Med 319: 1573–1578

Selnes OA, Miller E, McArthur JC, Gordon B, Munoz A, Sheridan K, Fox R, Saah AJ (1990) HIV-1 infection: no evidence of cognitive decline during the asymptomatic stages. Neurology 40: 204–208

Snider WD, Simpson DM, Nielsen S, Gold JW, Metroka CE, Posner JB (1983) Neurological complications of acquired immune deficiency syndrome: analysis of 50 patients. Ann Neurol 14: 403–418

Stern Y, Marder K, Bell K, Chen J, Dooneief G, Goldstein S, Mindry D, Richards M, Sano M, Williams J, Gorman J, Ehrhardt A, Mayeux R (1991) Multidisciplinary baseline assessment of homosexual men with and without human immunodeficiency virus infection. III. Neurologic and neuropsychological findings. Arch Gen Psychiatry 48: 131–138

Tross S, Price RW, Navia B, Thaler HT, Gold J, Sidtis JJ (1988) Neuropsychological characterization of the AIDS dementia complex: a preliminary report. AIDS 2: 81–88

van Gorp WG, Mandelkern M, Gee M, Hinken CH, Stern CE, Paz D, Dixon W, Evans G, Flynn F, Frederick CJ, Ropchan J, Blahd W (1992) Cerebral metabolic dysfunction in AIDS: findings in a sample with and without dementia. J Neuropsychiatry Clin Neurosci 4: 280–287

Velin RA, Grant I, Heaton RK, McCutchan JA, Chandler J (1992) Effects of advancing age on neurocognitive functioning in HIV-infection. Symposium presentation at the 45th annual scientific meeting of the Gerontological Society of America, Washington DC, Nov 22, 1992

White DA, Heaton RK, Monsch AU, HNRC Group (1995) Neuropsychological studies of asymptomatic human immunodeficiency virus-type 1 infected individuals. J Int Neuropsychol Soc 1: 304–315

Working Group of the American Academy of Neurology AIDS Task Force (1991) Nomenclature and research case definitions for neurologic manifestations of human immunodeficiency virus-type 1 (HIV-1) infection. Neurology 41: 778–785

World Health Organization (1988) Report of the consultation on the neuropsychiatric aspects on HIV infection. Paper presented at the Global Programme on AIDS, Geneva, Switzerland

AIDS Dementia Complex and HIV-1 Brain Infection: A Pathogenetic Framework for Treatment and Evaluation

R.W. Price

1 Introduction

This review focuses on three aspects of prevention and treatment of the AIDS dementia complex (ADC) and central nervous system (CNS) human immunodeficiency virus type one (HIV-1) infection: (1) a pathogenetic framework for classification and development of therapies, (2) derivative therapeutic strategies and (3) evaluation methodologies. While preventing initial exposure and systemic HIV-1 infection is clearly the most important approach to eliminating the morbidity and mortality of AIDS and its diverse manifestations, this review considers only approaches to treatment of neurologically symptomatic patients and prevention of neurological diseases in those already infected, so called secondary prophylaxis.

Department of Neurology, University of Minnesota Health Science Center, Box 295, 420 Delaware St SE, Minneapolis, MN 55455-0323, USA

In discussing therapy in these therapeutic and prophylactic settings, I will maintain a separation between ADC, which refers to a clinical neurological syndrome, on the one hand, and HIV-1 infection of the CNS, which describes a pathobiological process, on the other. While neurological dysfunction and brain infection are importantly linked, this dichotomy is useful in considerations of pathogenesis, therapy and evaluation. By way of general background, I will first consider these separate concepts and their linkage.

1.1 AIDS Dementia Complex

AIDS dementia complex is one of several terms used to designate a clinical syndrome complicating HIV-1 infection and affecting cognition, motor performance and behavior in a characteristic pattern (Navia et al. 1986b; Price and Brew 1988). The syndrome is classified among the subcortical dementias because of clinical similarities to other disorders, including hydrocephalus and Huntington's disease, in which dysfunction of deeper brain structures rather than the cortex appears to predominate (Benson 1987; Navia 1990). In the cognitive sphere, salient aspects of ADC include impairment in attention and concentration, and in the speed and agility of dealing with complex or unrelated sets. This also parallels the motor impairment characterized early by slowing of rapid and fine movements and later, in severe cases, by progression to more frank incoordination and weakness. Referring to the combined alteration of thinking and movement as involving "psychomotor slowing" is descriptively apt and points to major involvement of diencephalic motor systems that are also involved in cognitive agility. This is in contrast to the cortical dementias such as Alzheimer's or Creutzfeldt-Jakob diseases that are accompanied by amnesia, aphasia, apraxia and other focal cortical deficits. Behavioral abnormalities in ADC patients follow this same pattern and are most notable for blunting of animation, interest and volition. While sharing a commonality of features, ADC varies from patient to patient in several ways, including: clinical severity and course, the relative prominence of particular symptoms and signs, pathological substrate, and prevalence even among those who are severely immunosuppressed with low blood $CD4^+$ T lymphocyte counts (Navia et al. 1986b; McArthur et al. 1994).

1.2 HIV-1 Infection in the Central Nervous System

A number of observations have contributed to defining the course of CNS HIV-1 infection. It has been known for some time that the CNS, or at least the cerebrospinal fluid (CSF) surrounding the CNS, is frequently exposed to HIV-1 early in the course of systemic infection. Indeed, such early exposure is likely the rule. Evidence for this includes detection of early local immune response (including intrathecal antibody production) and more direct identification of the virus and its components in the CSF (Appleman et al. 1988; Buffet et al. 1991;

ELOVAARA et al. 1987; MCARTHUR et al. 1988; RESNICK et al. 1988). One case report also documents HIV-1 in the CNS shortly after iatrogenic virus inoculation (DAVIS et al. 1992). The promiscuous penetration of lymphocytes into the CNS after immune activation (HICKEY 1991) may underlie the early and persistent traffic into brain and choroid plexus of HIV-1-infected blood cells. However, at least in the short term, this early exposure is seemingly almost always benign in that no clinical neurological disease ensues for many years, with the exception of uncommon cases of aseptic meningitis or very rare instances of acute encephalitis in the context of the seroconversion-related illness (BREW et al. 1989b).

Nonetheless, persistent virus and host reactions in CSF indicate a continued biological process, even if the effects are subclinical. A major open question remains whether this early seeding of the CNS has long-term sequelae, either by producing early mild CNS dysfunction or in introducing virus that remains latent or indolent only to reactivate later and cause more overt disease, perhaps as it undergoes mutation and selection to replicate locally. However, one can also speculate that this early infection of the brain is far overshadowed by subsequent continuous viral exposure throughout the course of systemic infection, particularly when intense viremia reemerges late in disease in the setting of advanced immunosuppression. It is in this late stage that some patients clearly suffer overt HIV-1 encephalitis with local replication in brain macrophages and microglia, very often with characteristic histopathological changes of multinucleated cell encephalitis (BUDKA 1991; NAVIA 1986a; ROSENBLUM 1990; VAZEUX et al. 1987). This late brain infection has been shown to involve selected variants of the quasispecies that populates and "drifts" within each infected individual; these variants exhibit properties of macrophage tropism and limited heterogeneity (LI et al. 1991; LIU et al. 1990; O'BRIEN 1994; O'BRIEN et al. 1990; PANG et al. 1991; WESTERVELT et al. 1992). An important question regarding these variants relates to the timing and location of their origin and genetic change. It is uncertain whether the progenitors of these neuropathic strains, in fact, first populate the brain early in infection and then evolve in situ or whether they undergo mutation and selection outside of the brain only to enter and replicate later when conditions are favorable (POWER et al. 1994). If both of these pathways to brain infection are taken, which is more important?

1.3 Links Between AIDS Dementia Complex and Central Nervous System HIV-1 Infection

While most investigators feel that HIV-1 brain infection causes ADC, the connection between virus and disease is not fully understood (PRICE et al. 1988, 1991; PRICE 1994). Tab. 1 lists both some of the observations that suggest that infection is directly causative and findings that challenge this role, at least in the same manner as typical neurotropic viruses cause other encephalitides. These observations and their interpretation have been discussed elsewhere (PRICE 1994), and detailed

Table 1. Some observations bearing on the causative role of HIV-1 brain infection in AIDS dementia complex

Supporting the role of CNS HIV-1 infection in causing ADC
ADC is unique to HIV-1 infection and not reported in other states of immunosuppression.
Animal lentiviruses related to HIV-1, including visna and simian immunodeficiency virus, infect the CNS and cause neurological dysfunction.
HIV-1 infects the brain, at times intensely, and some studies suggest a correlation between the CNS "virus load" and neuronal changes, particularly in severe cases.
Severe brain infection involves strains of the virus that share macrophage tropism and perhaps other genetic and phenotypic features.
No alternative pathogen has been consistently correlated with ADC.
Neurotoxicity has been observed both in vitro and in vivo (including in a transgenic mouse model) related to HIV-1 gene products, most notably gp120.
Treatment with the antiretroviral nucleoside zidovudine can ameliorate and probably prevent ADC.
Challenging the role of CNS HIV-1 infection in causing ADC
Productive HIV-1 infection of the brain is confined (principally or exclusively) to cells of bone marrow origin, including monocyte-macrophages and microglia, rather than the "functional" cells of the brain, i.e., neurons, oligodendrocytes and astrocytes, of neuroectodermal lineage.
Neither multinucleated cell encephalitis nor immunocytochemical evidence of infection is invariably present in ADC patients, particularly in milder cases, and even when present may seen "less intense" than the extent of clinical dysfunction.
Immune activation, as revealed in CSF and brain, may correlate better with ADC severity than the CNS virus load

ADC, AIDS dementia complex.

discussion will not be repeated here. In the aggregate, they suggest that HIV-1 is responsible for ADC, but that pathogenesis of brain injury involves indirect effects of this infection in which host factors serve as important proximate mediators of pathology. These observations underlie the effort to develop models describing a more complex and circuitous relation between virus and disease.

2 Pathogenetic Models

As a first step in approaching prevention and treatment of ADC, it is useful to first consider a pathogenetic framework of the condition, and then, based on this framework, to define a set of therapeutic targets. Taking this approach, we have examined models at three levels: a general *systems level, a functional cell level* and a *molecular level* (Price 1994; Spencer and Price 1992).

2.1 System Interactions in AIDS Dementia Complex: Agonists and Target

Broken down very simply, ADC can be considered to relate to three "major players" or systems. The first is HIV-1 itself or, considered more essentially, its genome, which populates the host and pathologically disturbs the homeostasis of the immune system. This is the primary or initiating agonist, the driver of pathological processes, including those that affect the brain. This complex genetic unit, packaged in the virus particle, is well adapted to persist and replicate in the host. In addition to integration into host DNA, its replication shares regulatory mechanisms with those controlling events in the host cell. It also undergoes change within the infected individual, leading to the emergence of mutant strains that are altered in a number of ways, including their susceptibility to immune clearance, pathogenicity and tissue tropism (Cheng-Mayer et al. 1991; Cheseboro et al. 1991; Koyangi et al. 1987; Li et al. 1991; O'Brien 1994; Steuler et al. 1992). The viral component is thus both well-adapted and dynamic, undergoing important evolution over time. In addition to disturbing the behavior and viability of infected immunocytes, HIV-1 codes for some gene products, including the external glycoprotein gp 120, with signal properties that can alter the physiology of uninfected cells (Dreyer et al. 1990; Lipton 1994).

The second major pathogenetic component is the immune system which can be considered the secondary or reacting agonist and relates to the virus in several fundamental ways. The immune system is the primary target of infection. $CD4^+$ T lymphocytes and macrophages are the principal cells harboring and replicating HIV-1, thereby supporting its persistence and spread within the host. Over time this infection, in turn, has important effects on the function of the immune system which together can be considered dysregulatory. Most notably, infection causes loss of $CD4^+$ T lymphocytes and a distinct profile of impaired cell-mediated defenses. These defenses are involved not only in protecting against other intracellular pathogens, but also against HIV-1 itself; thus, their loss allows systemic infection to flourish once again late in the course of disease. While the relative importance of different defenses in suppressing, although never eliminating, HIV-1 infection have yet to be precisely defined (and need not be considered in detail here), suppression of the viremia associated with primary infection (Clark et al. 1991; Daar et al. 1991) provides clear evidence of the existence of such antiviral defenses, likely involving multiple arms of the immune system. With time, as a result of the progressive crippling of immune defenses, important changes occur. The immune defenses can no longer suppress HIV-1 infection and viremia reemerges, in turn accelerating immune decline, leading to a terminal crescendo of immunosuppression and opportunistic infections. Presumably the damaged immune system has increasing difficulty in controlling new virus mutations generated by the genetically evolving quasispecies. Thus, virus and immune defenses, the initiating and reacting agonists, are locked in a dynamic relationship. The combined genetic drift of the virus and changes in immune reactions confer an important dimension to the relationship between the two agonists, the dimension of time.

Importantly, the effects of HIV-1 infection on the immune system include not only simple loss of cell-mediated defenses, but also alteration of the character of responses, including the profile of cytokine production (Pantaleo et al. 1993b). Thus, the immune system exhibits two important types of dysimmunity, it is less effective in protecting against certain organisms, including itself, and the character of its response changes. Both of these changes appear to be important in the pathogenesis of ADC.

The third pathogenetic component of ADC is the target organ, the CNS, in which the physiology of certain cell components are altered by both the virus and the immune system, or more precisely by the gene products of both the virus and the host cells involved in immune reactions. Both agonists, the driving virus and the responding immune system, have the potential to perturb target cell function and hence to inflict damage on the CNS.

2.2 Cell Interactions in AIDS Dementia Complex: Pathogenic Cell Networks

The three way interaction among HIV-1 and the immune system (including their evolving relationship over time) and the CNS target can be extended by constructing a model of underlying cellular interactions. Figure 1 presents a cell model of infection involving four functional cell roles (but without assigning definite cell types to each of these roles) (Price 1994). This model considers brain injury to involve indirect mechanisms of cell dysfunction and the immune system to be involved in both defense and injury.

Unlike prototypes of productive-lytic viral infections of the nervous system in which the predominating pathological force involves direct damage to target cells from within (such as polioviruses killing motoneurons or JC virus destroying

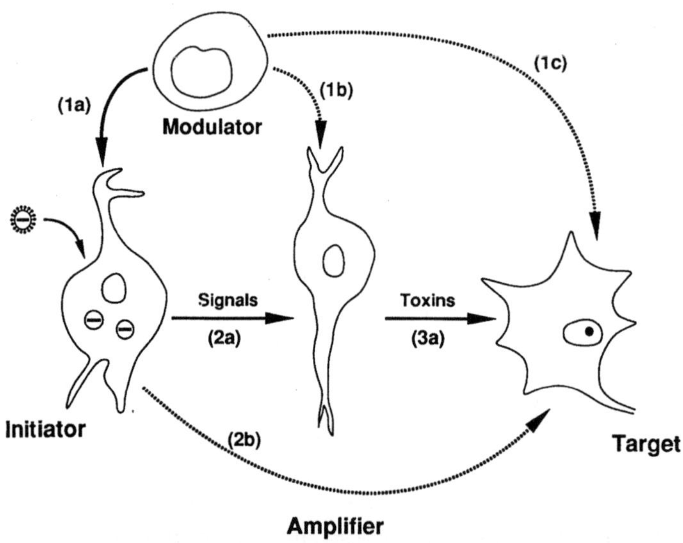

Fig. 1. The four-cell model of the pathogenesis of AIDS dementia complex. The cell types and their major interactions depicted by the *labeled arrows* are described in the text. The figure is a variant of that described elsewhere (Price 1994)

oligodendrocytes) as a sequela of viral gene expression in the infected target cell, HIV-1 does not appear to directly infect and kill neurons or other major functional elements derived from neuroectoderm. Rather HIV-1 infection appears to cause neuropathological damage from without by a process involving a pathogenic network of cellular interactions and chemical mediators. Among other features, this four-cell model attempts to take into account: (a) indirect mechanisms of injury, (b) an extent of injury more severe and more widely distributed than productive HIV-1 infection, (c) the importance of immune defenses in controlling infection early and releasing it later, and (d) the neurotoxic effects of cytokine circuits. In the context of this essay, the model has potential heuristic value in serving as a scaffold upon which to consider and organize a rational approach to prevention and therapy using several complimentary strategies. The model hypothesizes four functional cell types:

1. A *modulator cell* acts to control the extent of virus replication and gene expression by a variety of mechanisms, including eliminating infected cells or free virus and more directly modulating replication intracellularly (arrow 1a in Fig. 1). This cell represents the various immune reactions involved in defending against the virus. Early in the course of systemic infection these protective reactions are effective in containing HIV-1 replication and spread; later they are impaired. This cell's responses to the initiator cell and viral antigens might also result in the release of molecules that have secondary, bystander, effects upon the amplifier and target cells (arrows 1b,c).
2. The *initiator cell* is the virus-infected cell. This cell not only supports HIV-1 replication and spread, but is also the origin of signals that lead to neuropathology. These signals, in the simplest version of the model, are virus-coded gene products that are released into the surrounding tissue and act either on the amplifier (arrow 2a in Fig. 1) or more directly on the target (arrow 2b). Infection might also alter the profile of host gene expression in the initiator cell resulting in excessive production of cell-coded products (including cytokines) or underproduction of trophic factors (a negative toxic signal). It is within this cell that HIV-1 plays its primary agonist role.
3. The *amplifier cell* stands between the initiator and target cells. It transduces and amplifies the signals from the infected cell, producing, in turn, toxins that act on the target cells (arrow 3a in Fig. 1). While signals from the initiator could directly affect the target (simplifying the model), the amplifier is invoked to account for the seemingly excessive nervous system dysfunction in relation to the paucity of brain infection (Brew BJ, Rosenblum M, Cronin K, Price RW, unpublished; Glass et al. 1993). It also takes into account the importance and complexity of cytokine circuits, including their cascading release of multiple factors from uninfected cells. The amplifier might respond directly to virus-coded products that have specific signal functions or, less specifically, might respond to the byproducts of cell infection and injury. These reactions, along with alterations resulting from signals by the modulator cell reacting more specifically to particular antigens (arrows 1b,c), can be considered part of the

defensive response to infection and viewed in broad terms as immunopathological. Additionally, the profiles of modulator and amplifier cell reactions are likely perturbed as part of the dysregulation of cytokine responses alluded to earlier. A central aspect of many biological processes, including immune responses, is the involvement of regulatory feedback loops. In the case of late HIV-1 infection, one can hypothesize that such loops might be disturbed so that physiological inhibitory factors are damaged and what are normally phasic reactions become chronic and, as a result, more likely to cause tissue damage (PRICE et al. 1989).

4. The *target cell* is acted upon by toxic signals (or deprived of essential trophic substances) (arrows 3a, 2b and 1c in Fig. 1) leading to dysfunction or death and the characteristic symptoms and signs of ADC.

The model as illustrated only depicts the simplest, predominating interactions among the four cell types. Further variations could be added. The amplifier effect may well involve multiple reverberating cell-cell interactions. Likewise, damage to the target cell might alter the number and replicative state of the initiator cell, for example, by attracting and activating monocyte/macrophages and microglia; these are effects that reverse the direction of arrow 2b in Fig. 1. Indeed, one can envision feedback loops or other mechanisms in which signals and actions reverse the direction of each of the depicted arrows. However, for the present it is probably most useful to first consider only these elementary interactions as depicted. Table 2 lists some of the candidate cell types for each role along with some of the mechanisms possibly governing their interactions.

2.3 Molecular Basis of AIDS Dementia Complex: Pathogenetic Mediators

The molecular basis of ADC is now under increasing study and the subject of other contributions to this volume. Within the context of the cell model described

Table 2. Possible cell and molecular mediators involved in the four-cell model of AIDS dementia complex viral pathogenesis.

Functional type	Candidate cells	Examples of possible mechanisms and mediators
Modulator cell	CD8+,CD4+ T, cells; macrophage; B cell (antibody)	Direct cytotoxicity; IFN-γ; colony stimulating factors; TNF-α
Initiator cell	Macrophage-microglia	Viral signals: gp120, tat, nef
Amplifier cell	Microglia-macrophage; astrocyte; endothelial cell	Neurotoxins:TNF-α, quinolinic acid, interferons, eicosinoids, nitric oxide, unidentified NMDA agonists
Target cell	Neuron, Oligodendrocyte	Sites of action: NMDA receptors, Ca^{2+} channels

IFN, interferon; TNF, tumor necrosis factor.

above, understanding ADC pathogenesis at the molecular level entails defining the interactions among the different functional cell classes and the mechanisms controlling both the elaboration of and responses to the intercellular signals. This includes virus-coded and virus-regulated events as well as cell-coded mechanisms, the latter involving principally cytokines and other products of immune reactions. Some of these have been identified (Table 2), although defining their relative importance requires a great deal more work. However, as discussed below, adjunctive therapeutic strategies will rely on a more refined knowledge of the molecular basis of these interactions.

3 Therapeutic Strategies

The cell model of ADC pathogenesis provides a framework for categorizing prophylactic and therapeutic strategies into three major types of interventions: *immunoenhancing*, *antiviral*, and *adjunctive* therapies with the last of these divided into subcategories of *immunosuppressive*, *neuroprotective*, and *compensatory* approaches (Table 3).

3.1 Immunoenhancement: Correcting Modulator-Initiator Cell Interactions

This approach to intervention aims at preserving, restoring or enhancing anti-HIV-1 immunity. If one of the critical components of ADC pathogenesis is the loss of immune defenses that control replication of the AIDS virus itself, one therapeutic strategy is to correct this deficit. Early in infection, HIV-1 is present in the CNS, but its replication in the brain is limited and there is little evidence of clinical or pathological abnormality. This relates, at least in part, to immune defenses acting within the CNS that control replication. In this sense, immunosuppression has a

Table 3. Classification of therapeutic strategies with some examples

Immunoenhancing therapy
Biological response modifiers and antiviral cytokines (e.g., interferons, interleukins)
Cell or gene transplantation
Secondary immunization (vaccine)
Antiviral therapy
Nucleoside anti-reverse transcriptase inhibitors (e.g., zidovudine)
Nonnucleoside reverse transcriptase inhibitors
Protease inhibitors
Adjunctive therapy
Immunosuppressive therapy: corticosteroids, pentoxifylline
Neuroprotective therapy: nimodipine, memantine
Compensatory therapy: lithium, neuroleptics, antidepressants, amphetamines

permissive effect on CNS HIV-1 infection and, hence, on ADC pathogenesis just as it does on the development of important opportunistic infections; indeed, HIV-1 brain infection can be regarded as an opportunistic CNS infection that systemic infection by the same agent has prepared the way for. Importantly, preservation of effective immune reactions should prevent or ameliorate ADC. For example, if a TH-1 or other type of immune response is protective and a TH-2 or another type is not, then attempts to "tilt the balance" of responses to the former might be advantageous in ADC as in other sequelae of HIV-1 infection (CLERICI et al. 1994). In fact, preservation of immunity is one of the ways in which systemic antiviral therapy might prevent or treat ADC, since immunosuppression is induced by HIV-1 infection itself. However, other strategies which primarily target the interaction between immune defenses and the virus-infected cell might be pursued independently, for example, by the use of biologic response modifiers, antiviral cytokines or immune reconstitution through transplantation or various molecular genetic techniques.

3.2 Antiviral Therapy: Inhibiting Virus Replication and Gene Expression in the Initiator Cell

Antiviral strategies provide the principal focus of current efforts to combat established systemic infection and have similarly been directed at CNS infection (FISCHL 1995). The overall rationale of the antiviral approach to prevention and treatment of ADC is therefore straightforward. Reducing the viral burden systemically and in the CNS should, in turn, down-regulate the secondary processes induced by infection. If active HIV-1 infection is the prime mover in ADC pathogenesis, interfering with viral replication and viral gene expression should interrupt the sequence of subsequent reactions that eventually lead to brain injury, whether involving viral or host gene products as the final toxic pathway. Of all the approaches to therapy for ADC on the immediate horizon, this is likely to have the biggest payoff.

While considerations of ADC therapy generally direct major attention to the effects of infection within the brain, treatment of systemic infection may be important for two reasons: (1) it is the ultimate source of brain seeding and (2) it might itself lead to release of substances into the bloodstream that either directly injure the brain or trigger secondary pathogenic reactions in the brain. Chronicity of exposure, in concert with an altered blood-brain barrier, might result in an accumulated neurotoxic effect ultimately caused by systemic infection, perhaps involving steps similar to those outlined in the model of brain infection.

With respect to HIV-1 infection within the brain, antiviral therapy is aimed at: (1) reducing virus entry and exposure, (2) inhibiting replication in the brain, and, at least theoretically, (3) eliminating or inactivating the viral genome in brain. The first of these involves reduction of the systemic viral burden as discussed above, but also conceivably inhibiting passage of virus or infected cells into the brain more selectively. The second requires the additional consideration of access of the antiviral treatment to the CNS through the blood-brain barrier and is considered below in the context of evaluating antiviral treatment effect. The third involves strategies presently unachievable, yet within the realm of possibility, of molecular attack on the latent proviral DNA. A discussion of the specific strategies to achieve these ends is beyond the scope of this contribution.

3.3 Adjunctive Therapy: Interrupting/Compensating for Initiator-Amplifier-Target Cell Interactions

Rather than targeting the primary processes of infection, adjunctive therapies are directed downstream at the secondary reactions that lead to neurological dysfunction. Adjunctive approaches can be usefully subdivided into: (1) immunosuppressive treatments modifying interactions of cells participating in immune responses, (2) neuroprotective therapies aimed at interrupting neurotoxic reactions, and (3) compensatory treatments augmenting surviving functions or correcting symptoms. Division into three subcategories has heuristic value because each of these parallels a set of strategies applied to other neurological diseases sharing pathogenetic mechanisms or therapeutic needs. These other conditions therefore hold lessons for ADC and, vice versa, ADC may provide lessons for them.

3.3.1 Immunosuppressive Therapy

The connections between the infected initiator cell and the target, particularly those mediated through the amplifier cell, for the most part can be considered immunological in a broad sense. They involve cells and mechanisms that partake in immune responses and in teleological terms are part of the defensive reactions to foreign invaders, both outside and within the CNS. However, in excess, either in concentration or exposure time, they may also be pathogenic, particularly to the delicate nervous system. Measures to decrease these responses can, then, be considered in this same way to be immunosuppressive.

Additionally, as was noted above, an important aspect of the immune involvement in HIV-1 is dysregulation of cytokine responses with a change in the profile of these responses over the course of systemic infection. For example, cytokine profiles in blood late in infection tend to favor viral replication. These cytokines may likewise be neurotoxic, particularly if released locally in the brain or if an altered blood-brain barrier allows their unaccustomed penetration. In small amounts or over limited time, these cytokines may cause negligible and fully reversible neurological disturbance, but in larger concentrations over a more protracted time course such physiological responses may be neuropathogenic. Hence, efforts to combat these responses, to deamplify reactions to infection or shorten exposure times may be therapeutically beneficial.

Some of these efforts, if broad and unspecific, may oppose the immuno-enhancing efforts needed to defend against HIV-1. Virus-promoting and virus-inhibiting immune responses are not likely to fully segregate into CNS-injuring and CNS-innocent responses. This is one of the difficulties of broad measures; they may constitute a "two-edged sword" with combined favorable and unfavorable effects. Therapies following this general strategy are, of course, not novel. In the context of AIDS, this general rationale underlies the use of corticosteroids in patients with *Pneumocystis carinii* pneumonia in which pulmonary dysfunction relates more to local inflammatory responses than to the offending parasite which is, nonetheless, the initiator (MONTANER JSG et al. 1990). Outside of the AIDS context, this rationale is also the basis for using corticosteroids along with antibiotics for treatment of bacterial meningitis (LEBEL et al. 1988). In both of these therapeutic settings exogenous antibiotics replace the protective host responses, allowing their partial suppression without giving the invading pathogen an undue advantage. Current efforts to apply this rationale to ADC include clinical trials of pentoxifylline and thalidomide, both of which block tumor necrosis factor (TNF)-α, in systemic infection as well as ADC (DEZUBE et al. 1993).

3.3.2 Neuroprotective Therapy

This class of adjuvant therapy centers on interrupting the signals that act upon the target cell (and particularly the neuron) and the processes within the target that lead to and mediate cell dysfunction. Thus, neuroprotective therapy seeks to interrupt the reception of neurotoxins as well as the secondary harmful processes within the target cell. This same approach is currently being very actively pursued in other neurological diseases and hence brings ADC treatment into a larger context that includes efforts to reduce brain damage in stroke or to arrest the progression of genetic and sporadic neurodegenerative diseases. Segregating these components of ADC pathogenesis highlights consideration of shared interventions, for example, of applying measures now being tested to reduce infarct size to the problem of treating ADC.

This overall neuroprotective approach underlies current efforts to treat ADC with nimodipine, a Ca^{2+} channel blocker, and memantine, a *N*-methyl-D-aspartate (NMDA) receptor blocker (LIPTON 1994). It also encompasses one of the rationales of treatment with peptide-T, based on the latter's putative neurotrophic property (LIPTON 1994). Additional neurotoxin inhibitors (e.g., nitric oxide synthetase inhibitors), neurotrophic factors and neuroprotective agents will likely similarly be proposed for ADC in the future (DAWSON et al. 1993; EPSTEIN and GENDELMAN 1993; PRICE 1994; PULLIAM et al. 1991; WESSELINGH et al. 1993).

3.3.3 Compensatory Therapy

This class of treatment is aimed at substituting for or overcoming the physiological disturbance caused by the underlying disease. It includes therapies that follow the paradigm of L-dopa usage in Parkinson's disease, anticonvulsants for seizure disorders and neuroleptics in psychoses. These treatments do not alter the disease process but mitigate the abnormal physiology and, as a result, improve symptoms and functional capacity. They are an important part of palliation of neurological dysfunction. Indeed, they include some of the most useful measures currently available to the neurologist.

4 Evaluation of Therapies

Just as the goals of treatment of ADC and CNS HIV-1 infection can be considered separately, so too can their evaluation, although the two clearly converge to the extent that infection and neurological dysfunction are linked, and most intimately when the therapeutic strategy is antiviral treatment. However, the evaluation methods vary, with the success of ADC therapy defined by neurological status and that of antiviral treatment most directly by laboratory measures of the viral burden.

4.1 Assessing Treatment

The goals of ADC therapy are defined by clinical outcome. Secondary prevention is directed at curtailing the development of ADC in infected individuals, while success in treating clinically symptomatic patients can be considered at two levels: arresting progression and, more ambitiously, actually reversing the clinical deficits. At present, efficacy is judged principally using clinical measures of neurological function with some additional assistance from clinical neuroimaging studies and, occasionally, clinical electrophysiology.

4.1.1 Clinical Assessments of Neurological Function

Clinical evaluations are the mainstay of assessing ADC, since this is a syndrome without a specific laboratory marker to either confirm the diagnosis or accurately follow therapy. In practice, clinical assessments for ADC are of two complimentary types: qualitative bedside neurological examination and quantitative neurological testing. Each has an important place in evaluating the efficacy of prevention and therapy. The first is used principally to establish the presence or absence of an ADC diagnosis and its functional severity, while the second is used to assess change, either worsening (treatment failure) or improvement (success). Together these methods, as currently applied are in fact quite precise and useful.

Qualitative neurological evaluation of ADC involves a standardized bedside neurological history and examination with attention to functional status. This establishes the presence or absence of ADC and serves as the gold standard of diagnosis and outcome. It importantly encompasses both inclusion of characteristic features and exclusion of other disorders. In the research setting, accurate and consistent clinical diagnosis is critical for defining eligibility for trails of ADC therapy as well as for declaring ADC as an endpoint in more general therapeutic trials involving "AIDS-defining" clinical events as outcomes (PRICE and SIDTIS 1990a). In the clinical practice setting, diagnosis is essential in order to direct appropriate treatment and predict prognosis. For clinical trials we have developed standardized evaluations with more (macro evaluation) or less (micro evaluation) detail in the history and examination. Additionally, for both therapeutic trials and clinical practice, we have empirically defined an ADC staging scheme based on functional cognitive and motor status (PRICE and BREW 1988; PRICE and SIDTIS 1990b). This relatively simple staging provides a very useful common vocabulary.

Quantitative neurological evaluation of ADC uses the tools of neuropsychological testing to assess in a more precise fashion the cognitive and motor deficits characteristic of ADC. Again, complementary quantitative tests have been developed for macro- and microtesting to be used in different settings. The resultant scores can be used individually or normalized and averaged to obtain a combined neuropsychological Z-score (NPZ) (SIDTIS et al. 1993; SIDTIS 1994). The test components are chosen on the basis of their capacity to measure the characteristic deficits of ADC, their acceptability to subjects, repeatability and reproducibility without major confounding practice effect, and the simplicity of their administration. While such testing is not a replacement for clinical diagnosis and the results are not disease specific, occasionally it may be valuable in confirming whether the profile of abnormalities is typical of ADC. However, this type of quantitative testing is most useful for serial measurement, so that once a clinical diagnosis is established and confounding conditions eliminated, it provides an accurate and reproducible means of following either the natural history or treatment effects in an individual patient or in measuring changes in patient groups. This type of methodology has provided the principal way in which the efficacy of zidovudine has been demonstrated (PIZZO et al. 1988; SCHMITT et al. 1988; SIDTIS et al. 1993; YARCHOAN et al. 1987). In this sense quantitative neurological testing provides a close surrogate for standard neurological clinical evaluation, although it should always be interpreted in the context of accurate clinical diagnosis.

4.1.2 Neuroimaging

The role of clinical anatomic neuroimaging modalities, including magnetic resonance imaging (MRI) and computed tomography (CT) scanning, is chiefly for differential diagnosis, particularly for assessing the presence of alternative diagnosis such as primary CNS lymphoma, cerebral toxoplasmosis or progressive multifocal leukoencephalopathy (PML), rather than for specifically diagnosing ADC. However, these anatomic methods may also be more directly supportive of an ADC diagnosis by showing cerebral atrophy or increased water content in white matter or diencephalon (DAL PAN et al. 1992; GELMAN and GUINTO 1992; NAVIA et al. 1986b; POST et al. 1991). Functional neuroimaging methods, including positron emission tomography (PET), single photon computed emission tomography (SPECT) and MRI spectroscopy have not yet achieved a place in routine diagnosis although their utility is now being explored. Similarly, these modalities are being assessed with respect to their value in evaluating therapies, although they must be at present regarded as experimental and of uncertain value in providing surrogate measures of disease. The hope is that they might eventually provide more sensitive and objective markers for detecting early responses to therapy, thereby accelerating the assessment of new treatments.

4.1.3 Electrophysiology

Electrophysiological testing using electroencephalography (EEG), evoked potentials and other related methods are occasionally used in diagnosis, but also generally are of limited value. In selected cases they may be useful in differential diagnosis (e.g., in suggesting toxic or metabolic encephalopathy). However, like functional neuroimaging, their value in assessing therapy has not yet been shown to be widely applicable, being neither more sensitive nor more specific than clinical measures.

4.1.4 Pathological Assessment

The pathological findings in ADC are variable and their clinical implications somewhat controversial (Budka 1991; Price et al. 1991; Price 1994). For this reason, pathological diagnoses cannot be specifically translated to a clinical diagnosis or indicator of ADC severity. This, along with the fact that cerebral biopsy is not indicated for diagnosis or follow-up of ADC and the logistical difficulty of linking treatment protocols to autopsy, makes neuropathological findings of limited use in assessing treatment efficacy. While one of the pathologies associated with ADC, multinucleated-cell encephalitis, appears to be less common now than earlier in the epidemic and might reflect an effect of widespread use of zidovudine (Gray et al. 1993; Portegies et al. 1989; Vago et al. 1993), this reduction is perhaps better considered an indicator of an effect on CNS HIV-1 infection and will be considered below.

4.2 Evaluation of Antiviral Treatment

Assessment of the effectiveness of therapy on CNS HIV-1 infection involves strategies to measure the virus load in various compartments and, in the context of this review, particularly in the CNS. If infection and ADC are directly related and indeed HIV-1 is the prime mover in determining ADC, then the clinical evaluation methodologies discussed above are pertinent to antiviral therapeutic effect as well. Indeed, they are more important since the desired clinical result is prevention or amelioration of CNS dysfunction. However, not all treatment approaches involve suppression of virus infection and antiviral treatment may not always translate into an effect on clinical disease. Hence, it is worthwhile to consider how therapeutic effect on infection itself, and particularly on CNS infection, might be measured more directly. I will consider two overall approaches to assessing the viral load: the first is straightforward and involves measuring different components of the virus; the second involves studying markers of immune activation and is of less certain relevance.

4.2.1 Virological Studies

The antiviral strategy to prevent and treat ADC need to target both systemic and CNS infections, since the former gives rise to the latter and both may potentially contribute to CNS injury. While antiviral treatment will often have parallel effects in the two compartments, certain factors may lead to their divergence, generally with reduced efficacy in the CNS (Table 4). The most important factors contributing to such divergence relate to differences in susceptibility between systemic and CNS virus strains and to limitations in drug penetration of the blood-brain barrier. Among the aspects of infection that are of paramount importance in assessing treatment effect are: (1) quantitative measures of the virus load; (2) genotypic and phenotypic differences in virus strains isolated from different sources, and in particular, differences in antiviral resistance; and (3) the relation of fluid-phase (e.g., plasma or CSF) assays to tissue infection.

While assessment of antiviral therapy in the context of ADC and CNS infections needs to examine effects on both systemic and CNS virus infections, considerations of the latter are more central to the focus of this essay. However, because systemic infection has been so extensively studied as well as being more accessible for study, its methodology is more advanced than that which has been applied to the CNS and, hence, serves as a guide both conceptually and technically. Unfortunately, even in the case of systemic infection quantitative measurement of the virus burden has met with some difficulty, although several newer techniques appear to represent real and practical advance; hopefully, they will be similarly applied to CNS infection and its response to therapy. While simple culture isolation does not provide an index of the intensity of infection, quantitative culture of plasma has provided a clearer picture of the dynamic aspects of systemic infection and might be similarly useful in assessing treatment effect. However, because of the expense and difficulty of this technique, it is impractical for wide use and its utility in assessing CSF has not been tested. Measurement of p24 core antigen in CSF by immunoassay has been disappointing and is likely to

Table 4. Conditions leading to convergence or divergence of the effects of antiviral therapy on systemic and CNS infections and AIDS dementia complex

Parallel efficacy of systemic treatment on systemic and CNS infections should occur:
If systemic infection alone causes ADC (e.g., by blood-borne toxins) *or* if CNS infection causes ADC, *and either* CNS infection requires reseeding by systemic infection, *or* systemic and CNS strains share drug susceptibility profiles, *and* the drug penetrates the blood-brain barrier well enough to confer similarly effective systemic and CNS tissue levels.

Reduced efficacy of systemic treatment on CNS infection compared to systemic infection should occur:
If CNS infection alone causes ADC, *and* CNS infection continues independently from systemic infection, *and either* antiviral resistance of CNS strains diverges from that of systemic strains, *or* limited drug penetration across the blood barrier leads to inadequate CNS tissue levels.

ADC, AIDS dementia complex.

be of little use in assessing CNS infection; this relates to the relatively low concentration of p24 in CSF and the inconsistency of its detection in ADC patients. Indeed, this antigen, whether because of antibody binding or poor release from tissue (see below), is too often undetectable in CSF to serve as a consistent measure of treatment effect (BREW et al. 1994). Studies are now needed to see if newer quantitative techniques to measure HIV-1 nucleic acid, including quantitative PCR or branched chain DNA assays, when applied to CSF, will be useful in assessing the state of CNS infection and its response to therapy as it seems they are in blood.

It is clear that virus strains, or at least identifiable predominant strains, can differ across both the blood-CNS and blood-CSF barriers and between CNS and CSF with respect to genotype and phenotype (KOYANGI et al. 1987; O'BRIEN 1994). Studies are only beginning to address the relevance of such differences to therapy, including the possible clinical importance of resistant strains that may arise or remain sequestered in the CNS. If infection proceeds independently in the brain without substantial renewal from the blood, then these differences might be very important.

Central to assessing these quantitative and qualitative aspects of infection and treatment is the problem of sampling and the use of CSF to assess brain infection. This can be considered at two levels. The first is the practical difficulty of obtaining serial samples of CSF. Lumbar puncture is not accepted by patients in the same way as venipuncture chiefly because of the incidence and morbidity of postlumbar puncture headache. This understandably often lowers patients' willingness to participate in or comply with therapeutic trials requiring this procedure.

The second problem of CSF sampling is more fundamental and relates to how well measurements of lumbar subarachnoid fluid reflect the state of brain infection. This issue has its parallel in systemic infection also. Virus in whole blood or in its fluid-phase, plasma, may not accurately reflect tissue infections, including variations among different organs. The magnitude of this discrepancy is underscored by the observations of extensive lymph node infection during the clinically latent phase of infection (EMBERTSON et al. 1993; PANTALEO et al. 1993a). The case of the CSF is even more difficult than that of blood. The leptomeninges likely serve as a different replicative compartment than the brain. Its access from blood involves the choroid plexus with a different barrier from that of the brain parenchyma, and it is lined by a local population of tissue macrophages. This presumably explains why HIV-1 strains may differ between brain and CSF. Additionally, productive HIV-1 brain infection appears to preferentially infect deep brain structures, including basal ganglia and central white matter (KURE et al. 1990) far from the spinal subarachnoid space. Hence, what is detected in the CSF may poorly reflect infection in the brain.

This difficulty could be overcome by a noninvasive method of detecting brain infection, but none now exists. As discussed above, autopsy studies may detect treatment effects in patient groups, reflected in the detection of antigen or resultant pathology (e.g., multinucleated giant cell encephalitis) or even in more quantitative measurements of infection similar to those applied to blood.

However, this is clearly not practical for clinical trials. Further studies are therefore needed to examine correlations between CSF and brain with respect to virus type and amount and their respective responses to treatment, before drawing negative or positive conclusions about this issue. The capacity to assess the effects of antiviral and other treatments on brain infection itself, in addition to the clinical ADC functional outcomes discussed earlier, is a worthwhile objective. It might provide more direct and earlier answers regarding particular therapies in individual patients and might also help to test the hypothesis of ADC viral pathogenesis presented earlier.

4.2.2 Biochemical/Immunological Assessments

A second, less direct, way to possibly assess the viral load in response to antiviral therapy is founded on the pathogenetic scheme discussed earlier, or at least that component hypothesizing that immune activation is driven by HIV-1 infection. If this is the case, then measuring the levels of various cytokines that are up-regulated or cytokine markers that are similarly increased, such as neopterin and β2 microglobulin (Brew et al. 1989a, 1990, 1992), could be used to evaluate the effects of therapy. In fact, observations of decreasing CSF concentrations of neopterin, β2 microglobulin and quinolinic acid following zidovudine therapy are consistent with this hypothesis. While interpretation of these markers in CSF vis-a-vis brain infection is subject to the same caveats raised earlier regarding measurement of the virus and its components, some of the cytokine markers may both diffuse more widely and be more stable, thereby permitting CSF to more accurately reflect their synthesis in brain. While production of cytokines in response to infection may differ among patients, measurement of changes in individual patients in response to treatment may still be useful. Moreover, even if these cytokine responses in CSF reflect not only infection but also important host reactions that are involved in the pathogenesis of brain injury, their measurement might remain helpful in assessing treatments, including adjuvant-immunosuppressive treatments. While these relationships remain to be clearly defined and the interpretation of cytokine markers needs to be fortified, such measurements might prove very useful in following responses to treatment, including early responses before clinical or other changes are detectable. If this information proves of sufficient value as an early indicator of treatment effect, then even the risk of morbidity from repeated lumbar puncture would be acceptable.

5 Conclusions

ADC presents both a pathogenetic and therapeutic challenge. Its pathogenesis is complex and entails an evolving interplay of virus and host, and of various cell types and intra- and intercellular reactions. It is accompanied by considerable morbidity and at stage 2 level with a poor overall prognosis. Prevention and therapy, in parallel, raise important theoretical and practical issues. This chapter presents one view of pathogenesis and argues that while ADC and brain HIV-1 infection are intimately related, it is important to keep them separate in order to advance our understanding and to develop both diverse approaches to treatment and clear evaluation tools. It is hoped that these considerations will be useful in clarifying and stimulating the work that is needed to deal with this important complication of HIV-1 infection and AIDS. It is also hoped that this pathogenetic scheme might serve as a model for consideration of other infections and even some degenerative conditions afflicting the nervous system.

Acknowledgment. Our research program studying the AIDS Dementia Complex is supported by US Public Health Service Grant NS-25701.

References

Appleman ME, Marshall DW, Brey RL, Houk RW, Beatty DC, Winn RE, Melcher GP, Wise MG, Wumaya CV, Boswell RN (1988) Cerebrospinal fluid abnormalities in patients without AIDS who are seropositive for the human immunodeficiency virus. J Infect Dis 158: 193–199

Benson DF (1987) The spectrum of dementia: a comparison of the clinical features AIDS/dementia and dementia of the Alzheimer type. Alz Dis and Assoc Disorders 1(4): 217–220

Brew BJ, Paul MO, Nakajima G, Khan A, Gallardo H, Price RW (1994) Cerebrospinal fluid HIV-1 p24 antigen and culture: sensitivity and specificity for AIDS-dementia complex. J Neurol Neurosurg Psychiatry 57: 784–789

Brew BJ, Bhalla R, Paul M, Sidtis JJ, Keilp JJ, Sadler AE, Gallardo H, McArthur JC, Schwart MK, Price RW (1992) Cerebrospinal fluid B_2-microglobulin in patients with AIDS dementia complex: an expanded series including response to zidovudine treatment. AIDS 6: 461–465

Brew BJ, Bhalla RB, Paul M, Gallardo H, McArthur JC, Schwartz MK, Price RW (1990) Cerebrospinal fluid neopterin in human immunodeficiency virus type 1 infection. Ann Neurol 28: 556–560

Brew BJ, Bhalla RB, Fleisher M, Paul M, Khan A, Schwartz MK, Price RW (1989a) Cerebrospinal fluid B_2 microglobulin in patients infected with human immunodeficiency virus. Neurology 39: 830–834

Brew BJ, Perdices M, Darveniza P et al (1989b) The neurological features of early and "latent" human immunodeficiency virus infection. Aust NZ Med 19: 700–705

Budka H (1991) Neuropathology of HIV encephalitis. Brain Pathol 1: 163–175

Duffet R, Agut H, Chieze R, Katlama C, Bolgert F, Devillechabrolle A, Diquet B, Schuller E, Pierrot Deseilligny C, Gentilini M, Huraux J-M (1991) Virological markers in the cerebrospinal fluid from HIV-1 infected individuals. AIDS 5: 1419–1424

Cheng-Mayer C, Shioda T, Levy JA (1991) Host range, replicative, and cytopathic properties of human immunodeficiency virus type 1 are determined by very few amino acid changes in tat and gp120. J Virol 65(12): 6931–6941

Chesebro B, Nishio J, Perryman S, Cann A, O'Brien W, Chen ISY, Wehrly K (1991) Identification of human immunodeficiency virus envelope gene sequences influencing viral entry into CD4-positive HeLa cells, T-leukemia cells, and macrophages. J Virol 65(11): 5782–5789

Clark SJ, Saag MS, Decker WD, Campbell-Hill S, Roberson JL, Veldkamp MS, Kappes JC, Hahn BH and Shaw GM (1991) High titers of cytopathic virus in plasma of patients with symptomatic primary HIV-1 infection. N Engl Med 324: 954–960

Clerci M and Berzofsky JA (1994) Cellular immunity and cytokines in HIV infection. AIDS 8(suppl1): S175–S182

Daar ES, Mougkil T, Meyer RD, Ho DD (1991) Transient high levels of viremia in patients with primary human immunodeficiency virus type 1 infection. N Engl Med 324: 961–964

Dal Pan GJ, McArthur JH, Aylward E, Selnes OA, Nance-Sproson TE, Kumar AJ, Mellits ED, McArthur JC (1992) Patterns of cerebral atrophy in HIV-1-infected individuals: results of a quantitative MRI analysis. Neurology 42: 2125–2130

Davis LE, Hjelle BL, Miller VE, Plamer DL, Llewellyn AL, Merlin TL, Young SA, Mills RG, Wachsman W, Wiley CA (1992) Early brain invasion in iatrogenic human immunodeficiency virus infection. Neurology 42: 1736–1739

Dawson VL, Dawson TM, Uhl GR, Snyder SH (1993) Human immunodeficiency virus type 1 coat protein neurotoxicity mediated by nitric oxide in primary cortical cultures. Proc Natl Acad Sci USA 90: 3256–3259

Dezube BJ, Pardee AB, Chapman B, Beckett LA, Korvick WJ, Chiurco J, Kasdan P, Ahlers CM, Ecto LT, Curmpacker CS, NIAID AIDS Clinical Trials Group (1993) Pentoxifylline decreases tumor necrosis factor expression and serum triglycerides in patients with AIDS. J Acq Immun Defic Synd 6a: 7787–794

Dreyer EB, Kaiser PK, Offermann JT, Lipton SA (1990) HIV-1 coat protein neurotoxicity prevented by calcium channel antagonists. Science 248: 364–367

Elovaara I, Iivanainen M, Valle SL, Suni J, Tervo T, Lahdevirta J (1987) CSF protein and cellular profiles in various stages of HIV infection related to neurological manifestations. J Neurol Sci 78: 331–342

Embertson J, Zupancic M, Ribas JL, Burke A, Racz P, Tenner-Racz K, Haase A (1993) Massive covert infection of helper T lymphocytes and macrophages by HIV during the incubation period of AIDS. Nature 362: 359–362

Epstein LG, Gendelman HE (1993) Human immunodeficiency virus type 1 infection of the nervous system: pathogenetic mechanisms. Ann Neurol 33: 429–436

Fischl MA (1995) Treatment of HIV infection. In: Sande ME, Volberding PA (eds) Medical management of AIDS, 4th edn. Saunders, Philadelphia, pp 141–160

Gelman BB, Guinto FC Jr (1992) Morphometry, histopathology and tomography of cerebral atrophy in acquired immunodeficiency syndrome. Ann Neurol 31: 32–40

Glass JD, Wesselingh SL, Selnes OA and McArthur JC (1993) Clinical-neuropathological correlation in the HIV-1 associated dementia. Neurology 43: 2230–2237

Gray F, Belec L, Keohane C, De Truchis P, Clair B, Durigon M, Sobel A, Gherardi R (1994) Zidovudine therapy and HIV encephalitis: a 10-year neuropathological survey. AIDS 8: 489–493

Hickey WF (1991) Migration of hematogenous cells through the blood-brain barrier and the initiation of CNS inflammation. Brain Pathol 1: 97–105

Koyangi Y, Miles S, Mitsuyasu RT, Merrill JE, Vinters HV, Chen ISY (1987) Dual infection of the central nervous system by AIDS viruses with distinct cellular tropisms. Science 236: 819–822

Kure K, Weidenheim KM, Lyman WD, Dickson DW (1990) Morphology and distribution of HIV-1 gp41-positive microglia in subacute AIDS encephalitis. Pattern of involvement resembling a multisystem degeneration. Acta Neuropathol 80: 393–400

Lebel MH, Freij BJ, Syrongiannopoulos GA, Chrane DF, Hoyt MJ, Stewart SM, Kennard BD, Olsen KD, McCraken GH (1988) Dexamethasone therapy for bacterial meningitis: results of two double blind, placebo controlled trials. N Engl J Med 319: 964–971

Li Y, Kappes JC, Conway JA, Price RW, Shaw GM, Hahn BH (1991) Molecular characterization of human immunodeficiency virus type 1 cloned directly from uncultured brain tissue: identification of replication-competent and -defective viral genomes. J Virol 65(8): 3973–3985

Lipton SA (1994) Laboratory basis of novel therapeutic strategies to prevent HIV-related neuronal injury. In: Price RW, Perry SW (eds) HIV, AIDS and the brain. Raven, New York, pp 183–202

Liu, Z-Q, Wood C, Levy JA, Cheng-Meyer C (1990) The viral envelope gene is involved in macrophage tropism of a HIV-1 strain isolated from brain tissue. J Virol 64(12): 6148–6153

McArthur JC, Selnes OA, Glass JD, Hoover DR and Bacellar H (1994) HIV dementia, Incidence and risk factors. In: Price RW, Perry SW (eds) HIV, AIDS and the brain. Raven, New York, pp 251–272

McArthur JC, Cohen BA, Farzadegan H, Cornblath DR, Selnes OA, Ostrow D, Johnson RT, Phair J, Polk BF (1988) Cerebrospinal fluid abnormalities in homosexual men with and without neuropsychiatric findings. Ann Neurol 23(suppl): S34–S37

Montaner JSG, Lawson LM, Letitt N, Belzberg A, Schechter MT, Reudy J (1990) Oral corticosteroids prevent early deterioration in patients with moderately severe AIDS-related *Pneumocystis carinii* pneumonia. Ann Intern Med 113: 14–20

O'Brien WA, Koyangi Y, Namazie A, Zhao J-Q, Diagne A, Idler K, Zack JA, Chen ISY (1990) HIV-1 tropism for mononuclear phagocytes can be determined by regions of gp120 outside the CD4-binding domain. Nature 348: 69–73

O'Brien WA (1994) Genetic and biologic basis of HIV-1 neurotropism. In: Price RW, Perry SW (eds) HIV, AIDS and the brain. Raven, New York, pp 47–70

Navia BA (1990) The AIDS dementia complex. In: Cummings JL (ed) Subcortical dementia. Oxford University, New York, pp 181–198

Navia BA, Cho E-W, Petito CK, Price RW (1986a) The AIDS dementia complex: II. Neuropathology. Ann Neurol 19: 525–535

Navia BA, Jordan BD, Price RW (1986b) The AIDS dementia complex: I. Clinical features. Ann Neurol 19: 517–524
Pang S, Vinters HV, Akashi T, O'Brien WA, Chen ISY (1991) HIV-1 env sequence variation in brain tissue of patients with AIDS-related neurological disease. J Acq Immun Defic Syn 4: 1082–1092
Pantaleo G, Graziosi C, Demarest JF, Butini L, Montroni M, Fox CH, Orenstein JM, Kotler DP, Fauci AS (1993a) HIV infection is active and progressive in lymphoid tissue during the clinically latent stage of disease. Nature 362: 355–358
Pantaleo G, Graziosi C, Fauci AS (1993b) The immunopathogenesis of human immunodeficiency virus infection. N Engl J Med 328: 327–335
Pizzo PA, Eddy J, Falloon J, Balis FM, Murphy RF, Moss H, Wolters P, Brouwers P, Jarosinski P, Rubin M, Broder S, Yarchoan R, Brunetti A, Maha M, Nusinoff Lehrman S, Poplack DG (1988) Effect of continuous intravenous infusion of zidovudine (AZT) in children with symptomatic HIV infection. N Engl J Med 319: 889–896
Post MJD, Berger JR, Quencer RM (1991) Asymptomatic and neurologically symptomatic HIV- seropositive individuals. Prospective evaluation with clinical MR imaging. Radiology 178: 131–139
Portegies P, de Gans J, Lange JM, Derix MA, Speelman H, Bakker M, Danner SA, Goudsmit J (1989) Declining incidence of AIDS dementia complex after introduction of zidovudine treatment. Br Med J 299: 819–821
Price RW (1994) Understanding the AIDS dementia complex (ADC). The challenge of HIV and its effects on the central nervous system. In: Price RW, Perry SW (eds) HIV, AIDS and the brain. Raven, New York, pp 1–45
Price RW, Sidtis JJ, Brew BJ (1991) AIDS dementia complex and HIV-1 infection: a view from the clinic. Brain Pathology 1: 155–162
Price RW, Sidtis JJ (1990a) Evaluation of the AIDS dementia complex in clinical trials. J Acquir Immune Defic Syndr 3 (Suppl 2): S51–S60
Price RW, Sidtis JJ (1990b) Early HIV infection and the AIDS dementia complex. Neurology 40: 323–326
Price RW, Brew BJ, Rosenblum (1989) The AIDS dementia complex and HIV-1 infection: a pathogenetic model of virus-immune interaction. In: BH Waksman (ed) Immunologic mechanisms in neurologic and psychiatric disease. Raven, New York, pp 269–290
Price RW, Brew BJ (1988) The AIDS dementia complex. J Infect Dis 158: 1079–1083
Price RW, Brew BJ, Sidtis J, Rosenblum M, Scheck AC, Cleary P (1988) The brain in AIDS: central nervous system HIV-1 infection and AIDS dementia complex. Science 1988; 239: 586–592
Power C, McArthur JC, Johnson RT, Griffin DE, Glass JD, Perryman S, Chesebro (1994) J Virol 68: 4643–4649
Pulliam L, Herndier BG, Tang HM, McGrath MS (1991) Human immunodeficiency virus-infected macrophages produce soluble factors that cause histological and neurochemical alterations in cultured human brains. J Clin Invest 87: 503–512
Resnick L, Berger JR, Shapshak P, Tourtellotte WW (1988) Early penetration of the blood-brain-barrier by HIV. Nourology 38: 9–11
Rosenblum MK (1990) Infection of the central nervous system by the human immunodeficiency virus type 1: Morphology and relation to syndromes of progressive encephalopathy and myelopathy in patients with AIDS. Pathol Ann 25: 117–169
Schmitt FA, Bigley JW, McKinnis R, Logue PE, Evans RW, Drucker JL, AZT Collaborative working Group (1988) Neuropsychological outcome of zidovudine (AZT) treatment of patients with AIDS and AIDS-related complex. N Engl J Med 319: 1573–1578
Sidtis JJ (1994) Evaluation of the AIDS dementia complex in adults In Price RW and Perry SW (eds) HIV, AIDS and the brain. Raven, New York, pp 273–287
Sidtis JJ, Gatsonis C, Price RW, Singer EJ, Collier AC, Richman DD, HIrsch MS, Schaerf FW, Fischl MA, Kieburtz K, Simpson D, Koch MA, Feinberg J, and the AIDS Clinical Trials Group (1993) Zidovudine treatment of the AIDS dementia complex: results of a placebo-controlled trial. Ann Neurol 33: 343–349
Spencer DC, Price RW (1992) Human immunodeficiency virus and the central nervous system. Ann Rev Microbiol 46: 655–693
Steular H, Storch-Hagenlocher B, Wildemann B (1992) Distinct populations of human immunodeficiency virus type 1 in blood and cerebrospinal fluid. AIDS Res Hum Ret 8(1): 53–59
Vago L, Castagna A, Lazzarin A, Trabattoni G, Cinque P, Costanzi G (1993) Reduced frequency of HIV-infected brain lesions in AIDS patients treated with zidovudine. J Acq Immun Defic Synd 6: 42–45

Vazeux R, Brousse N, Jarry A et al. (1987) AIDS subacute encephalitis: identification of HIV-infected cells. Am J Pathol 126: 403–410

Wesselingh SL, Power C, Glass JD, Tyor WR, McArthur JC, Farber JM, Griffen JW, Griffen DE (1993) Intracerebral cytokine messenger RNA expression in acquired immunodeficiency syndrome dementia. Ann Neurol 33: 576–582

Westervelt P, Trowbridge DB, Epstein LG, Li Y, Hahn BH, Shaw GM, Price RW, Ratner L (1992) Macrophage tropism determinants of human immunodeficiency virus type 1 in vivo. J Virol 66: 2577–2582

Yarchoan R, Berg G, Brouwers P et al. (1987) Response of human immunodeficiency virus associated neurological disease to 3' -azido-3'- deoxythymidine. Lancet 1: 132–135

Quantitative Neuropathologic Assessment of HIV-1 Encephalitis

C.A. WILEY

1 Introduction

Since the beginning of the AIDS epidemic, morbidity and mortality associated with opportunistic infections of the nervous system have been well appreciated (LEVY et al. 1985; ANDERS et al. 1986). In 1985, Dr. GEORGE SHAW and colleagues noted that HIV was present in the brain at concentrations that exceeded that of the spleen. The cases that Dr. SHAW studied consisted of a small group of individuals exhibiting what became known as the AIDS dementia complex (NAVIA et al. 1986). These patients had no evidence of opportunistic infections within the brain, but a Southern blot of brain tissue showed abundant HIV DNA. It is now clear that approximately a quarter of all terminally ill AIDS patients develop the clinical symptomatology initially described as AIDS dementia complex and more recently renamed HIV-1 associated cognitive motor complex (JANSSEN et al. 1991). The neurologic symptoms associated with this complex emanate from an overall slowing of mentation. In addition to slowness of thought and motor activity, a variety of behavioral abnormalities have been noted (GRANT et al. 1987). The dementia associated with AIDS was thus unlike the well described cortical dementias such as Alzheimer's disease, notable for aphasia, apraxia and agnosia (NAVIA et al. 1986; PORTEGIES et al. 1993). Dementia caused by slowing of mentation had been previously described in neurologic diseases associated with damage to subcortical structures (e.g. Huntington's disease or Parkinson's

Presbyterian University Hospital, Neuropathology Division, A-506, 200 Lothrop Street, Pittsburgh, PA 15213-2582, USA disease). This form of dementia implied a disproportionate damage to the basal ganglia structures in AIDS.

What I would like to describe in this chapter is the pathologic substrate of AIDS dementia complex, HIV encephalitis (MASLIAH et al. 1992a). It is my bias as a pathologist that knowledge of the pathologic substrate is critical to understanding the pathogenesis of the dementing illness. I have four objectives for this chapter. This first is to define HIV encephalitis from a clinical and pathologic perspective. The second objective is to demonstrate the localization and quantitation of HIV. Here I will introduce the concept of viral burden in the central nervous system (CNS). How much virus and how long it has been present in the CNS appear to determine the severity of the neuropathology. AIDS dementia complex is the result of months of uncontrolled replication of HIV within the CNS. As with other neurodegenerative diseases, our glimpse into the pathogenesis of this disease emanates from study of human autopsy tissue. It could be readily inferred that studies of chronic infection by random single time point analysis would be quite difficult. The third objective is to introduce new neuropathologic techniques used in the assessment and quantitation of neurologic damage. These new quantitative techniques now allow us to assay, at the neuropathologic level, subtle cognitive deficits. Finally, I will briefly describe hypothetical mechanisms of neurodegeneration in HIV encephalitis.

2 Definition of HIV Encephalitis

One of the most important features to recognize about HIV encephalitis is that it is a chronic encephalitis. The chronicity of the disease makes it conceptually more difficult to understand than the better understood acute encephalitides such as herpes simplex encephalitis. Herpes simplex encephalitis is an acute limbic encephalitis in which the clinical symptoms are readily attributable to the regions of the brain affected by herpes simplex virus (HSV). In HSV encephalitis acute neuronal and glial infection resulting in lysis of these cells readily implies the pathologic damage and the clinical symptomatology. In contrast, HIV encephalitis, while clearly infecting and affecting subcortical structures, damages this region without significant neuronal or glial infection. Dementia results from a chronic encephalitis whose complexity reflects the life cycle of a retrovirus within the human CNS.

The neuropathology of HIV encephalitis is distinct from that of most previously described encephalitides. In fact, the noninflammatory nature of the neuropathology made many individuals uncomfortable with calling the disease an encephalitis at all (BUDKA et al. 1991). Nevertheless, based upon the fact that virus replicates to very high levels within the CNS, the presence of microglial nodules and multinucleated giant cells was considered histologic evidence for the entity HIV encephalitis. With the advent of improved immunohistochemical and immunocytochemical probes for HIV, the criteria to diagnose HIV encephalitis has been refined to include not only those cases with multinucleated giant cells, but also those cases with demonstrable HIV infection. Unfortunately, new technology permits detection of HIV at a high level of sensitivity such that almost all CNS tissues (and possibly systemic tissues) subjected to polymerase chain reaction (PCR) reveal some HIV. Clearly presence or absence of virus is inadequate in assessment of the presence or absence of HIV encephalitis and some form of quantitation is required.

3 Quantitation of HIV in CNS Tissues

To quantitate the amount of HIV in any particular region of the CNS, we developed a semiquantitative immunocytochemical assay for the HIV envelope protein gp41 (Masliah et al. 1992a). Using cortical gray matter, cerebral white matter and deep gray matter histologic sections immunostained for gp41, we developed a scale ranging from 0–6 for abundance of virus. Comparison of the semiquantitative immunocytochemical assay with other quantitative assays of viral burden showed comparable results. Antigen capture detection of HIV gag protein p24 proved to be less sensitive than immunocytochemical analysis of viral envelope protein, while PCR detection of HIV DNA showed a spectrum of infection similar to the immunocytochemistry. Assessing approximately 60 AIDS autopsies in which there was no evidence of significant opportunistic infection within the CNS, we noted that approximately one third had no evidence of HIV in the CNS, another third had some HIV, while the final third had abundant HIV within the three CNS regions (Achim et al. 1994).

When we compared these immunocytochemical measurements of HIV with the H & E histology, we noted that multinucleated giant cells were present in approximately half of the cases with a high viral burden measured by immunocytochemistry (Wiley and Achim, submitted 1994). We would conclude that multinucleated giant cells are not a very sensitive marker of the presence of HIV. When patients were assessed retrospectively for presence or absence of dementia, approximately one half of the demented patients had multinucleated giant cells while the other half did not. All demented patients had abundant HIV as assessed by immunocytochemistry. We also observed that 28 out of 44 patients had abundant virus, but no clinical chart summary indicating dementia. These latter cases either represent under reporting of dementia or, more likely, the fact that virus needs to be present within the CNS for a period of time in order to mediate neurologic damage.

4 Quantitation of Neurologic Damage

There are a variety of techniques available for quantifying damage of the human brain. We have used three-dimensional reconstruction of serial magnetic resonance images to assess ratios of various CNS tissue volumes (Heindel et al., in press). In these studies digitized images of the CNS are analyzed for ratios of neocortical gray matter, neocortical white matter, abnormal white matter and subcortical structures. These ratios have then been compared to viral burden measurements and the nonspecific measurement of brain damage, gliosis. In a small study comparing these ratios we identified significant correlations between loss of gray matter within the caudate, lenticular nuclei, thalamus and cortical gray and the abundance of virus. A similar but positive correlation was noted between the abundance of virus and the abundance of abnormal white matter. No such correlation was evident when compared to the amount of gliosis.

We have taken these imaging studies of gross brain atrophy and extended them to include histologic quantification of neuronal loss. Comparing different regions of the neocortex for abundance of neurons (defined on the basis of cell size), we identified a loss of the large neuronal cells in patients with HIV encephalitis compared to those without (Wiley et al. 1991). We have extended the neuronal count studies to include quantitative measurements of synaptic and dendritic complexity in the cerebral cortex (Masliah et al. 1992b). Using the laser confocal microscope and immunofluorescent tags for the postsynaptic marker microtubule associated protein and the presynaptic marker synaptophysin, we have identified a strong correlation between severity of HIV encephalitis and the abundance of the pre- and postsynaptic markers. The quantifications have further suggested that the postsynaptic elements appear to be disproportionately involved in HIV encephalitis. This is in contradistinction to what is seen in Alzheimer's disease, in which the presynaptic element appears to be more severely involved.

Golgi impregnations are excellent techniques for studying postsynaptic spines. Using this technique on patients with or without HIV encephalitis we noted a strong correlation between the loss of dendritic spines on neocortical apical dendrites and the abundance of virus (Masliah et al. 1992b). These results are consistent with and complimentary to the laser confocal microscope studies. We have assembled these observations into a hypothesis that, at some point during the course of infection, the control of HIV replication is lost. Uncontrolled HIV replication within the CNS is followed at some finite time by cumulative neurologic damage. When that damage exceeds a specific threshold the clinical symptomatology of AIDS dementia complex is evident.

5 Hypothetical Mechanisms of HIV Encephalitis Associated Neurodegeneration

The fact that HIV does not directly infect nervous system cells leads to the paradox of explaining how neurons degenerate. Several theories have been put forth. The first consists of a direct neurotoxic effect of HIV (Dreyer et al. 1990; Dawson et al. 1993; Giulian et al. 1993; Lipton 1993). In this model HIV replication within CNS macrophages results in the secretion of HIV or HIV envelope proteins. Retroviral proteins are notorious for cytotoxicity and there are now numerous studies suggesting that the envelope protein gp120 may be neurotoxic. Alternatively, HIV replication within activated macrophages leads to the production of macrophage derived neurotoxic compounds (DREYER et al. 1990; LIPTON et al. 1991; DAWSON et al. 1993; EPSTEIN and GENDELMAN 1993; GIULIAN et al. 1993). There are a variety of potential neurotoxins that macrophages are known to elaborate. We and others have studied the production of quinolinic acid and in both the human and simian cases there does appear to be a correlation between neurologic symptomatology and the abundance of quinolinic acid within the cerebrospinal fluid (HEYES et al. 1991; BROUWERS et al. 1993). We have extended these studies to look at the presence and quantitation of quinolinic acid within CNS tissue and have noted that patients with severe HIV encephalitis have high levels of quinolinic acid, particularly within the deep gray structures (WILEY et al. 1992). These concentrations are at many fold that known to be neurotoxic in vitro.

One other class of potential neurotoxins including cytokines elaborated by macrophages. Perhaps the most studied of all monokines in HIV infection is tumor necrosis factor-α (TNF-α). Many have noted an abundance of TNF-α within the cerebrospinal fluid of AIDS patients (LAHDEVIRTA et al. 1988; SELMAJ and RAINE 1988; GALLO et al. 1989; MINTZ et al. 1989; MERRILL and CHEN 1991; PEUDENIER et al. 1991; GENIS et al. 1992; HOFMAN and HINTON 1992). We have additionally observed high concentrations of TNF-α within the tissue particularly of patients with severe HIV encephalitis (WILEY et al. 1992; ACHIM et al. 1993). Overall there are a variety of studies suggesting immune activation within the CNS of patients with HIV encephalitis. Curiously this immune activation, unlike any other viral encephalitis, occurs in the absence of T cells.

6 Conclusion

Why a fraction of patients develop HIV encephalitis is unclear. It would be reasonable to hypothesize that an increase of severity of neurologic damage is a reflection of the time of death relative to debilitation of the immune system during which HIV has grown within the CNS. Patients must survive numerous opportunistic infections before the immune system destruction has resulted in uncontrolled HIV replication. It appears that unmitigated viral replication within the CNS leads to much damage. It will be the subject of future investigation to find the molecular site and mechanisms of the CNS damage.

References

Achim AL, Heyes MP, Wiley CA (1993) Quantitation of human immunodeficiency virus, immune activation factors, and quinolinic acid in AIDS brains. J Clin Invest 91: 2769–2775

Achim C, Wang R, Miners D, Wiley C (1994) Brain viral burden in HIV infection. J Neuropathol Exp Neurol 53: 284–294

Anders KH, Guerra WF, Tomiyasu U, Verity MA, Vinters HV (1986) The neuropathology of AIDS. UCLA experience and review. Am J Pathol 124: 537–558

Brouwers P, Heyes MP, Moss HA, Wolters PL, Poplack DG, Markey SP, Pizzo PA (1993) Quinolinic acid in the cerebrospinal fluid of children with symptomatic human immunodeficiency virus type 1 disease: relationships to clinical status and therapeutic response. J Infect Dis 168: 1380–1386

Budka H, Wiley CA, KP Artigas J, Asbury AK, Cho E-S, Cornblath DR, Dal Canto MC, DeGirolami U, Dickson D, Epstein LG, Esiri MM, Giangaspero F, Gosztonyi G, Gray F, Griffin JW, Henin D, Iwasaki Y, Janssen RS, Johnson RT, LPL, Lyman WD, MJC, Nagashima K, Peres N, Petito CK, Price RW, Rhodes RH, Rosenblum M, Said G, Scaravilli F, Sharer LR, Vinters HV (1991) HIV-associated disease of the nervous system: review of nomenclature and proposal for neuropathology-based terminology. Brain Pathol 1: 143–152

Dawson VL, Dawson TM, Uhl GR, Snyder SH (1993) Human immunodeficiency virus type 1 coat protein neurotoxicity mediated by nitric oxide in primary cortical cultures. Proc Natl Acad Sci USA 90: 3256–3259

Dreyer EB, Kaiser PK, Offermann JT, Lipton SA (1990) HIV-1 coat protein neurotoxicity prevented by calcium channel antagonists (see comments). Science 248: 364–367

Epstein LG, Gendelman HE (1993) Human immunodeficiency virus type 1 infection of the nervous system: pathogenetic mechanisms (review). Ann Neurol 33: 429–436

Gallo P, Piccinno MG, Krzalic L, Tavolato B (1989) Tumor necrosis factor alpha (TNF alpha) and neurological diseases. Failure in detecting TNF alpha in the cerebrospinal fluid from patients with multiple sclerosis, AIDS dementia complex, and brain tumours. J Neuroimmunol 23: 41–44

Genis P, Jett M, Bernton EW, Boyle T, Gelbard HA, Dzenko K, Keane RW, Resnick L, Mizrachi Y, Volsky DJ et al. (1992) Cytokines and arachidonic metabolites produced during human immunodeficiency virus (HIV)-infected macrophage-astroglia interactions: implications for the neuropathogenesis of HIV disease. J Exp Med 176: 1703–1718

Giulian D, Wendt E, Vaca K, Noonan CA (1993) The envelope glycoprotein of human immunodeficiency virus type 1 stimulates release of neurotoxins from monocytes. Proc Natl Acad Sci USA 90: 2769–2773

Grant I, Atkinson JH, Hesselink JR, Kennedy C, Richman D, Spector S, McCutchan JA (1987) Evidence for early central nervous system involvement in the acquired immunodeficiency syndrome (AIDS) and other human immunodeficiency virus (HIV) infections. Studies with neuropsychologic testing and magnetic resonance imaging (published erratum appears in Ann Intern Med 108(3): 496, 1988). Ann Intern Med 107: 828–836

Heindel WC, Jernigan TL, Archibald S, Achim CL, Masliah E, Wiley CA (in press) The relationship of quantitative brain MRI measures to neuropathological indices of HIV infection. Arch Neurol

Heyes MP, Lackner A, Kaufman S, Milstien S (1991) Cerebrospinal fluid and serum neopterin and biopterin in D-retrovirus-infected rhesus macaques (Macaca mulatta): relationship to clinical and viral status. AIDS 5: 555–560

Hofman FM, Hinton DR (1992) Tumor necrosis factor-alpha in the retina in acquired immune deficiency syndrome. Invest Ophthalmol Vis Sci 33: 1829–1835

Janssen RS, Cornblath DR, Epstein LG, Foa RP, McArthur JC, Price RW, Asbury AK, Beckett A, Benson DF, Bridge TP, Leventhal CM, Satz P, Saykin AJ, Sidtis JJ, Tross S (1991) Nomenclature and research case definitions for neurological manifestations of Human Immunodeficiency Virus Type-1 (HIV-1) infection. Neurology 41(6): 778–785

Lahdevirta J, Maury CP, Teppo AM, Repo H (1988) Elevated levels of circulating cachectin/tumor necrosis factor in patients with acquired immunodeficiency syndrome. Am J Med 85: 289–291

Levy RM, Bredesen DE, Rosenblum M (1985) Neurological manifestations of the acquired immunodeficiency syndrome (AIDS): experience at UCSF and review of the literature. J Neurosurg 62: 475–495

Lipton SA (1993) Human immunodeficiency virus-infected macrophages, gp120, and N-methyl-D-aspartate receptor-mediated neurotoxicity (letter). Ann Neurol 33: 227–228

Lipton SA, Sucher NJ, Kaiser PK, Dreyer EB (1991) Synergistic effects of HIV coat protein and NMDA receptor-mediated neurotoxicity. Neuron 7: 111–118

Masliah E, Achim CL, Ge N, DeTeresa R, Terry RD, Wiley CA (1992a) Spectrum of human immunodeficiency virus-associated neocortical damage. Ann Neurol 32: 321–329

Masliah E, Ge N, Morey M, DeTeresa RM, Terry RD, Wiley C (1992b) Cortical dendritic pathology in HIV encephalitis. Lab Invest 66: 285–291

Merrill JE, Chen JS (1991) HIV-1, macrophages, glial cells, and cytokines in AIDS nervous system disease. FASEB J 5: 2391–2397

Mintz M, Rapaport R, Oleske JM, Connor EM, Koenigsberger MR, Denny T, Epstein LG (1989) Elevated serum levels of tumor necrosis factor are associated with progressive encephalopathy in children with acquired immunodeficiency syndrome. Am J Dis Child 143: 771–774

Navia BA, Jordan BD, Price RW (1986) The AIDS dementia complex: I. Clinical features. Ann Neurol 19: 517–524

Peudenier S, Hery C, Montagnier L, Tardieu M (1991) Human microglial cells: characterization in cerebral tissue and in primary culture, and study of their susceptibility to HIV–1 infection. Ann Neurol 29: 152–161

Portegies P, Enting RH, de Gans J, Algra PR, Derix MM, Lange JM, Goudsmit J (1993) Presentation and course of AIDS dementia complex: 10 years of follow-up in Amsterdam, The Netherlands. AIDS 7: 669–675

Selmaj KW, Raine CS (1988) Tumor necrosis factor mediates myelin and oligodendrocyte damage in vitro. Ann Neurol 23: 339–346

Shaw GM, Harper ME, Hahn BH, Epstein LG, Gajdusek DC, Price RW, Navia BA, Petito CK, O'Hara CJ, Groopman JE et al (1985) HTLV-III infection in brains of children and adults with AIDS encephalopathy. Science 227: 177–182

Wiley CA, Achim CL (1994) HIV encephalitis is the pathologic correlate of dementia in AIDS. Ann Neurol 36: 673–676

Wiley C, Masliah E, Morey M, Lemere C, DeTeresa RM, Grafe MR, Hansen LA, Terry RD (1991) Neocortical damage during HIV infection. Ann Neurol 29: 651–657

Wiley CA, Achim CL, Schrier RD, Heyes MP, McCutchan JA, Grant I (1992) Relationship of cerebrospinal fluid immune activation associated factors to HIV encephalitis. AIDS 6: 1299–1307

Penetration of Solutes, Viruses, and Cells Across the Blood-Brain Barrier

M.W. Brightman, S. Ishihara, and L. Chang

1 Solute Passage

1.1 Barrier Site

The endothelial barrier between the blood and interstitial or extracellular fluid of the central nervous system (CNS) closely regulates the composition of the immediate extracellular environment of brain cells. This regulation ensures that the ionic composition of the CNS interstitial fluid, so vital to the normal functioning of brain cells, remains stable in the face of fluctuations in the blood circulating through brain and cord (Bradbury 1979). The means by which the regulation is brought about is the endothelium's restriction in the movement of solutes between blood and its selective transfer of certain solutes. When a bolus of the glycoprotein horseradish peroxidase (HRP, ~40 kDa) is infused into the systemic blood, the HRP is excluded from the brain's interstitial fluid by a continuous belt of tight junctions between adjacent endothelial cells and by the paucity of endocytotic pits and vesicles (Reese and Karnovsky 1967). The junctions, being zonular, seal the intercellular cleft, thereby preventing movement of solutes from blood

Laboratory of Neurobiology, Bld. 36 Rm. 2A-21, National Institutes of Health, 9000 Rockville Pike, Bethesda, MD 20892, USA into CNS extracellular fluid and from interstitial fluid into blood (Bodenheimer and Brightman 1968). Smaller hydrophilic molecules such as heme-peptide, about 2 kDa in size, are also excluded from brain by the same mechanism (Brightman et al. 1970). The passive flooding of the interstitium with serum that would occur were the junctions patent is prevented by the junctional belts which thereby enable the endothelium to act on a solute. The endothelium can then selectively regulate a solute's passage by, e.g., metabolizing it or by endocytosing it after it binds to its receptor on the luminal face of the endothelium. Numerous chemically unrelated substances are, in this manner, largely excluded from the brain interstitial clefts. There are, however, means by which blood-borne solutes can gain entry into the intercellular fluid of the normal intact brain and cord by way of the contact sites between axon terminals and constitutively fenestrated, permeable, nonbarrier vessels. The following account compares the ways in which solutes, viruses and cells can cross the blood-brain barrier (BBB).

1.2 Entry Sites

The permeated sites where axon terminals come into contact with an interstitial space are: (a) where peripheral and cranial nerves project to muscle, both skeletal and visceral, the capillaries of which are normally permeable, (b) where neurites of neuroendocrine areas abut fenestrated permeable capillaries such as those of circumventricular organs, and (c) a few subpial regions that contain leaky vessels (Broadwell and Sofroniew 1993). When the concentration of HRP is maintained at a high level for a sufficient period, the glycoprotein crosses the permeable blood vessels of muscle and circumventricular organs of the CNS in amounts that can be visualized by light- and electron microscopy in a number of brain stem nuclei (Broadwell and Brightman 1976). Some of the neuronal groups, the dorsal motor nucleus of the vagus, nucleus of the tractus solitarius and ambiguus (Fig. 1), and the hypoglossal nucleus (Fig. 2), project their axons to viscera. The blood-borne protein exudes from the permeable vessels of these organs, enters the interstitial clefts, including the synaptic gutter at myoneural junctions, and is then incorporated by vesicles that bud from the terminal axolemma. The vesicles involved in this fluid phase endocytosis communicate with tubular cisterns within the axoplasm that eventually connect with lysosomes in the neuron's cell body (Broadwell and Brightman 1979). The neurons become tagged when these solutes accumulate in the lysosomes (Fig. 1 and 3), where the protein is eventually degraded, presumably by.cathepsins. Proteins and peptides move retrogradely from neuronal periphery to the soma via this vesiculotubular route. Autonomic neurons, like the A-1 group of the medulla, are likewise labeled because their neurites, too, presumably lie in the vicinity of permeable vessels. The vast majority of neurons within the CNS are not associated with permeable vessels and are not labeled by this nonspecific, fluid-phase vesicular uptake of a foreign glycoprotein.

Another, more specific type of vesicular incorporation of macromolecules involves adsorptive endocytosis, defined here as specific binding of a lectin to

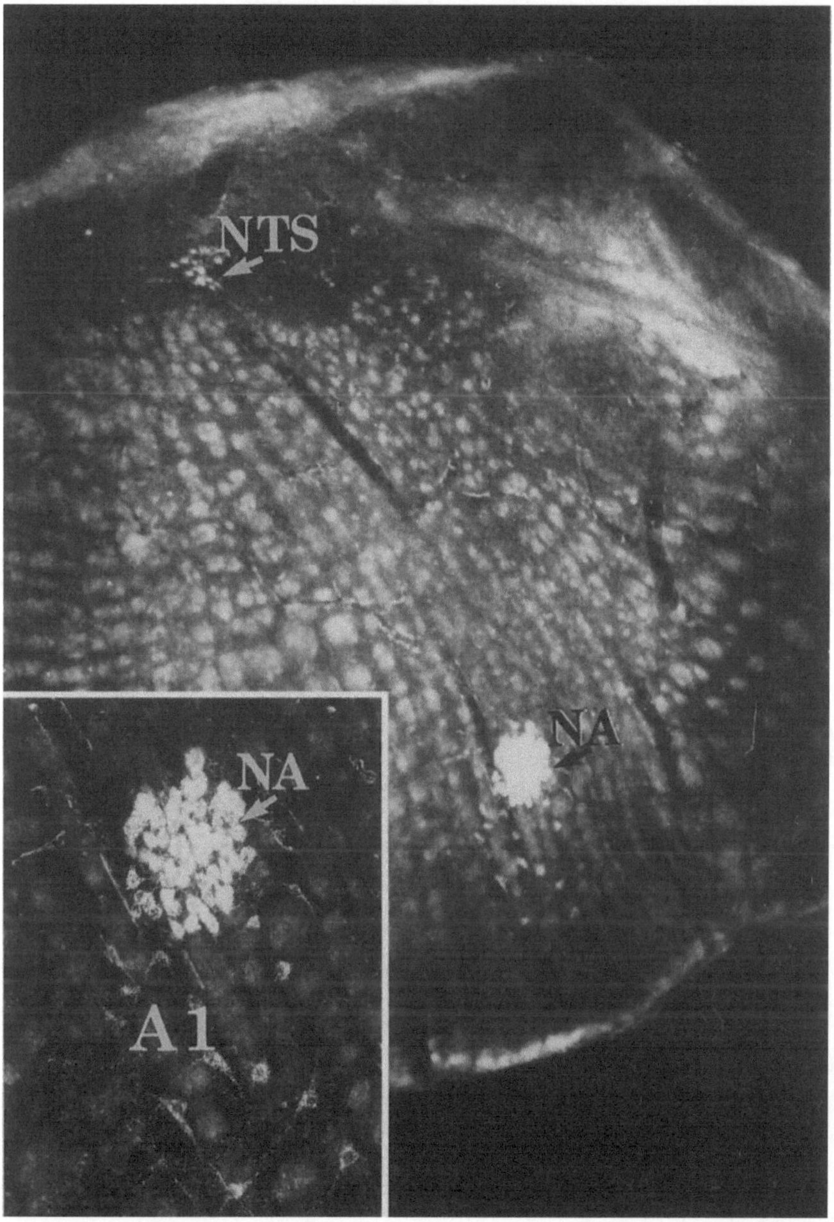

Fig. 1. Transverse section of the medulla from a mouse 24 h after the intravenous infusion of horseradish peroxidase (HRP). The nucleus of the tractus solitarius (*NTS*) was labeled, presumably by retrograde transport of the HRP along NTS axons projecting to the area postrema, which has no barrier and is rapidly flooded with peroxidase; x 65. *Inset,* The nucleus ambiguus (*NA*) and *A1* adrenergic neurons of the medulla also incorporated peroxidase from blood. Dark-field micrograph, x 175

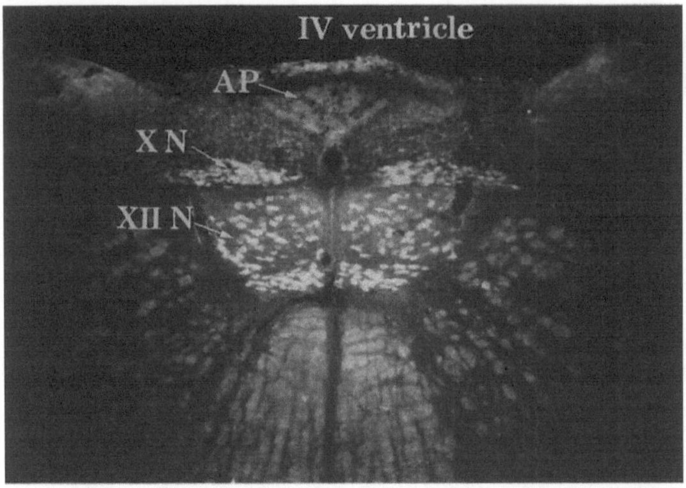

Fig. 2. Frontal section through a circumventricular organ, the area postrema (*AP*) of a mouse, 12 h after being given horseradish peroxidase (HRP) intravenously. Neuronal cell bodies in the dorsal motor nucleus (*XN*) of the vagus nerve and the hypoglossal nerve nucleus (*XII N*) have become labeled with peroxidase, taken up by their nerve terminals exposed to extracellular HRP that had escaped from the permeable vessel of viscera and muscle as well as from the AP. Dark-field micrograph, x 49

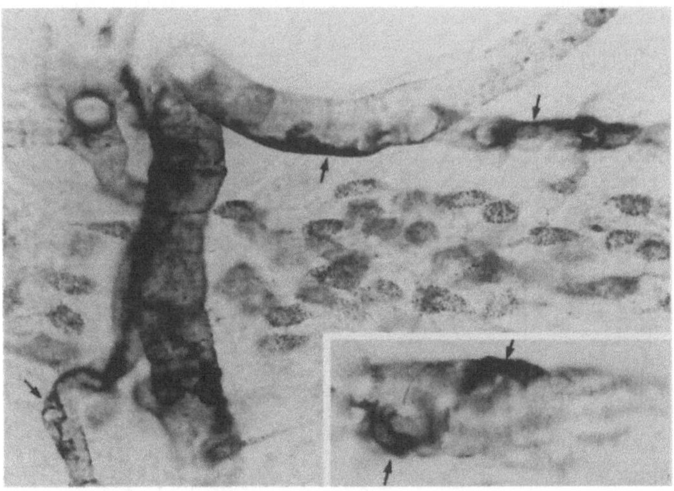

Fig. 3. Labeling of the neuronal cell bodies of the hypoglossal nucleus, depicted in Fig. 2, can be resolved here as a punctate labeling of lysosomes. Perivascular phagocytes (*arrows*) have endocytosed horseradish peroxidase (HRP) that had entered the interstitial clefts of the nucleus, presumably from the nearby area postrema. Bright-field micrograph, x 180

particular oligosaccharides on the cell membrane. When a lectin, such as wheat germ agglutinin, is conjugated to HRP and placed on the nasal mucosa, the complex is incorporated, as in fluid-phase uptake, by vesicles that pinch off the axon terminals of olfactory nerves. A crucial difference, however, is that only some of the lectin-protein complex is destined for lysosomes. The remainder of the conjugate is transported within the axon's cytoplasm and out, transynaptically, to a chain of several neurons (BAKER and SPENCER 1984, 1986; BROADWELL and BALIN 1985; SHIPLEY 1985).

The cytoplasmic route taken by solutes after adsorptive endocytosis is thus distinctly different from the route taken after fluid phase uptake. A third means of incorporating substances is by receptor mediated endocytosis. A circulating ligand, such as the protein transferrin, binds to its specific receptor on brain endothelium and is transcytosed across the endothelial cell and released at its abluminal side fronting the perivascular interstitial compartment (JEFFRIES et al. 1984).

Viral particles that are incorporated by receptor mediated endocytosis enter the cytoplasmic pathway used by some of the lectin-associated HRP. Virus and conjugate are transported retrogradely or anterogradely in vesicles to the endoplasmic reticulum, then to the Golgi apparatus and thence through the axoplasmic cytosol (CARD et al. 1993). A difference in the utilization of this pathway is that the conjugate is contained in membrane bounded vesicles and tubules throughout its journey, whereas herpes viral particles may move through axoplasm as free nucleocapsids, unenclosed by an organelle membrane (MARCHAND and SCHWAB 1986) to be exocytosed from neurites near where they abut synaptic clefts. Virions, like the lectin-HRP conjugate, are exocytosed at the neurite terminal into the extracellular and confluent synaptic clefts from which both HRP-lectin complex and virus can be incorporated by the dendrite of the next—order neuron (BAKER and SPENCER 1986; BROADWELL and BALIN 1985; SHIPLEY 1985; CARD et al. 1993).

2 Viral Passage

2.1 Viremia

A virus rendered noninfective does not enter the CNS. When HSV-1 is made nonvirulent by ultraviolet irradiation and infused into systemic or cerebral blood, it is, like smaller substances such as protein and peptide, excluded from the normal CNS. The inactivated virus, labeled with [^{35}S] methionine, can only enter the brain when the BBB is first opened hyperosmotically with the intracarotid infusion of 25% mannitol. This type of BBB opening is reversible and allows the passive entry of detectable amounts of virus from blood into the brain tissue (NEUWELT et al. 1991). Hyperosmotic opening of the barrier apparently shrinks CNS endothelium to the extent of deforming the tight junctions between adjacent endothelial cells; the structurally altered junctions then permit the extracellular passage of solutes from blood into the interstitial clefts on the side of the brain supplied by the carotid artery that had been infused with the hypertonic agent (BRIGHTMAN et al. 1973). Virus presumably enters the brain through patent tight junctions, although this route has yet to be actually demonstrated. One hour following barrier opening, the viral concentration on that side of the cerebrum is about fourfold greater than the radioactivity detected in brains without barrier modification.

No such imposed opening of the BBB is necessary for an infective replicating virus to enter the CNS. Viruses can enter the body across all of its surfaces (MORRISON and FIELDS 1991). Having entered an organ, a virus may replicate, bind selectively to receptors on the terminals of neurons innervating the organ and may also be shed into the vasculature supplying the organ. Entry into the CNS can, accordingly, be effected by two routes: neural and circulatory. Free virus may get into brain by way of peripheral neurites that lie outside of the brain and spinal cord or it may enter directly across CNS endothelium. The viremia that follows infection of an organ must be sustained and be of sufficient magnitude for appreciable amount of virus to penetrate the endothelium of brain and spinal cord. Having entered cerebral endothelium, enteroviruses, arboviruses and retroviruses may have to replicate there and then be extruded from the endothelium into the brain (TYLER and MCPHEE 1987).

The permeable fenestrated endothelium of circumferential organs, such as the choroid plexus and area postrema (Fig. 2) to name only two out of seven such organs, may provide the same entry portal for some viruses (TYLER and FIELDS 1990) as it does for solutes (BRIGHTMAN 1968). HRP, infused as a bolus into systemic blood, rapidly crosses the fenestrated vessels of the choroid plexus into its stromal extracellular space, then into the clefts between ependymal cells and from there into the ventricular cerebrospinal fluid (CSF) and the brain parenchyma subjacent to the ependyma (BRIGHTMAN and RESSE 1969). The same route may be followed by viruses moving from blood to CSF. Additionally, however, some viruses eventually infect the ependyma and may then continue to be shed into the underlying brain parenchyma (TYLER and FIELDS 1990).

2.2 Transneuronal Entry of Virus

Pseudorabies virus, injected into the pinna of the ear, is transported retrogradely at a minimum rate of about 1.7 mm/h into dorsal root ganglia (FIELD and HILL 1975). Rabies virus can directly enter both motor and sensory neurites of mature mice (COULON et al. 1989), as can certain other viruses, if the host tissue is sufficiently immature. When three strains of mammalian reovirus were inoculated into the foot pad of neonate mice, the viruses were quickly and directly incorporated by sensory and motor fibers near the site of inoculation. They were retrogradely transported within the axoplasm to the somata of sensory, dorsal root ganglia and motor neurons within the ventral horn of the spinal cord and interneurons as well (FLAMAND et al. 1991). The extent to which motor or sensory neurons were penetrated depended on the amount of viral replication in either the skin, where most of the sensory nerves lie, or the muscle tissue, where the motor nerve terminals are situated. One of the serotypes also entered the blood and endothelium of the limb muscle. There was apparently no entry of viruses into CNS endothelium.

A physiological neural entry is by way of the intestine (MORRISON et al. 1991). Both poliovirus and a strain of reovirus can enter the brain stem from nerve terminals in the wall of the intestine. Moreover, brain endothelium, meninges and three circumventricular organs supplied by permeable vessels: choroid plexus, hypothalamus (presumably, the median eminence) and area postrema may be free of the reovirus used. Like the barrier vessels of the brain, the fenestrated, permeable vessels of the circumventricular organs are apparently impervious to viral particles although the same fenestrated capillaries are permeable to protein (BRIGHTMAN 1968; WEINDL 1973) and peptide (BRIGHTMAN et al. 1970). Only the dorsal motor nucleus of the vagus nerve, the axons of which supply the intestine, may be labeled by reovirus (MORRISON et al. 1991). Significantly, no mention is made of the nucleus of the solitary tract and the nucleus ambiguus which contribute fibers to the vagus nerve and which are labeled by circulating HRP (BROADWELL and BRIGHTMAN 1976). The lack of labeling by virus of the solitarius and ambiguus nuclei suggest that these particular branches of the X, VII, IX, and XI cranial nerves do not have receptors for the particular reovirus examined.

When placed on the corneal surface or into the vitreous of the eye, virions can bind selectively to the plasma membrane of distal axonal terminals and can be anterogradely transported or, like lectin-conjugated protein, be retrogradely transported to the neuronal soma through the axoplasm. The viral particles can then be released near synaptic contacts into the interstitial clefts of the CNS and then be transferred intracellularly to the next neuron with which it makes contact. So specific is viral entry into neurons and their neurites that consequent labeling has been used as a means for the tracing of axons belonging to the same functional network (KUYPERS and UGOLINI 1990).

The high specificity of neuronal infection by viruses involves interaction of the glycoprotein belonging to the viral envelope and specific receptors on the cell membrane. The α-herpes virus primarily selects, as do some other neurotropic viruses, sensory neurons and, when injected into the vitreous of the eye, travels through functionally discrete neural circuits. The predilection for specific fiber tracts is so high that mutant strains of this virus are reproducibly transported in the axons of distinctly different neural pathways of the visual system (CARD et al. 1991). Although virus receptors on neurons have not been fully characterized, at least some of the receptors may include an oligosaccharide domain, in as much as lectins can inhibit the binding of virus to sensory neurons (ZIEGLER and POZOS 1981; MARCHAND and SCHWAB 1986).

The oligosaccharide domain that is apparently shared by some lectins and viruses in their binding to the cell membrane raises the question of whether the same intracellular pathway is utilized by both lectin and viruses in their progression through the neuronal cytoplasm. When superior cervical ganglia were first exposed, in vitro, to the lectin concanavalin-A, there was a 60% reduction in the amount of ^{35}S-labeled herpes virus that accumulated in the ganglion cell bodies. Concanavalin-A does not affect axoplasmic transport and the reduction in viral labeling was interpreted as the result of a decrement in the binding of virus to its receptor (MARCHAND and SCHWAB 1986). The binding was followed by the retrograde axoplasmic transport of free nucleocapsids, without an enclosing membrane.

2.3 Defective Barrier

It is likely that the BBB must be compromised in some way and at some site in order for virus to enter the cerebral interstitium. Even during the early stages of human immunodeficiency virus-1 (HIV-1) infection, the barrier appears to be affected. Diffuse breakdown of the BBB, manifested by extravasation of serum fibrinogen and immunoglobulin G, detected immunohistochemically in the brain, is an accompaniment of the infection (Petito and Cash 1992; Rhodes 1991). The diffuse pallor of subcortical myelin in such brains is not due to demylination but rather to edema. The pallor accompanies a slowly progressive, chronic, wide spread opening of the barrier and a subsequent passive entry of fluid and serum proteins into the white matter (Power et al. 1993). A breached barrier is also manifested by an increase in the CSF/serum ratio of serum proteins (Marshall et al. 1991; McArthur et al. 1992) and by uptake of albumin and haptoglobin by reactive astrocytes and some neurons (Power et al. 1993). The two serum proteins are rarely found in brain cells of individuals with an intact BBB but they do enter the brain during inflammation, including HIV infection. As the proteins are exuded in nondemented AIDS patients, it has been inferred that the barrier has been opened before the onset of dementia. Diffuse opening of the barrier has been regarded as an initial "priming" event that permits entry of HIV or HIV-infected cells and of cytokines such as tumor necrosis factor-α (TNF-α) that are neurotoxic (Power et al. 1993).

The lowering of the barrier that leads to the exudation of serum proteins is unlikely to be the consequence of HIV infection of endothelium; the virus has yet to be detected in vivo within endothelial cells. One possible mechanism is the effect on endothelium of circulating cytokines (Power et al. 1993), known to be elevated in the serum of AIDS patients (Rosenberg and Fauci 1990). The inflammatory cytokines interferon-β (IFN-β) and TNF-α, when infused intracisternally, open the barrier to ^{125}I-labeled bovine serum albumin in rats (Quagliarello et al. 1991). The increase in the number of perivasular macrophages and their hypertrophy in the brains of HIV-infected people has suggested that these cells may be a source of cytokines that lower the BBB (Power et al. 1993).

Another role of cytokines is as attractant for monocytes/macrophages that could bring virus into the brain. Sources of such cytokines intrinsic to CNS are glia and endothelium. Astrocytes, oligodendrocytes and microglial cells all contain the mRNA for the cytokine transforming growth factor-β1 (TGF-β1) and constitutively produce it (da Cunha et al. 1993). The synthesis of TGF-β1 is activated by another cytokine, interleukin-1 (IL-1), in vitro and within human brains (da Cunha et al. 1993). As an intrinsic source of cytokines, the perivascular astrocyte, being the endothelium's nearest and most extensive neighbor, is in an optimal position to affect capillary activity. The suggestion has been made (Vitković et al. 1994) that, as a potent attractant for monocytes (WAHL et al. 1987), TGF-β1 may promote the entry of viral-containing monocytes/macrophages into brain parenchyma. It is possible that astroglial and endothelial sources of TGF-β1 reinforce each other. Having been attracted to endothelium, subsequent adhesion of leukocytes to endothelium may be initiated by IL-1 which, in turn, enhances the production of other cytokines that induce production of cell adhesion molecules (CAMs) such as VCAM-1 and ELAM-1 (SPRINGER 1990). So begins a series of events culminating in the entry of viral bearing monocytes and macrophages into brain.

3 Leukocyte Passage

The otherwise immunological isolation of the normal CNS is due, in part, to the absence of specialized dendritic antigen presenting cells, common to many other organs, and to the absence of a lymphatic drainage system (METLAY et al. 1989). There is no evidence to suggest that the resident macrophage-like cells, or any other brain cells, can pick up antigen and migrate out of the CNS to lymph nodes from which they could sensitize the immune system to that antigen (SEDGWICK and DÖRRIES 1991).

3.1 Immune System Arousal

There are, nevertheless, several ways in which the immune system can be informed of a pathological event within the CNS (SEDGWICK and DÖRRIES 1991). First, free antigen, produced by the event, may escape the confines of the BBB by the bulk flow of interstitial fluid that would carry it to the CSF, through the arachnoid villi, across the permeable vessels of the dura and into the systemic circulation to reach the spleen and lymph nodes.

A second way in which the immune system is alerted to events within the CNS is through the passage of antigen out of the brain along cranial nerves. Lectin-conjugated HRP can enter brain through the axoplasm of the olfactory nerves and tract, as previously discussed. Antigen, generated within or injected into the brain substance, can escape by passing alongside but external to the olfactory nerves. Antigen within the cerebral parenchyma is carried by the interstitial fluid into the CSF and thence into the subarachnoid space surrounding the olfactory nerves. At the emergence of these cranial nerves from the cribriform plate of the ethmoid bone, antigen would accumulate and enter lymphatics that drain into the cervical lymph nodes (HARLING-BERG et al. 1989) from where it would lead to the activation of appropriate lymphocytes.

Thirdly, antigen may be presented by CNS endothelial cells at their luminal face where circulating cells can be activated. A fourth way in which the immune system can be aroused is by surveillance lymphocytes that randomly cross the BBB in small numbers within apparently healthy individuals (HICKEY 1991).

Activated lymphocytes traverse CNS endothelium, where they are immediately confronted by an endothelial neighbor, which is also potentially an antigen presenting cell: the perivascular macrophage (HICKEY and KAMURA 1988) or astrocytes. If perivascular phagocytes prove to be antigen processing cells, they too are strategically placed. They readily endocytose foreign protein (Fig. 3) and peptide as they would endogenous antigen and could then present the processed antigen to competent lymphocytes.

3.2 Migration of Leukocytes Across the Barrier

How circulating cells of the immune system are induced to enter the CNS parenchyma is unknown. One of the first steps in the extravasation of sensitized immune cells is their nonspecific adhesion to brain endothelium. The entry of macrophages from blood to a brain that has a focal sterile disturbance is given by the following example. In otherwise normal brains, a focal, chronic cellular interaction can be created by placing a piece of skeletal muscle, as an auto- or isograft, onto the surface of the brain (Fig. 4) with minimal disturbance to the rest of the brain (ISHIHARA et al. 1993). The graft acts as a permanent focal opening of the BBB to solutes. It also creates an entry portal into brain for circulating macrophages. Whereas solutes enter the brain directly from the graft's interstitial compartment, macrophages enter across vessels in the brain near the graft rather than directly from the graft. The graft induces an elevated expression of intercellular adhesion molecule-1 (ICAM-1) on endothelial cells of capillaries in the medulla near the graft. The elevation of endothelial ICAM-1 in vitro is mediated by the cytokines TNF-α and IFN-γ (POBER et al. 1986). Peritoneal macrophages are harvested then activated with lipopolysaccharide or, more effectively, by phorbol myristate acetate (ISHIHARA et al. 1993). The cells are then stained with a fluorescent dye and infused, in rats, as a bolus into the axillary artery which eventually communicates with the vertebral artery. The activated macrophages not only adhere to the endothelial cells (Fig. 5) that, very likely, express such adhesion molecules as E-selectin in addition to ICAM-1, but also cross the endothelial cells to enter and migrate for a short distance through the parenchyma of the medulla. (Fig. 6) (ISHIHARA et al. 1993).

This passage into the brain parenchyma of macrophages, including virus containing ones, appears to take place across an endothelium that had been made receptive by a preceding, focal, pathological event. Normal monocytes and macrophages, derived from administered bone marrow, are called forth from the peripheral circulation by an inherited demyelinating process within the CNS (HOOGEBRUGGE et al. 1988). Adhesion may also be brought about by stimulation of brain endothelium with cytokines or by a peripheral activation of monocytes (LUSCINSKAS et al. 1991). For example, IFN-γ is one cytokine that augments the passage of nonactivated T-11 lymphocytes across a monolayer of endothelial cells derived from umbilical vein, in a dose- and time-dependent manner (OPPENHEIMER-

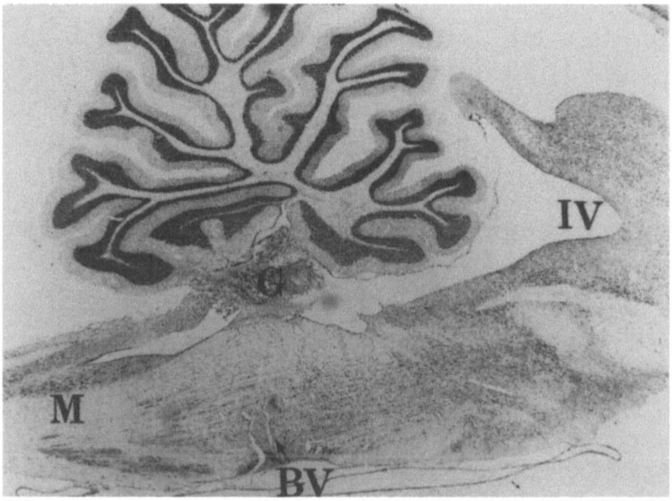

Fig. 4. Autograft (*G*) of skeletal muscle inserted into the IV ventricle (*IV*) between the medulla oblongata (*M*) and overlying cerebellum of a rat 5 weeks prior to aldehyde fixation. Radial medullary blood vessels (*BV*) course dorsoventrally, x 11

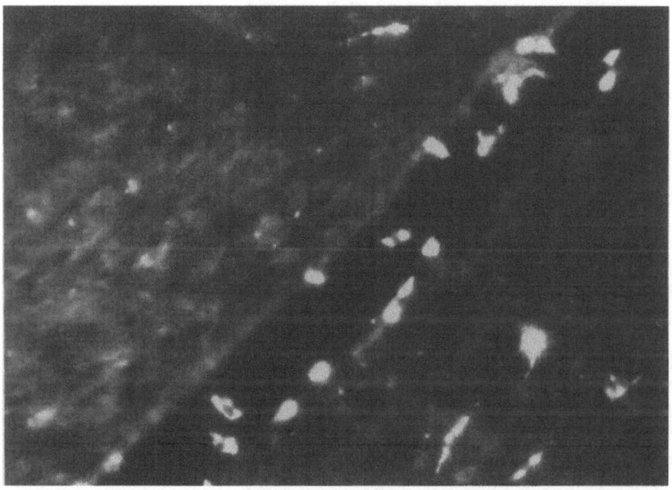

Fig. 5. Exogenous, isogeneic macrophages that had been activated with phorbol myristate acetate, labeled with a lipid-soluble fluorescent dye, and then infused into the axillary artery of a graft-bearing rat were permitted to circulate for 2 h before fixation by perfusion of aldehydes through the vascular tree. The macrophages (*bright bodies*) adhered to the wall of blood vessels to such a degree that they were not dislodged by the perfusion of fixative. Fluorescence micrograph, x 570

Marks and Ziff 1988). In the absence of a pathological event, activated macrophages would still have to adhere to endothelium before its cytokines could act directly and locally on the capillary wall. Only then would the macrophages migrate across endothelium from blood to brain.

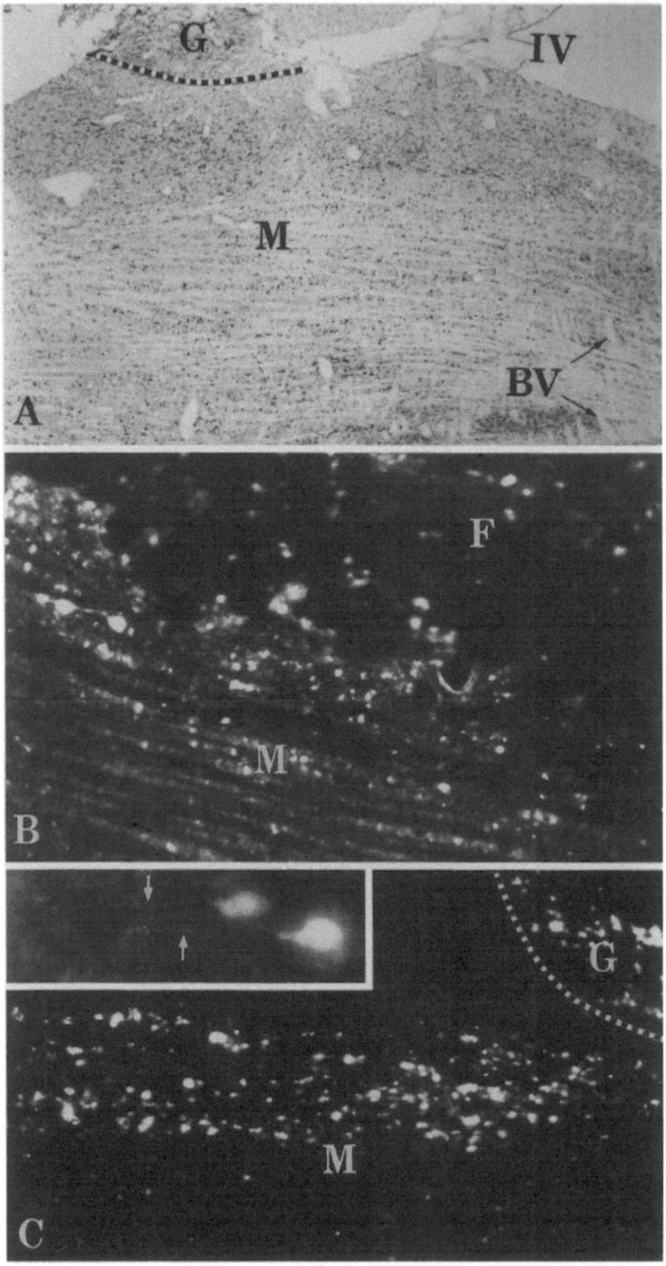

Fig. 6A–C. Entry of macrophages, treated as were those in Fig.5, into the medulla of a graft-bearing rat. **A** Bright-field micrograph of graft's(*G*) typical position and relation to medulla (*M*), where longitudinal axons lie parallel to each other. Blood vessels (*BV*) in the medulla parenchyma and the IV ventricle (*IV*) are also shown; x 23. **B** Gelfoam (*F*) inserted into IV ventricle 5 weeks prior to intravascular infusion of activated macrophages, a few of which have entered the medulla (*M*) near the acellular graft to lie parallel with the longitudinal axons. Fluorescence micrograph, x 220. **C** Isogeneic, activated macrophages have migrated longitudinally through the medullary parenchyma (*M*) for approximately 1 mm in a brain that had received a muscle graft but that was otherwise undisturbed. Fluorescence micrograph, x 225. *Inset*, Higher magnification of C. Some of the exogenous macrophages have long processes (*arrows*) that lie parallel to the axons; x 780

It would be advantageous to the stability of the interstitial fluid composition for the BBB to remain intact during the passage of leukocytes across brain endothelium into the CNS parenchyma. The advantage is realized, at least in vitro. The impermeability imposed by zonular belts of tight junctions on a barrier layer of epithelial or endothelial cells is maintained during the passage of motile cells across the layer (MILKS et al. 1986). The permeability of the monolayers can be estimated in vitro by measuring the electrical resistance across it; the more complete the belts of tight junctions between the cells, the less the paracellular flow of ions and the higher the resistance across the layer (e.g., MARCIAL et al. 1984). Leukocytes can traverse a monolayer of endothelial cells from human umbilical vein without increasing the permeability to protein, such as albumin, or to ions, as manifested by an unchanged electrical resistance (HUANG et al. 1988). It is likely that the even more impermeable endothelium of cerebral vessels may also retain its barrier level during the passage of leukocytes across it in vivo. Circulating leukocytes leave cerebral vessels when attracted by α-bungarotoxin applied to the pial surface of the brain. The passage of the leukocytes appears to be transcytotic rather than paracellular in the inflammed region's endothelium, the junctions of which do not appear to be separated by the migrating cells. The lumenal face of the endothelium evaginates as arm-like projections that engulf the leukocytes, drawing them through the endothelial cells and out to the periendothelial basal lamina (FAUSTMANN and DERMIETZEL 1985).

The signals coming from CNS that attract circulating leukocytes are largely unknown, although cytokines may be one type. A ready source of extrinsic cytokines that act directly on brain endothelium is the perivascular macrophage. The mRNA for expression of IL-1 and its receptor also resides in endothelium, astroglia and neurons and choroid plexus (CUNNINGHAM and DESOUZA 1993). The number of macrophages increases within HIV-1 infected patients and these cells frequently contain the virus itself (POWER et al. 1993). When activated, such macrophages themselves synthesize and release cytokines (TYOR et al. 1992) which could then affect endothelium. In the brain with an intact BBB, resident perivascular phagocytes incorporate HRP that had been infused systemically (Fig. 3). The labeling by HRP of these phagocytes is a sign that the HRP had penetrated interstitial clefts. The sources of the HRP that had crossed the BBB to enter the cerebrum are probably focal leaks from a few subpial, cortical vessels, whereas the sources of HRP in more ventral regions may be the circumventricular organs (BROADWELL and SOFRONIEW 1993). It is possible that these resident perivascular phagocytes could be a vector in the spread of HIV-1 virus in brain parenchyma.

4 Summary

The aspects presented here of how solutes, viruses and cells are able to cross the BBB indicate that there must be an active interaction of endothelium with viruses and immune system cells before they can penetrate the brain and spinal cord. The axoplasmic pathway taken by lectin-solute conjugates is similar but not identical to that followed by viral particles during their retrograde or anterograde transit through the axoplasm. Both the conjugates and virus are transferred to other neurons transsynaptically but the receptor mediated transfer utilized by viruses is far more specific. Cranial nerves are involved in both the entry and egress of antigens into and out of the brain. Antigen, generated within the CNS, may be able to escape from the brain to lymphoid tissue by passing into the fluid around a cranial nerve, thence via the lymph into lymph nodes to initiate an immune response involving the CNS.

References

Baker H, Spencer RF (1984) Olfactory receptor neurons are a peripheral conduit for access of foreign substances to the central nervous system. Soc Neurosci Abstr 10: 859

Baker H, Spencer RF (1986) Transneuronal transport of peroxidase-conjugated wheat germ agglutinin (WGA-HRP) from the olfactory epithelium to the brain of adult rat. Exp Brain Res 63: 461–473

Bodenheimer TS, Brightman MW (1968) A blood-brain barrier to peroxidase in capillaries surrounded by perivascular spaces. Am J Anat 122: 249–267

Bradbury MWB (1979) The concept of the blood-brain barrier. Wiley, New York

Brightman MW (1968) The intracerebral movement of proteins injected into blood and cerebrospinal fluid of mice. Prog Brain Res 29: 19–37

Brightman MW, Reese TS (1969) Junctions between intimately apposed cell membranes in the vertebrate brain. J Cell Biol 40: 648–677

Brightman MW, Reese TS, Feder N (1970) Assessment with the electronmicroscope of the permeability to peroxidase of cerebral endothelium and epithelium in mice and sharks. In: Crone C, Lassen NA (eds) Capillary permeability. Munksgaard, Copenhagen, pp 468–476

Brightman MW, Hori M, Rapoport SI, Reese TS, Westergard E (1973) Osmotic opening of tight junctions in cerebra; endothelium. J Comp Neurol 152: 317–326

Broadwell RB, Brightman MW (1976) Entry of peroxidase into neurons of the central and peripheral nervous system from extracerebral and cerebral blood. J Comp Neurol 166: 257–284

Broadwell RB, Brightman MW (1979) Cytochemistry of undamaged neurons transporting exogenous protein in vivo. J Comp Neurol 185: 31–74

Broadwell RB, Balin BJ (1985) Olfactory labeling from mucosa. Endocytotic and exocytotic pathways of the neuronal secretory process in transynaptic transfer of wheat germ agglutinin-horseradish peroxidase in vivo. J Comp Neurol 242: 632–650

Broadwell RB, Sofroniew MV (1993) Serum proteins bypass the blood-brain fluid barriers for extracellular entry to the central nervous system. Exp Neurol 120: 245–263

Card JP, Whealy ME, Robbins AK, Moore RY, Enquist LW (1991) Two α-herpes virus strains are transported differentially in the rodent visual system. Neuron 6: 957–969

Card JP, Rinaman L, Lynn RB, Lee B-H, Meade RP, Miselis RR, Enquist LW (1993) Pseudorabies virus infection of the rat central nervous system: ultrastructural characterization of viral replication, transport and pathogenesis. J Neurosci 13: 2515–2539

Coulon P, Derbin C, Kucera P, Lafay F, Prehaud C, Flamand A (1989) Invasion of the peripheral nervous systems of adult mice by the CVS strain of rabies and its avirulent derivative Av01. J Virol 63: 3550–3554

Cunningham ET, DeSouza EB (1993) Interleukin 1 receptors in the brain and endocrine tissues. Immunol Today 14: 171–176

Da Cunha A, Jefferson JA, Jackson RW, Vitkovic L (1993) Glial cell-specific mechanisms of TGF-b1 induction by IL-1 in cerebral cortex. J Neuroimmunol 42: 71–86

Faustmann PM, Dermietzel R (1985) Extravasation of polymorophonuclear leukocytes from the cerebral microvasculature. Inflammatory response induced by alpha-bungarotoxin. Cell Tissue Res 242: 399–407

Field HJ, Hill TJ (1975) The pathogenesis of pseudorabies in mice: virus replication at the inoculation site and axonal uptake. J Gen Virol 26: 145–148

Flamand A, Gagner J-P, Morrison LA, Fields BN (1991) Penetration of the nervous systems of suckling mice by mammalian reoviruses. J Virol 65: 123–131

Harling-Berg C, Knopf PM, Merriam J, Cserr HF (1989) Role of cervical lymph nodes in the systemic humoral response to human serum albumin microinfused into rat cerebrospinal fluid. J Neuroimmunol 25: 185–193

Hickey WF (1991) T-lymphocyte entry and antigen recognition of the central nervous system. In: Ader R, Felten DL, Cohen N (eds) Psychoneuroimmunology, 2nd edn. Academic, New York, pp 149–175

Hickey WF, Kimura H (1988) Perivascular microglial cells of the CNS are bone marrow derived and present antigen in vivo. Science 239: 290–292

Huang AJ, Furie MB, Nicholson SC, Fischsbarg J, Kiebovitch LS, Silverstein SC (1988) Effects of human neutrophil chemotaxis across human endothelial cell monolayers on the permeability of these monolayers to ions and macromolecules. J Cell Physiol 135: 355–366

Hoogebrugge PM, Suzuki K, Suzuki K, Poorthuis BJHM, Kobayashi T, Wagamaker G, vanBekkum DW (1988) Donor-derived cells in the central nervous system of twitcher mice after bone marrow transplantation. Science 239: 1035–1038

Ishihara S, Sawada M, Chang L, Kim J-M, Brightman MW (1993) Brain vessels near muscle autografts are sites for entry of isogeneic macrophages into brain. Exp Neurol 124: 219–230

Jeffries WA, Brandon MR, Hunt SV (1984) Transferrin receptor on endothelium of brain capillaries. Nature 312: 162–163

Kuypers HGJM, Ugolini G (1990) Viruses as transneuronal tracers. Trends Neurosci 1s3: 71–75

Lucinskas FW, Cybulski MI, Kiely J-M, Pekind CS, Davis VM, Gimbrone MA (1991) Cytokine-activated human endothelial monolayers support enhanced neutrophil transmigration via a mechanism involving both endothelial-leukocyte adhesion molecule-1 and intercellular adhesion molecule. J Immunol 146: 1617–11625

Marchand CF, Schwab ME (1986) Binding, uptake and retrograde axonal transport of herpes virus suis in sympathetic neurons. Brain Res 383: 262–270

Marcial MA, Carlson SL, Madara JL (1984) Partitioning of paracellular conductance along the ileal crypt-villus axis: a hypothesis based on structural analysis with detailed consideration of tight junction structure-function relationships. J Membr Biol 80: 59–70

Marshall DW, Brey RL, Butzin CA, Lucey DR, Abbadessa M, Boswell RN (1991) Changes in a longitudinal study of 124 neurologically normal HIV-1 infected United States Airforce personnel. J Acquir Immuno Defic Syndr 4: 777–781

McArthur JC, Nance-Sporson TE, Griffin DE, Hoover D, Selnes OA, Miller EN, Margolick JB, Cohen BA, Farzadegan H, Saah A (1992) The diagnostic utility of elevation in cerebrospinal fluid beta-microglobulin in HIV–1 dementia. Neurology 42: 1707–1712

Metlay JP, Pure E, Steinman R (1989) Control of the immune response at the level of antigen-presenting cells: a comparison of the function of dendritic cells and B lymphocytes. Adv Immunol 47: 45–116

Milks LC, Conyers GP, Cramer EB (1986) The effect of neutrophil migration on epithelial permeability. J Cell Biol 103: 2729–2738

Morrison LA, Fields BN (1991) Viral pathogenesis and central nervous system infection. Semin Neurosci (3): 83–91

Morrison LA, Sidman RL, Fields BN (1991) Direct spread of reovirus from the intestinal lumen to the central nervous system through vagal autonomic nerve fibers. Proc Natl Acad Sci USA 88: 3852–3856

Neuwelt EA, Pagel MA, Dix RD (1991) Delivery of ultraviolet-inactivated ^{35}S-herpesvirus across an osmotically modified blood-brain barrier. J Neurosurg 74: 475–479

Oppenheimer-Marks N, Ziff M (1988) Migration of lymphocytes through endothelial cell monolayers: augmentation by interferon-γ. Cell Immunol 114: 307–323

Petito CK, Cash KS (1992) Blood-brain barrier abnormalities in the acquired immunodeficiency syndrome: immunohistochemical localization of serum proteins in postmortem brain. Ann Neurol 32: 658–666

Pober JS, Gimbrone MA, Lapierre LA, Mendrick DL, Fiers W, Rothlein R, Springer T (1986) Overlapping patterns of activation of human endothelial cells by interleukin 1, tumor necrosis factor and immune interferon. J Immunol 137: 1893–1896

Power C, Kong P-A, Crawford TO, Wesselingh S, Glass JD, McArthur JC (1993) Cerebral white matter changes in acquired immunodeficiency syndrome dementia: alterations of the blood-brain barrier. Ann Neurol 34: 339–350

Quagliarello VJ, Wispelwey B, Long WJ Jr, Scheld WM (1991) Recombinant interleukin-1 induces mioningitis and blood-brain barrier injury in the rat. J Clin Invest 87: 1360–1366

Reese TS, Karnovsky MJ (1967) Fine structural localization of a blood-brain barrier to exogenous peroxidase. J Cell Biol 34: 207–217
Rhodes RH (1991) Evidence of serum-protein leakage across the blood-brain barrier in the acquired immunodeficiency syndrome. J Neuropathol Exp Neurol 50: 171–183
Rosenberg ZF, Fauci AS (1990) Immunopathogenic mechanisms of HIV infection: cytokine induction of HIV expression. Immunol Today 11: 176–180
Sedgwick JD, Dörries R (1991) The immune system response to viral infection of the CNS. Semin Neurosci 3: 93–100
Shipley MT (1985) Transport of molecules from nose to brain: transneuronal anterograde and retrograde labeling in the rat olfactory system by wheat germ agglutinin-horseradish peroxidase applied to the nasal epithelium. Brain Res Bull 15: 129–142
Springer TA (1990) Adhesion receptors of the immune system. Nature 346: 425–434
Tyler KL, Fields BN (1990) Pathogenesis of viral infection. In: Fields BN, Knipe DM, Chanock RM, Hirsch MS, Melnick JL, Monath TP, Roizman B (eds) Virology, 2nd edn. Raven, New York, pp 1911–1240
Tyler KL, McPhee DA (1987) Molecular and genetic aspects of the pathogenesis of viral infections of the central nervous system. CRC Crit Rev Neurobiol 3: 221–243
Tyor WR, Glass JD, Griffin JW, Becker PS, McArthur JC, Bezman L, Griffin D (1992) Cytokine expression in the brain during the acquired immunodeficiency syndrome. Ann Neurol 31: 349–360
Vitković L, da Cunha A, Tyor WR (1994) Cytokine expression and pathogenesis in AIDS brain In: Price RW, Perry SW (eds) HIV, AIDS and the brain. Raven, New York, pp 203–222
Wahl S, Hunt DA, Wakefield LM, Francis NM, Wahl LM, Roberts AB, Sporn MB (1987) Transforming growth factor type β induces monocyte chemotaxis and growth factor production. Proc Natl Acad Sci USA 84: 5788–5792
Weindl A (1973) Neuroendocrine aspects of circumventricular organs. In: Ganong WF, Martini L (eds) Frontiers in neuroendocrinology. Oxford University Press, London, pp 1–32
Ziegler RJ, Pozos RS (1981) Effects of lectins on peripheral infection by herpes simplex virus of rat sensory neurons in culture. Infect Immunol 34: 588–595

Cytotoxic T Lymphocytes and HIV-1-Related Neurologic Disorders

S.A. KALAMS and B.D. WALKER

1 Introduction

Human immunodeficiency virus (HIV)-1 infection is characterized by progressive and ultimately profound immune suppression, but a growing body of data indicate that this disease is also associated with extreme levels of virus-specific immune activation during both asymptomatic and symptomatic phases of the illness. An extremely vigorous cytotoxic T lymphocyte (CTL) response has been detected in infected persons (reviewed in KALAMS and WALKER 1994), although the potential contribution of this response in retarding disease progression has not been firmly established. Alternatively, CTLs or cytokines produced by these cells upon activation in vivo have been postulated to contribute to the immunopathogenesis of disease (PLATA et al. 1987; JASSOY et al. 1992, 1993). By analogy to animal models of virus infection involving the central nervous system, a better understanding of the CTL response to HIV-1 may be particularly relevant in understanding neuropathogenesis associated with this infection. The review will focus on the current understanding of the CTL response to HIV-1 in infected persons, including the CTL response in the central nervous system.

Infectious Disease Unit, Harvard Medical School and Massachusetts General Hospital, 55, Fruit Street, Boston, MA 02114, USA

2 Cytotoxic T Lymphocytes in HIV-1-1nfected Persons

Infection with HIV-1 is relatively unique among chronic viral infections in that this virus is associated with an extremely vigorous, activated, circulating CTL response. Although detection of antiviral CTL among peripheral blood mononuclear cells (PBMCs) typically requires a period of in vitro activation and expansion in order to boost this activity to the level of detection, the magnitude of the circulating CTL response in HIV-1-infected persons is often of sufficient magnitude that this activity can be detected using freshly isolated PBMCs (WALKER et al. 1987,1988; KOENIG et al. 1988; RIVIERE et al. 1989; KOUP et al. 1989). The ability to detect these cells in the circulation provides ample evidence that this infection is associated with marked immune activation.

A number of laboratories have contributed to an expanding knowledge not only of the viral proteins targeted by this response, but also the precise viral epitopes which are recognized by these cells (reviewed in MCMICHAEL and WALKER 1994). CTLs recognize viral proteins which are presented at the cell surface as a complex with an HLA molecule. In HIV-1 infection, the cells which mediate cytotoxicity are predominantly CD8-positive lymphocytes that recognize viral peptides presented by HLA class I molecules, although minor components of cytotoxicity mediated by class II restricted CD4-positive CTLs have also been reported (LITTAUA et al. 1992; KUNDU and MERIGAN 1992; CURIEL et al. 1993). CD8-positive, HLA class I restricted CTL responses to most of the HIV-1 proteins have been reported, including the gag, pol, env, nef, vif, tat and rev proteins (KALAMS and WALKER 1994). A number of studies have suggested that the gag protein may be the most frequently targeted protein, with up to 80% of infected persons demonstrating CTL responses to this structural protein (BUSEYNE et al. 1993). In some individuals, the levels of lysis directed at this protein may be extremely high, when fresh circulating PBMCs are evaluated (JOHNSON et al. 1991). Other studies have addressed the frequency of CTL precursors (CTLp) in the circulation in infected persons and again underscore the vigor of the CTL response, with frequencies of gag-specific CTL as high as 1/1700 cells (KOUP et al. 1991). Comparably high frequencies of CTL and of CTLp compared to other proteins such as the envelope protein have also been reported (KALAMS et al. 1994).

The CTL response to HIV-1 has also been evaluated with respect to the precise epitopes targeted by this effector cell population. For the most part, CTL clones have been used to make these determinations. The actual CTL epitopes are defined using synthetic viral peptides to sensitize uninfected autologous cells, using B lymphoblastoid cell lines (B-LCL) (WALKER et al. 1989). Using progressively truncated peptides, one can identify the minimum peptide which results in sensitization of target cells. The magnitude of the immune response even to a single epitope may be sufficient to detect cytolytic activity using fresh PBMCs derived directly from the peripheral circulation, as is shown for an envelope gp41 epitope in Fig. 1 . The HIV-1 epitopes which have been defined correspond very well with the predicted motifs for peptides based on elution of peptides from class I

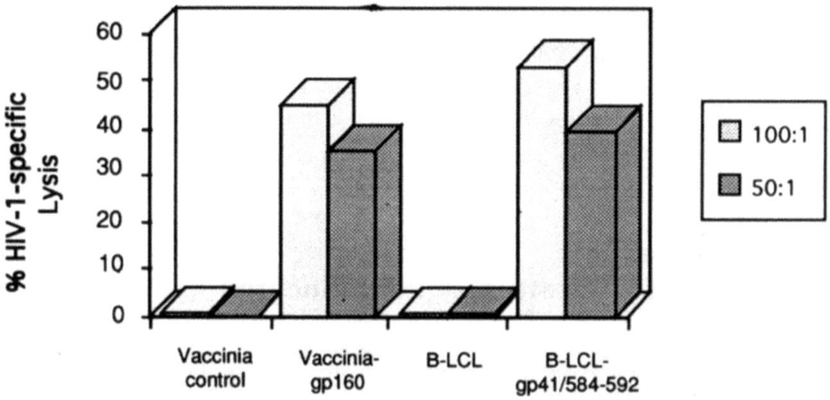

Fig. 1. Lysis of autologous B lymphoblastoid cell lines (B-LCL) infected with vaccinia expressing gp160 or incubated with peptide gp41/584-592. Fresh peripheral blood mononuclear cells (PBMC) from a patient were isolated over a Ficoll gradient and incubated with 1×10^4 target cells at effector: target ratios of 100:1 and 50:1. Target cells were B-LCL infected with a control vaccinia virus, a vaccinia-gp 160 vector or incubated with peptide gp41/584-592 or no peptide. Lysis of target cells incubated with irrelevant peptide was always < 5%

molecules (FALK et al. 1991). Such detailed studies of epitopes have revealed a number of features which are worthy of comment. The vast majority if not all epitopes identified thus far lie in regions of the virus in which sequence variation is tolerated. This may have important implications for disease pathogenesis in that even single amino acid changes may abrogate recognition by an established CTL response. This raises the question as to whether immune escape may play a pivotal role in disease progression. However, detailed studies in some individuals have revealed that as many as 14 distinct CTL epitopes may be targeted by the CTL response in a single individual (JOHNSON and WALKER 1994). Although the breadth of the CTL response is likely to vary among different individuals, such a broadly directed CTL response may be expected to decrease the likelihood that immune escape will occur. Another important feature of the CTL response to HIV-1 is that the epitopes recognized vary considerably depending on the HLA type of the infected individual. Even in persons sharing common HLA class I alleles, different class I restricted epitopes have been identified (JOHNSON et al. 1991).

Although the majority of studies of HIV-1-specific CTLs have relied on PBMCs as the source of effector CTLs, these cells have also been detected at specific tissue sites. In some of the earliest studies of HIV-1-specific CTLs reported, a vigorous CTL response was detected in the lungs of persons with lymphocytic alveolitis (PLATA et al. 1987). Other studies have demonstrated CTLs within the lymph nodes of infected persons (HADIDA et al. 1992), which may be particularly relevant since lymph nodes are thought to represent a major reservoir for virus even in the early stages of illness (PANTALEO et al. 1993). CTLs have also been detected within the central nervous system of infected persons. In the first published studies, CTLs were detected after multiple rounds of in vitro stimulation with viral antigen (SETHI et al. 1988). More recently, a vigorous CTL response has been detected in the CSF of persons with AIDS dementia complex (JASSOY et al. 1992), raising the issue whether the virus-specific immune response may contribute to the neurologic disease.

3 Functional Role of HIV-1-Specific Cytotoxic T Lymphocytes

The potential role of CTLs as an important correlate of protective immunity has been suggested by a number of studies, but definitive proof is still lacking. Some of the supportive data are inferential and derive from studies of persons in various stages of illness. Recent studies of persons in the acute phase of seroconversion have shown a temporal relation between decreasing viral load and the development of CTL responses. Although detectable neutralizing antibody responses were absent in all five subjects evaluated in one study, CTL precursors specific for gag, env and pol were present in four of five persons at or within 3 weeks of seroconversion and were present in all five at six months (KOUP et al. 1994). Frequencies of these cells ranged from 30 to 3800 per million PBMCs. Only the person with the absent initial response experienced prolonged viremia, persistent symptoms and a low CD4 count. Similar associations with CTL of responses and clearance of viremia have been seen in simian immunodeficiency virus (SIV)-infected macaques (YASUTOMI et al. 1993).

Other studies have focused on the CTL response in certain defined cohorts at various stages of illness. One group of individuals which has been evaluated for CTL responses consists of persons who have been infected for 10 years or more who have remained asymptomatic with no evidence of CD4 lymphocyte decline. Studies in our laboratory have revealed the presence of a vigorous CTL response in the majority of these individuals, with detectable circulating activated CTLs in approximately half of these individuals (T. HARRER, E. HARRER, S. KALAMS et al. unpublished observations). These studies indicate that immune activation is also present in persons in whom no disease progression has been observed. Other studies have characterized the CTL response in persons in various stages of CD4 decline, suggesting a correlation between decline in CTLs and disease progression. CARMICHAEL et al . (1993) have demonstrated a selective loss of H IV-1 specific CTLp as CD4 cell counts decline, while the Epstein-Barr virus EBV-specific CTLp remain stable, again suggesting that CTLs may contribute to the prolonged asymptomatic phase of the illness. The precise mechanism whereby CTL might exert a protective effect has not been fully determined. Studies by Levy and coworkers have shown that CD8 cells from infected persons are able to inhibit HIV-1 replication in vitro in PBMCs (WALKER et al. 1986), and have suggested that this effect is mediated by the release of a soluble factor (WALKER and LEVY 1989).

Others have suggest that this inhibitory activity is mediated by CTLs, and our own studies with cloned HIV-1-specific CTLs indicate that these cells are capable of inhibiting viral replication in vitro (R.P. Johnson et al., unpublished observations).

The persistence of virus replication and ultimate disease progression with HIV-1 infection despite a vigorous, virus-specific CTL response has prompted investigation into mechanisms that allow HIV-1 to evade suppression from the host immune response. Theories that have been proposed include HIV-1 sequence variation that allows escape from CTL recognition and a decline of HIV-1-specific CTL activity over the course of disease (CARMICHAEL et al. 1993; HOFFENBACH et al. 1989). Results from our laboratory indicate that in certain instances the CTL response against an immunodominant epitope may be due to the clonal expansion of a single progenitor cell (KALAMS et al. 1994). CTL clones specific for an epitope in gp41 were isolated from an infected individual at different timepoints over a 31 month period, and, over the first 27 months of the study, all of the CTL clones isolated expressed the identical T cell receptor (TCR) as defined by nucleotide sequence. In addition to showing a limited TCR repertoire by CTL able to recognize this epitope, this study emphasizes the longevity of CTL clones in vivo (KALAMS et al. 1994). Subsequent studies have extended the period of time we have been able to isolate this particular CTL clone out to almost 4 years (unpublished observations). Interestingly, this CTL clone is not able to recognize the dominant HIV-1 genotype present in the PBMCs of this subject (unpublished observations). Continued longitudinal study of the TCR repertoire of HIV-1-specific CTLs will allow us to address the ultimate fate of these cells.

Several groups have investigated the potential role HIV-1 variation leading to escape from CTL recognition. The extensive sequence variation of HIV-1 has led to the term "quasispecies" to describe the mixture of related viruses found in infected individuals (GOODENOW et al. 1989). It has been postulated that variation within CTL epitopes that allows viral escape from CTL recognition could confer a selective advantage in vivo. In SIV-infected macaques that express the MAMU A*01 MHC type, gag-specific CTLs are directed against a single immunodominant epitope (amino acids 182-190 in p24) (MILLER et al. 1991). CHEN et al. (1992) examined sequence variation in this epitope in three infected MAMU A*01 monkeys. Although sequence variation was present within this epitope, none of the amino acid substitutions resulted in loss of recognition by previously isolated CTLs. This study did not directly address the issue of immune escape because the CTLs used were not derived from the monkeys from whom the sequence data were derived. In a cohort of HIV-1-seropositive hemophiliacs, variant peptides corresponding to deduced amino acid sequences in HLA-B8 and HLA-B27 restricted epitopes were tested for recognition by CTL (PHILLIPS et al. 1991). Although there were shifts in the proportion of viruses containing sequences that were not recognized by autologous CTLs over time, there was no clear selection for these variants (PHILLIPS et al. 1991). MEYERHANS et al. (1991) evaluated sequence variation in the same HLA-B27 epitope in four patients and none of the amino acid changes led to loss of recognition by CTLs. These data do not unequivocally establish that sequence variation leading to escape from CTL recognition is a major mechanism for HIV-1 disease progression, and there is as yet no clear evidence that such variants have a selective advantage in vivo.

Other potential mechanisms that allow HIV-1 to evade the CTL response include down-regulation of MHC class I molecules (Kerkau et al. 1989), decreased expression of LFA-1 (Amiel et al. 1988), or a diminished ability to present antigen due to impaired dendritic cell function and thereby respond to virus mutation (Macatonia et al. 1990). The relative contribution of each of these to disease progression remains to be determined.

4 Cytotoxic T Lymphocytes and the Central Nervous System in HIV-1 Infection

The majority of all AIDS patients show mild to severe signs of AIDS dementia complex (ADC), characterized by varying degrees of motor, cognitive and behavioral impairment (Price et al. 1988; Navia et al. 1986 a,b). Pathologic changes in advanced stages of ADC tend to be diffuse and include lymphocytes, mainly of the CD8 phenotype, in perivascular regions and in microglial nodules (Achim et al. 1991; Vazeux et al. 1987). In murine lymphocytic choriomeningitis virus (LCMV) infection, as few as 100–1000 virus-specific class I restricted CTLs can induce fatal immunopathology when injected intracerebrally into infected mice (Klavinskis et al. 1989). Death occurs despite the absence of morphologic changes such as infiltrating lymphocytes, suggesting that soluble factors released by CTLs may contribute to the observed pathology. Two independent studies have shown that HIV-1-specific CTLs can be detected within the central nervous system in persons infected with HIV-1. In the first published study, Sethi et al. (1988) reported the detection of CTLs of both the CD4 and CD8 phenotype, following in vitro stimulation of CSF-derived lymphocytes with viral antigen or whole virus. The restriction patterns for these cells indicated marked phenotypic heterogeneity among the effector cells.

In our own studies of CTLs in the central nervous system, we have identified HIV-1-specific, class I restricted CTL in five of six subjects studied (Jassoy et al. 1992). All subjects had mild to moderate ADC, with CSF cell counts of up to 8/mm^3. The methodology employed involves the in vitro cloning of lymphocytes using an anti-CD3 monoclonal antibody as a stimulus for T cell proliferation. CTL directed at the gag, pol, env and nef protein were detected, and frequency analysis revealed that as many as one in 333 lymphocytes from the CSF were HIV-1-specific. In four of the subjects evaluated, the frequency of HIV-1-specific CTLs was higher in the CSF than in the peripheral blood, suggesting a local recruitment to or induction within the central nervous system. The specific targets of these cells within the CNS are yet to be determined, but are likely to include infected monocyte/macrophages, microglial cells, endothelial cells, astrocytes, and oligodendrocytes.

Although these cells may serve to inhibit HIV-1 replication, they may also contribute to disease pathogenesis. Histologic studies of persons with ADC do not indicate the presence of large numbers of infiltrating lymphocytes, but release of cytokines or other products by activated CTLs could potentially contribute to the observed changes characteristic of this illness. Elevated levels of interferon (IFN)-α have been detected in the peripheral blood (RINALDO et al. 1990; TYOR et al. 1992) and CSF (GRIFFIN et al. 1991) of infected persons. In addition, serum, brain tissue and CSF have been shown to contain elevated levels of TNF alpha (TYOR et al. 1992; KOBAYASHI et al . 1990). A potential source of these cytokines is activated CTL, although few studies have directly addressed the patterns of cytokine release by virus-specific, class I restricted human CTLs. We have recently demonstrated that HIV-1-specific CTLs derived either from the CSF or the peripheral blood release IFN-α, tumor necrosis factor (TNF)-α and TNF-β in an epitope-specific fashion when they encounter the cognate epitope in the proper HLA setting (JASSOY et al. 1983).

Several potential mechanisms involving CTL-related cytokines may contribute to observed neuropathologic changes seen in ADC. The ability of IFN-γ to induce metabolic changes in macrophages and microglial cells, including alterations in the metabolic pathways for kinurinine (HEYES et al. 1991) and tryptophan (WERNER et al. 1987), could lead to increased production of neopterin and quinolinic acid. Release of TNF-α or β may stimulate astrocytes to proliferate and may thereby contribute to reactive gliosis, a characteristic histopathologic change seen in the brains of persons with HIV-related encephalitis (SELMAJ et al. 1990). TNF-α has also been associated with demyelination in the central nervous system (SELMAJ and RAINE 1988). The release of these cytokines by HIV-1-specific CTLs provides a potential physiologic link between an activated immune system and the destructive neuropathic changes associated with this illness. However, it must be stressed that the contribution of CTL-associated cytokines to either protection from disease progression or to the neurologic consequences of this disease remains unclear.

5 Summary

In summary, one can conclude that infected persons exhibit an extremely vigorous, virus-specific CTL response, and in at least some individuals this response is broadly directed at multiple epitopes. These cells are present at the time or initial control of viremia and can also be detected after more than a decade of asymptomatic infection. These cells can also be found in the central nervous system in persons with ADC, and one can envision pathways in which the inflammatory cytokines released by these cells upon activation could contribute to the neurologic sequelae of infection. However, the precise role of these cells as a protective host defense and the possible contribution of these cells, or products released by these cells, to tissue damage at sites such as the lung and brain remain to be determined. Further delineation of the role played by CTLs in the pathogenesis of disease should be extremely useful in helping to understand the disease itself and to guide intervention strategies.

References

Achim CL, Morey M, Wiley C (1991) Expression of major histocompatibility complex and HIV antigens within the brain of AIDS patients. AIDS 5: 535–541

Amiel C, Bene MC, May T, Canton P, Faure GC (1988) LFA1 expression in HIV infection. AIDS 2(3): 211–214

Buseyne F, McChesney M, Porrot F, Kovarik S, Guy B, Riviere Y (1993) GagSpecific cytotoxic T lymphocytes from HIV-1 infected individuals: gag epitopes are clustered in three regions of the p24 gag protein. J Virol 67: 694

Carmichael A, Jin X, Sissons P, Borysiewicz L (1993) Quantitative analysis of the human immunodeficiency virus type 1 (HIV-1)-specific cytotoxic T lymphocyte (CTL) response at different stages of HIV-1 infection: differential CTL responses to HIV-1 and Epstein-Barr virus in late disease. J Exp Med 177(2): 249 256

Chen ZW, Shen L, Miller MD, Ghim SH, Hughes AL, Letvin NL (1992) Cytotoxic T lymphocytes do not appear to select for mutations in an immunodominant epitope of simian immunodeficiency virus gag. J Immunol 149(12): 4060–4066

Curiel TJ, Wong JT, Gorczyca PF, Schooley RT, Walker BD (1993) CD4+ human immunodeficiency virus type 1 (HIV-1) envelope-specific cytotoxic T lymphocytes derived from the peripheral blood cells of an HIV-1-infected individual. AIDS Res Hum Retroviruses 9(1): 61–68

Falk K, Rotzschke O, Stevanovic S, Jung G, Rammensee HG (1991) Allelespecific motifs revealed by sequencing of self-peptides eluted from MHC molecules. Nature 351(6324): 290–296

Goodenow M, Huet T, Saurin W, Kwok S, Sninsky J, Wain-Hobson S (1989) HIV-1 isolates are rapidly evolving quasispecies: evidence for viral mixtures and preferred nucleotide substitutions. J Acquir Immune Defic Syndr 2(4): 344–352

Griffin D, McArthur J, Cornblath D (1991) Neopterin and interferon-gamma in serum and cerebrospinal fluid of patients with HIV-associated neurologic disease. Neurology 41: 69–74

Hadida F, Parrot A, Kieny MP, Sadat SB, Mayaud C, Debre P, Autran B (1992) Carboxyl-terminal and central regions of human immunodeficiency virus-1 NEF recognized by Cytotoxic T lymphocytes from lymphoid organs. An in vitro limiting dilution analysis. J Clin Invest 89(1): 53–60

Heyes M, Brew B, Martin A, Price R, Salazar A, Sidtis J, Yergey J, Mouradian M, Sadler A, Keilp J, Rubinow D, Markey S (1991) Quinolonic acid in cerebrospinal fluid and serum in HIV-1 infection: relationship to clinical and neurological status. Ann Neurol 29: 202–209

Hoffenbach A, Langlade DP, Dadaglio G, Vilmer E, Michel F, Mayaud C, Autran B, Plata F (1989) Unusually high frequencies of HIV-specific cytotoxic T lymphocytes in humans. J Immunol 142(2): 452–462

Jassoy C, Johnson RP, Navia BA, Worth J, Walker BD (1992) Detection of a vigorous HIV-1-specific cytotoxic T lymphocyte response in cerebrospinal fluid from infected persons with AIDS dementia complex. J Immunol 149: 3113–3119

Jassoy CJ, Harrer T, Rosenthal T, Navia BA, Worth J, Johnson RP, Walker BD (1993) HIV-1-specific cytotoxic T cells release interferon gamma, tumor necrosis factor (TNF)-alpha and TNF-beta when they encounter their target antigens. J Virol 67:2844–2852

Johnson R, Walker B (1994) HIV-1-specific CTL: responses to structural proteins. Curr Opin Microbiol Immunol 189: 35–63

Johnson RP, Trocha A, Yang L, Mazzara GP, Panicali DL, Buchanan TM, Walker BD (1991) HIV-1 gag-specific cytotoxic T lymphocytes recognize multiple highly conserved epitopes. Fine specificity of the gag-specific response defined by using unstimulated peripheral blood mononuclear cells and cloned effector cells. J Immunol 147(5): 1512–1521

Kalams SA, Walker BD (1994) The cytotoxic T-lymphocyte response in HIV-1 infection. Clin Lab Med 14(2): 271–299

Kalams SA, Johnson RP, Trocha AK, Dynan MJ, Ngo HS, DAquila RT, Kurnick JT, Walker BD (1994) Longitudinal analysis of T cell receptor (TCR) gene usage by human immunodeficiency virus 1 envelope-specific cytotoxic T lymphocyte clones reveals a limited TCR repertoire. J Exp Med 179(4): 1261–1271

Kerkau T, Schmitt LR, Schimpl A, Wecker E (1989) Downregulation of HLA class I antigens in HIV-1-infected cells. AIDS Res Hum Retroviruses 5(6): 613–620

Klavinkis L, Tishon A, Oldstone M (1989) Efficiency and effectiveness of cloned virus-specific cytotoxic T lymphocytes in vivo. J Immunol 143(6): 2013–2016

Kobayashi N, Hamamoto Y, Yamamoto N (1990) Production of tumor necrosis factors by human T cell lines infected with HTLV-1 may cause their high susceptibility to human immunodeficiency virus infection. Med Microbiol Immunol (Berl) 179(2): 115–122

Koenig S, Earl P, Powell D, Pantaleo G, Merli S, Moss B, Fauci AS(1988) Group-specific, major histocompatibility complex class I-restricted cytotoxic responses to human immunodeficiency virus 1 (HIV-1) envelope proteins by cloned peripheral blood T cells from an HIV-1-infected individual. Proc Natl Acad Sci USA 85(22): 8638–8642

Koup RA, Sullivan JL, Levine PH, Brettler D, Mahr A, Mazzara G, McKenzie S, Panicali D (1989) Detection of major histocompatibility complex class I-restricted, HIV-specific cytotoxic T lymphocytes in the blood of infected hemophiliacs. Blood 73(7): 1909–1914

Koup RA, Pikora CA, Luzuriaga K, Brettler DB, Day ES, Mazzara GP, Sullivan JL (1991) Limiting dilution analysis of cytotoxic T lymphocytes to human immunodeficiency virus gag antigens in infected persons: in vitro quantitation of effector cell populations with p17 and p24 specificities. J Exp Med 174(6): 1593–1600

Koup RA, Safrit JT, Cao Y, Andrews CA, McCloud G, Borkowsky W, Farthing C, Ho DD (1994) Temporal association of cellular immune responses with the initial control of viremia in primary HIV-1 syndrome. J Virol 68(7): 4650–4655

Kundu SK, Merigan TC (1992) Equivalent recognition of HIV proteins, Env, Gag and Pol, by CD4+ and CD8+ cytotoxic T-lymphocytes AIDS 6(7): 643–649

Littaua RA, Oldstone MB, Takeda A, Ennis FA (1992) A CD4+ cytotoxic T-lymphocyte clone to a conserved epitope on human immunodeficiency virus type 1 p24: cytotoxic activity and secretion of interleukin-2 and interleukin-6. J Virol 66(1): 608–611

Macatonia SE, Lau R, Patterson S, Pinching AJ, Knight SC (1990) Dendritic cell infection, depletion and dysfunction in HIV infected individuals. Immunology 71: 38–45

McMichael AJ, Walker BD (1994) Cytotoxic T lymphocyte epitopes: implications for HIV vaccines. AIDS 8 [Suppl 1]: S 155–S 173

Meyerhans A, Dadaglio G, Varatanian JP, Langlade DP, Frank R, Asjo B, Plata F, S. W-H (1991) In vivo persistence of a HIV-1-encoded HLA-B27-restricted cytotoxic T lymphocyte epitope despite specific in vitro reactivity. Eur J Immunol 21(10): 2637–2640

Miller MD, Yamamoto H, Hughes AL, Watkins DI, Letvin NL (1991) Definition of an epitope and MHC class I molecule recognized by gag-specific cytotoxic T lymphocytes in SIVmac-infected rhesus monkeys. J Immunol 147(1): 320–329

Navia B, Cho E-S, Petito C, Price R (1986a) The AIDS dementia complex: II. neuropathology. Ann Neurol 19: 525–535

Navia B, Jordan B, Price R (1986b) The AIDS dementia complex: I. Clinical features. Ann Neurol 19: 517–524

Pantaleo G, Graziosi C, Demarest JF, Butini L, Montroni M, Fox CH, Orenstein JM, Kotler DP, Fauci AS (1993) HIV infection is active and progressive in lymphoid tissue during the clinically latent stage of disease. Nature 362 (6418): 355–358

Phillips RE, Rowland JS, Nixon DF, Gotch FM, Edwards JP, Ogunlesi AO, Elvin JG, Rothbard JA, Bangham CR, Rizza CR et al. (1991) Human immunodeficiency virus genetic variation that can escape cytotoxic T cell recognition. Nature 354(6353): 453–459

Plata F, Autran B, Martins LP, Wain HS, Raphael M, Mayaud C, Denis M, Guillon JM, Debre P (1987) AIDS virus-specific cytotoxic T lymphocytes in lung disorders. Nature 328(6128): 348–351

Price R, Brew B, Sidtis J, Rosenblum M, Scheck A, Cleary P (1988) The brain in AIDS: central nervous system HIV-1 infection and AIDS dementia complex. Science 239: 586–592

Rinaldo C, Armstrong J, Kingsley L, Zhou S, Ho M (1990) Relation of alpha and gamma interferon levels to development of AIDS in homosexual men. J Exp Pathol 5(3):127–132

Riviere Y, Tanneau SF, Regnault A, Lopez O, Sansonetti P, Guy B, Kieny MP, Fournel JJ, Montagnier L (1989) Multiple cytotoxic effector are induced by infection with the human immunodeficiency virus. Res Immunol 140(1): 110–115

Selmaj KW, Raine CS (1988) Tumor necrosis factor mediates myelin and oligodendrocyte damage in vitro. Ann Neurol 23(4): 339–346

Selmaj KW, Farooq M, Norton WT, Raine CS, Brosnan CF (1990) Proliferation of astrocytes in vitro in response to cytokines. A primary role for tumor necrosis factor. J Immunol 144(1): 129–135

Sethi KK, Naher H, Stroehmann I (1988) Phenotypic heterogeneity of cerebrospinal fluid-derived HIV-specific and HLA-restricted cytotoxic T-cell clones. Nature 335(6186): 178–181

Tyor W, Glass J, Griffin J, Becker P, McArthur J, Bezman L, Griffin D (1992) Cytokine expression in the brain during acquired immunodeficiency syndrome. Ann Neurol 31: 349–360

Vazeux R, Brousse N, Jarry A, Henin D, Marche C, Vedrenne C, Mikol J, Wolf M, Michon C, Rozenbaum W, Bureau J-F, Montagnier L, Brahic M (1987) AIDS subacute encephalitis: identification of HIV-infected cells. Am J Pathol 126: 403–410

Walker CM, Levy JA (1989) A diffusible lymphokine produced by CD8+T lymphocytes suppresses HIV replication. Immunology 66(4): 628–630

Walker CM, Moody DJ, Stites DP, Levy JA (1986) CD8+ lymphocytes can control HIV infection in vitro by suppressing virus replication. Science 234(4783):1563–1566

Walker BD, Chakrabarti S, Moss B, Paradis TJ, Flynn T, Durno AG, Blumberg RS, Kaplan JC, Hirsch MS, Schooley RT (1987) HIV-specific cytotoxic T lymphocytes in seropositive individuals. Nature 328 (6128): 345–348

Walker BD, Flexner C, Paradis TJ, Fuller TC, Hirsch MS, Schooley RT, Moss B (1988) HIV-1 reverse transcriptase is a target for cytotoxic T lymphocytes in infected individuals. Science 240(4848): 64–66

Walker BD, Flexner C, Birch LK, Fisher L, Paradis TJ, Aldovini A, Young R, Moss B, Schooley RT (1989) Long-term culture and fine specificity of human cytotoxic T-lymphocyte clones reactive with human immunodeficiency virus type 1. Proc Natl Acad Sci USA 86(23): 9514–9518

Werner E, Bitterlich G, Fuchs D, Hausen A, Reibnegger G, Szabo G, Dierich M, Wachter H (1987) Human macrophages degrade tryptophan upon induction by interferon-gamma. Life Sci 41: 273–280

Yasutomi Y, Reimann KA, Lord CI, Miller MD, Letvin NL (1993) Simian immunodeficiency virus-specific CD8+ lymphocyte response in acutely infected rhesus monkeys. J Virol 67(3):1707–1711

Distinct HIV-1 *env* Sequences Are Associated with Neurotropism and Neurovirulence

C. Power[1,6,7], J.C. McArthur[1,2], R.T. Johnson[1], D.E. Griffin[1,3], J.D. Glass[1,4], R. Dewey[5], and B. Chesebro[6]

1 Introduction

Retroviral infections in many species are commonly associated with different forms of neurological illness. These neurological diseases differ widely in their clinical manifestations and in the extent of pathological changes such as severity of inflammation and sites of tissue destruction in the nervous system. For example, the human oncovirus HTLV-1 causes a myelopathy in less than 1% of infected adults (Roman 1988). Spinal cord pathology shows intense inflammation with T cell infiltrates and minimal virus detectable within the lesions (Power et al.

[1]Department of Neurology, Johns Hopkins University, 600 North Wolfe Street, Baltimore, MD 21287, USA

[2]Department of Epidemiology, Johns Hopkins University, 600 North Wolfe Street, Baltimore, MD 21287, USA

[3]Department of Medicine, Johns Hopkins University, 600 North Wolfe Street, Baltimore, MD 21287, USA

[4]Department of Pathology, Johns Hopkins University, 600 North Wolfe Street, Baltimore, MD 21287, USA

[5]Department of Neurosurgery, St. Patrick's Hospital, 601 West Spruce Street, Missoula, MT 59801, USA

[6]Laboratory of Persistent Viral Diseases, Rocky Mountain Laboratories, National Institute of Allergy and Infectious Diseases, 903 South Fourth Street, Hamilton, MT 59840, USA

[7]Present address: Section of Neurology, University of Manitoba, GF-543-700 William Ave.,Winnipeg, MB R3E 0Z3, Canada

1991; Roman 1988). The type C murine retroviruses can induce a variety of CNS diseases. Cas Br E (Lynch et al. 1991; Gardner et al. 1973; Jolicoeur et al. 1992a), Friend PVC 211 (Masuda et al. 1992, 1993), and Moloney-ts 1 murine leukemia viruses (Szurek et al. 1990) cause a disease with spastic paralysis associated with noninflammatory spongiform degeneration involving primarily the gray matter of the spinal cord and brain stem. In contrast, FMCF-98D murine retrovirus (Buller et al. 1990) causes an ataxic disease associated with infection of white matter tracts of the cerebellum but showing minimal pathology. Friend TR 1.3 induces a stroke-like syndrome secondary to infection of brain capillary endothelial cells (Park et al. 1993). The lentiviruses including visna, CAEV, SIV, FIV, and HIV also cause a number of differing clinical syndromes with variable pathology. SIV, CAEV, and visna may be characterized by a progressive downhill neurological course of varying rapidity accompanied by widespread inflammation of the CNS (Sharma et al. 1992; Georgsson et al. 1989; Lackner et al. 1991; Narayan and Clements 1989). In contrast, FIV encephalopathy is slowly progressive with minimal inflammation (Lackner et al. 1991). HIV is associated with a wide spectrum of neurological syndromes (Johnson et al. 1988). Neurological diseases such as aseptic meningitis and acute demyelinating peripheral neuropathy occur in the early stages of HIV infection while HIV dementia, vacuolar myelopathy and painful sensory neuropathy arise in the later stages of HIV infection. All of these conditions are thought to be primary HIV-induced neurological syndromes, in contrast to opportunistic infections such as cerebral toxoplasmosis and cryptococcal meningitis which are secondary to prolonged immune suppression.

HIV dementia (HIVD) is one of the most devastating clinical syndromes befalling patients with AIDS because of its poor prognosis and incapacitating consequences (McArthur 1987; McArthur et al. 1993). HIVD occurs in 15%–20% of AIDS patients although HIV proteins or nucleic acids can be detected in the brains of nearly all AIDS patients. HIVD is associated with a triad of clinical features including motor impairment, cognitive decline and behavioral abnormalities (Navia et al. 1986). The severity of HIVD varies widely from patient to patient, and it is unclear what determines this diversity. Although a variety of histopathological abnormalities have been observed in CNS tissues of AIDS patients (Budka et al. 1987; Budka 1992), precise correlation of clinical deficits with pathology has been difficult, and it is apparent that many patients with severe HIVD have no easily detectable abnormalities on routine histological analysis at autopsy (Budka 1991; Glass et al. 1993).

Multiple hypotheses have emerged to explain the pathogenesis of HIVD. These include both direct and indirect neurotoxicity of both viral and host molecules. For example, viral proteins such as gp120 have been shown to be toxic in vitro (Dreyer et al. 1990; Lipton 1991). Host agents including quinolinic acid and nitric oxide may also play an excitotoxic role in HIV-related disease (Heyes et al. 1991; Dawson et al.1993). Immunological molecules including cytokines and arachidonic acid metabolites have been shown to be elevated in patients with HIVD (Tyor et al. 1992; Wesselingh et al. 1993; Griffin et al. 1994). The consequences of increased cytokine levels such as altered blood-brain barrier permeability and abnormal myelination have also been reported to occur in HIV-infected patients (POWER et al. 1993; PETITO and CASH 1992).

Many of the virological parameters of HIV dementia have been examined but no precise correlation with dementia has been found. HIV antigens and nucleic acids have been identified in the brains of AIDS patients both with and without HIVD (WILEY et al. 1986; TYOR et al. 1992; EPSTEIN et al. 1991; KOENIG et al. 1986). Viral localization is primarily in the deep gray and white matter which is noteworthy because HIVD is a subcortical form of dementia (KURE et al. 1990). Quantitation of HIV p24 i n CSF indicates that HIVD patients have higher levels of viral protein than in nondemented patients (ROYAL et al. 1994), although similar analyses of viral quantity in the brain in relation to clinical neurological status have not been performed to date.

HIV sequences derived from brain are comparatively homogeneous and display marked homology to HIV macrophage-tropic strains, and the few brain-derived viral isolates examined to date appear to be macrophage-tropic in vitro (SHARPLESS et al. 1992). This is in contrast to HIV blood-derived sequences which show a much broader range of sequence diversity (CHESEBRO et al. 1992; PANG et al. 1991). CSF-derived HIV sequences and isolates appear to mirror the matched blood isolates (KEYS et al. 1993). However, the question whether HIVD might be related to specific HIV strains remains unanswered by the above studies. Therefore, we designed the present study in which HIV *env* sequences from HIVD patients were compared to a control group comprised of nondemented AIDS patients. The results of this study indicate that brain-derived HIV sequences from HIVD patients differed from brain-derived HIV sequences from nondemented AIDS patients suggesting that there may be distinct strains of HIV that participate in the development of HIVD (POWER et al. 1994).

2 Study Design and Results

2.1 Methods and Materials

Brain- and spleen-derived HIV *env* sequences were analyzed in demented and nondemented AIDS patients followed prospectively by the Neurology Department at Johns Hopkins University (POWER et al. 1993; TYOR et al. 1992; WESSELINGH et al. 1993; GLASS et al. 1993; MCARTHUR 1987). Patients were defined as either AIDS nondemented (ND; $n = 8$) or AIDS with HIV dementia (HIVD; $n = 14$). Diagnostic criteria for HIV dementia included all of the following: (1) HIV-1 seropositivity, (2) history of progressive cognitive decline, (3) neurological and/or neuropsychological findings consistent with a decline from the premorbid baseline, (4) exclusion of CNS opportunistic infections by computed tomography (CT) or magnetic resonance imaging (MRI) and CSF analysis during the phase of clinical dementia (JANSSEN et al. 1991). The dementia observed in the HIVD group was a primary HIV-induced syndrome and unlikely to be caused by other AIDS-associated CNS diseases such as CNS lymphoma and CNS opportunistic infections which were excluded during the long clinical observation period before death. The severity of HIV dementia was assessed by using a dementia rating scale (Memorial Sloan-Kettering, MSK) (PRICE and BREW 1988). The MSK scale ranges from mild dementia or minimal cognitive and motor deficits (MSK = 1) to severe dementia or end-stage vegetative state (MSK = 4). The patients' mental status was assessed over the course of disease based on MSK scores (Fig. 1). The HIVD and ND groups were well matched for mean ages, CD4 counts, neuropathological findings and antiretroviral drug therapy without statistically significant differences between groups (Table 1). Spleens and subcortical white matter from the midfrontal gyrus of brains of 22 patients were selected from the AIDS Brain Bank at the Johns Hopkins University based on clinical status. Multinucleated giant cells, typical of HIV encephalitis, were identified in only three of 14 HIVD patients and in none of the ND patients. Thus, the 11 demented patients who lacked pathological abnormalities detectable by routine histopathology were typical of many HIV dementia cases which show a marked discrepancy between the severe clinical defects and mild or absent pathological findings (BUDKA 1991; GLASS et al. 1993). Our patients differ from other patients with HIV dementia or encephalopathy in whom severe encephalitis was described and HIV DNA levels were unusually high, allowing direct molecular cloning without PCR amplification (LI et al. 1991). This diversity in pathological findings may imply that the pathogenic mechanisms involved in the development of CNS disease differ between patients with and without prominent HIV encephalitis.

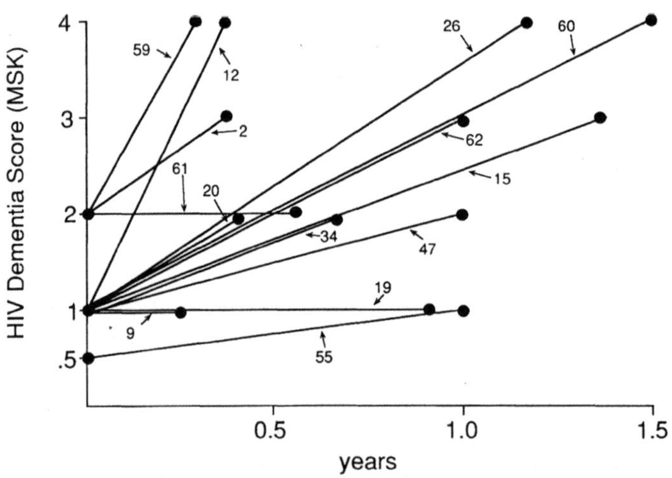

Fig. 1. Memorial Sloan-Kettering (MSK) scores of HIV dementia (HIVD) patients are plotted vs time for initial and final evaluations. The numbers indicate individual patients whose sequences are shown in Fig. 3. The clinical status of each patient was assessed at the time of initial diagnosis of HIVD and the final clinical evaluation before death by the same clinician (J.C.M.) during the course of disease progression

Because of previous data suggesting that the V3 region of the envelope influence infectivity for macrophages, microglia, T lymphocytes and CD4-positive HeLa cells (CHESEBRO et al. 1991; HWANG et al. 1991; O'BRIEN et al. 1990; PANG et al. 1991; SHIODA et al. 1991; WATKINS et al. 1990; WESTERVELT et al. 1991), we focused our initial efforts on sequence analysis of this region of the HIV genome. To avoid selection bias by in vitro viral isolation, we used DNA obtained directly from infected brains and spleens to examine the V3 and flanking regions. Sequences were amplified by using a nested polymerase chain reaction (PCR) protocol, and the first and second PCR reactions were carried out as described previously (CHESEBRO et al. 1992). This amplification yielded a fragment of 430 bases that included both the C2 and V3 regions of the HIV envelope gene. The DNA product was cloned and the sequence was analyzed by dideoxy sequencing.

2.2 Comparison of Brain- and Spleen-Derived Clones

In four patients, sequences from paired brain- and spleen-derived clones were compared to ascertain the diversity of HIV sequences between lymphoid tissue and brain (Tab. 2). In the 143 amino acid fragment including the C2 and V3 regions of HIV envelope, spleen-derived HIV sequences appeared to differ significantly from brain-derived sequences in each of four patients analyzed, both within and outside the V3 region. Regardless of clinical diagnosis there were four to five differing amino acids between the sequences of paired brain-derived clones from each patient, whereas the brain to spleen comparison from the same individual

Table 1. Clinical features in demented and nondemented AIDS patients

Clinical group	Age[a]	CD4[b]	HIV encephalitis[c]	Opportunistic infections[d]	Antiretroviral drug therapy[e]
HIV dementia	38±8.8	15±12	3/14	8/14	8/14
Nondemented	38±13.3	44±74	0/8	3/8	2/8

Groups did not differ statistically ($p > 0.05$) for any of the above features (ages and CD4 counts: Mann-Whitney *U* test; HIV encephalitis, opportunistic infections and antiretroviral therapy frequencies: Fisher's exact test).

[a]Mean age (±SD) (years) for each clinical group.

[b]Mean CD4 positive blood cell count (±SD) (cells/mm^3) for each clinical group.

[c]Number of patients with HIV encephalitis per group. The diagnosis of HIV encephalitis was made based on the presence of multinucleated giant cells in at least one of multiple sections taken from the brain of each patient.

[d]Number of patients with CNS opportunistic infections. HIV dementia: cytomegalovirus (CMV) ventriculitis ($n = 1$), progressive multifocal leukoencephalopathy PML ($n = 1$), cryptococcal meaningitis ($n = 1$), and cortical microabscesses ($n = 1$); CMV encephalitis ($n = 2$), CMV myelitis ($n = 1$), and toxoplasma cerebritis ($n = 1$). Nondemented: toxoplasma cerebritis ($n = 2$), cryptococcus meningtis, and PML ($n = 1$). Opportunistic infections were not observed in any adjacent sections to the tissues used for extraction of DNA. Based on repeated clinical evaluations, these opportunistic infections developed between the last clinical assessment and death and thus did not contribute to the signs and symptoms of HIV dementia.

[e]Number of patients receiving antiretroviral therapy i.e., zidovudine (AZT), dideoxyinosine (DDI), and/or dideoxycytidine (DDC), within a year of death.

Table 2. Comparison of the number of sequence differences between paired brain- and spleen-derived clones

Group (*n*)	Number of differing amino acids in the 143 residue fragment (C2 and V3)	
	Brain vs brain[a]	Brain vs spleen[b]
ND(10)	4	16(16)
ND(14)	5	24(22)
HIVD(9)	4	21(21)
HIVD(15)	4	16(16)

[a]Two brain-derived clones from each patient were aligned and compared.
[b]A single spleen-derived clone was compared to each of the brain-derived clones from the same patient. The number in parentheses represents the comparison for the second brain clone.

showed from 16 to 24 differing amino acids (Table 2). These results indicated that the similarity among the brain-derived clones was much greater than between the brains and spleen-derived clones. This finding was similar to previous reports comparing multiple brain- and blood-derived HIV clones from two patients (PANG et al. 1991). Therefore, we concentrated on the brain-derived HIV sequences to study differences between the HIVD and ND groups in a larger group of AIDS patients.

2.3 Comparison of Brain-Derived V3 Sequences to Macrophage-Tropic Strains

The V3 region of all the brain-derived sequences from the 22 patients studied were grouped by clinical diagnosis and compared to blood-derived HIV isolates of known cell tropism (Fig. 2). A striking similarity was observed between a previously described macrophage-tropic consensus sequence derived from blood-derived viruses and all the brain-derived sequences of both HIVD and ND clinical groups (CHESEBRO et al. 1992). In contrast, the brain-derived sequences differed extensively from sequences of blood-derived nonmacrophage-tropic viruses. The similarity of brain-derived V3 sequences from both demented and nondemented patients to the macrophage-tropic V3 consensus sequence and to the blood-derived macrophage tropic isolates was consistent with the previous findings of in vitro macrophage-tropism of most brain-derived HIV isolates (KOYANAGI et al. 1987; LI et al. 1991, 1992) and suggested that macrophage tropism might be necessary for CNS infection but not sufficient for the induction of HIVD.

2.4 Tropism of Recombinant Infectious HIV Clones

To examine the influence of the brain-derived sequences on HIV tropism, the *Stu* I to *Nhe* I fragments of the *env* gene cloned from each patient were ligated into an infectious molecular clone, NL4-3, which is known to be T cell tropic (CHESEBRO et al. 1992). Infectivity was tested in four of human cell types including CD4-positive HeLa cells (1022 cells), peripheral blood mononuclear cells (PBMCs),

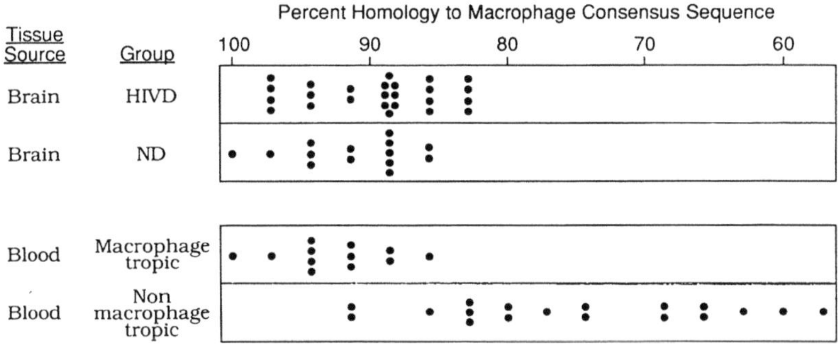

Fig. 2. Comparison of the homology of V3 sequences of brain-derived clones and previously reported blood-derived clones and isolates of known tropism with the macrophage-tropic consensus sequence (CHESEBRO et al. 1992). Each *dot* represents a differ clone. Regardless of clinical group, HIV dementia (HIVD) or not demented (ND), brain-derived isolates were similar to blood-derived macrophage-tropic isolates and differed significantly from blood-derived nonmacrophage-tropic isolates. Macrophage-tropic isolates included previously described macrophage-tropic clones (CHESEBRO et al. 1992) and NSI isolates defined as "nonsyncytia inducing" in phytohemagglutinin (PHA)-stimulated T lymphocyte cultures (FOUCHIER et al. 1992). Nonmacrophage-tropic strains included isolates capable of syncytia induction in PHA-stimulated lymphocytes (FOUCHIER et al. 1992) and clones capable of infecting 1022 CD4-positive HeLa cells (CHESEBRO et al. 1992)

blood-derived macrophages using previously reported procedures from our laboratory. In addition, human mixed glial cultures (MGC) were derived from surgical specimens obtained during operative procedures for deep CNS lesions requiring removal of healthy tissue from adults. To prepare the MGC, human brain tissue in DMEM (20% FCS) with the meninges removed was passed through a sieve, triturated, and centrifuged followed by seeding the cells on poly-L-lysine coated dishes at 2 x 10^6 per well. MGC were infected 2 weeks after seeding. Immunocytochemical staining of MGC indicated that approximately 80% of cells were CD68-positive microglia, 20% of cells were GFAP-positive astrocytes and none were positive for the oligodendroglial marker 2',3'-cyclic nucleotide 3'-phosphodiesterase. The infectivity of the brain-derived recombinant clones was compared to several previously characterized HIV strains as shown in Tab. 3. The infectivity of all viruses was measured by end-point dilution in PBMCs, indicating that all viruses could infect and replicate in PBMCs. Consistent with previous results (CHESEBRO et al. 1992), confirmed T cell tropic strains including NL4-3, 40-15, 55-17, and 56-22 were able to infect 1022 cells; however, none of the macrophage-tropic strains infected 1022 cells. In contrast, T cell tropic HIV strains infected and replicated minimally in macrophages, whereas established macrophage-tropic HIV strains including Ba-L, JRFL and the recombinant viruses: 31-3, 49-5, 32-5, and 51-9 infected macrophages at a wide range of infectivity measured by p24 level and foci. All of the brain-derived recombinant viruses infected and replicated in macrophages to variable levels but were not infectious for 1022 cells. This result confirmed the suggestion based on V3 region sequencing (POWER et al. 1994), that all the brain-derived sequences from both HIVD and ND groups conferred tropism for macrophages. Although the infectivity of brain-derived recombinants for macrophages was lower than infectivity of the Ba-L strain,

Table 3. Infectivity of various HIV strains in PBMC, 1022 cells, primary macrophages and mixed glial cultures

Group	Clone[a]	PBMC[b]	1022[c]	MØ[g] p24[d] pg/ml	MØ[g] Foci[e]	MGC[h] p24[4] pg/ml	MGC[h] Foci[f]
Blood-derived	NI4-3	10^3	2900	70	0	105	–
1022-tropic	40-15	10^3	2400	117	1	nt	nt
strains	55-17	10^3	600	87	0	nt	nt
	56-22	10^3	1300	36	0	nt	nt
	Ba-L	10^2	<3	27500	>250	6780	+++
Blood-derived	JRFL	10^3	<3	1995	200	2700	+
MØ-tropic	31-3	10^3	<3	6606	>250	nt	nt
strains	32-5	10^3	<3	213	3	nt	nt
	51-9	10^3	<3	309	17	nt	nt
	49-5	10^3	<3	748	14	nt	nt
	26-1	10^3	<3	912	5	295	+
Brain-derived	59-1	10^3	<3	1240	23	891	+
HIVD	15-2	10^3	<3	281	9	nt	nt
clones	62-1	10^3	<3	957	4	1995	++
	2-4	10^3	<3	501	9	1023	+
	19-1	10^3	<3	3981	122	nt	nt
	17-2	10^3	<3	7941	4	nt	nt
Brain-derived	10-1	10^3	<3	2198	104	nt	nt
ND	48-1	10^3	<3	724	5	750	+
clones	54-1	10^3	<3	416	7	977	++
	14-2	10^3	<3	4466	59	630	+

PBMC, peripheral blood monuclear cells; MØ macrophage; MGC,mixed glial cultures; HIVD, HIV Dementia; ND, nondemented.

[a]Clone numbers correspond to previously reported bood-derived reported clones (Chesebro et al. 1992) and brain-derived clone numbers in Fig. 3.

[b]Values represent median tissue culture infective dose ($TCID_{50}$) levels in PBMC obtained by end-point dilution per 0.1 ml of tissue culture supernatant.

[c]Foci/0.2 ml in clone 1022 (HeLa-CD4) cells at 3 days postinoculation indicated by FIA (Chesebro et al. 1992).

[d]p24 levels at 10 days postinoculation measured by ELISA.

[e]Number of foci per 0.2 ml measured by flourescent immunoassay (FIA) at 14 days postinoculation.

[f]Number of cells positive for HIV p24 by indirect immunofluorescence per well inoculated at the viral titer indicated in PBMC column per 0.1 ml– (0 cells); + (1–10); ++ (>10–20); +++ (>20): and nt (not tested).

[g]Primary macrophage cultures prepared as previously reported (Chesebro et al. 1992).

[h]Primary mixed glial cultures prepared as described in text.

comparison with blood-derived macrophage-tropic recombinant viruses 51-9 and 49-5, containing a *Stu* I-*Nhe* I fragment in the NL4-3 background, indicated that the brain-derived recombinant viruses in many cases were more infectious at the same input titer. Brain-derived recombinant clones also infected mixed glial cultures, and infectivity in macrophage and mixed glial cultures did not differ between HIVD and ND derived clones.

Simultaneous immunofluorescent detection of HIV p24 antigens and cellular antigens, CD68 for microglia, and glial fibrillary acidic protein (GFAP) for astrocytes, revealed that only microglia, and not astrocytes, were infected by the brain-derived HIV clones from both HIVD and ND groups. Thus, macrophage/microglial-tropism appeared to be necessary for brain infection but not sufficient for the development of HIVD.

2.5 Envelope Sequence Differences Between Clinical Groups

Although some clones in the HIVD group were very homologous to clones in the ND group, the development of HIVD might be dependent on a few amino acid residue differences, as has been found for CNS diseases caused by Moloney murine retrovirus, rabies, Sindbis, and dengue viruses (DIETZSCHOLD et al. 1983; KAWANO et al. 1993; SZUREK et al. 1990; TUCKER et al. 1993). To look for specific differences, sequences in each group were analyzed to see whether conserved amino acids were found at any position in either group (Fig. 3). Between clinical groups, significant differences in the amino acid residues at two positions were identified (Table 3). A histidine at position 305 was identified in 17 of 25 HIVD clones and in none of the ND clones. At the same position a proline was identified in four of eight ND clones but not in any of the HIVD clones. At position 329, a leucine was found in 15 of 25 HIVD clones and in only one of the ND clones, whereas an isoleucine was present in 11 of 14 ND patients and in nine HIVD clones. Although a histidine at position 305 and leucine at position 329 were most strongly correlated with the HIVD clinical group, these substitutions have been frequently found in blood-derived HIV isolates (Table 4) including strains with differing tropisms (CHESEBRO et al. 1992; FOUCHIER et al. 1992; HWANG et al. 1991; MYERS et al. 1993; WESTERVELT et al. 1991). Since many of these isolates probably came from nondemented patients, it is clear that presence of His-305 or Leu-329 in virus derived from blood or lymphoid tissues is not sufficient to cause dementia. Viral strains bearing the His-305 and Leu-329 may be pathogenic in the CNS but not pathogenic elsewhere in the body, as has been reported for neurotropic murine retroviruses which also infect various cells in the male and female reproductive tracts without any apparent pathogenic sequelae (PORTIS et al. 1987).

In our previous studies on in vitro tropism of HIV we found that no single V3 amino acid controlled cell tropism, but instead the entire V3 sequence context was important (CHESEBRO et al. 1992). Other *env* regions besides V3 can also influence tropism (GROENINK et al. 1992, 1993; SHIODA et al. 1991; WESTERVELT et al. 1992) which suggests that the 3-dimensional folding of the entire envelope protein might be important in determining tropism. Similarly, other envelope amino acids might account for the lack of absolute correlation between HIVD and His-305 or Leu-329. For example, in the HIVD group there were 21 amino acids at specific positions that were unique relative to the ND group, whereas only two residues at specific positions were unique to the ND group (Fig. 3). Furthermore, the most severely demented patients (MSK = 4) had the highest frequency of unique amino acids (KRUSKAL-WALLIS, $p < 0.027$; Fig. 3). The findings of highly

Clone	201					251		
MØ	ACPKVSFEPI	PIHYCAPAGF	AILKCNDKKF	NGKGPCTNVS	TVQCTHGIRP	VVSTQLLLNG	SLAEEEVVIR	SENFTNNAKT
HIVD								
26-1	....	....	......N.T.	..T.......	....	....	....(D).....	.(A)..D....
26-2	....	....	......N.T.	..T.......	....	....	....(D).....	.(A)..D....
59-2	........A.	....	......N.T.	..T.......	....	....	...(K)......	.(A)(L)D...I
59-3	....	....	......N.T.	..T.......	....	....	...(K)......	.(A)(L)D...I
60-1	....	....	....	..T.......	....	....	...(K)......	.(A)(L)D...I
60-2	......S...	....	....	..T.......	....	....	...(K)......	.(A)(L)D...I
12-1	....	..(F).......	......N.T.	....	....	....	.(V).(D)....	.D...S....
15-1	.....T....	.....T....	....	....	....	....	....	.D...D.G.N
15-2	.....T....	....	....	....	....	....	....	.D...D.G..
62-1	....	....	....(K).....	....	....	....	....	I
62-2	....	....	....(K).....	....	....	....	....	I
2-1	...A......	....	....	..T..K....	....	....	....	....
2-4	....	....	....	..T.Q.K...	....	....	....	....A....N
61-1	....	....	....(K).....	....	....	....	....	I
61-2	....	....	....(K).....	....	....	....	....	I
34-1	.....T....	.....T....	.L..(K).N..	..T..K....	L	....	.....K....	....
20-1	....	..(F).......	......N.T.	....	........K.	....	.(V).(D)....	.D...D.V..
20-2	....	..(F).......	......N.T.	....	........K.	....	.(V).(D)....	.D........
47-1	....	....	....(K).....	..T..K....	....A.....	....	......I...	.R..SD...I
47-2	....	....	....(K).....	..T..K....	....	....	......I...	.K..SD..K.
9-1	....	..(F).......	......N.T.	D...A.....	....	....	......I...	.........S
9-2	....	..(F).......	......N.T.	D...A.....	.......V..	....	......I...	.........S
19-1	....	....	....	..T..K....	....	....	....	.K...D....
19-2	....	....	....	..T..K....	....	....	....	.K...D....
55-1	.....T....	....	.L...K..N.	..T..K....	L	....	.....K....	....
ND								
17-2	....	....	......N.T.	..T.S.R...	....	....	......(I)..	.....D...N
1-2	.....T....	....	....	....	....	......P...	....	.D...D.G..
1-4	.....T....	.......V..	......N.T.	....	....	....	....	.D...D.G..
10-1	.....T....	....	.L...R....	..T..K....	....	....	......I...	....
10-2	.....T...T	....	.L...R....	..T..K....	V.........	....	....	.....D....
14-1	.......T..	.........S	.L........	E.R.E.N...	....	....	.........G	...IK..I..
14-2	.......T..	.....T....	.L........	E.R.E.N...	....	....	....	...IK..I..
48-1	.....T....	....	....	.E........	....	....	....	.K......E.
48-2	.....TS...	....	V.........	....	....	....	....	.K......E.
49-1	....	....Y.....	......N...	..T.......	........K.	....	......(I)..	..D.......
49-2	....	....	....	..T.......	........K.	....	......(I)..	..D.......
54-1	......W...	.....P....	....	..T.......	....	....	....G.....	....M....I
54-2	......W...	.....T....	....	..T.......	....	....	....G.....	....M....I
36-1	.....T....	....	......N.T.	....	....	....	....	.D...D.G..

conserved amino acids at two specific positions and the high number of unique amino acids in the HIVD group indicated that the brain-derived HIV sequences differed between the HIVD patients and the AIDS nondemented patients.

3 Discussion

3.1 The Role of HIV Cell Tropism in the Development of HIV Dementia

Unlike earlier studies of brain-derived HIV sequences, the present analysis utilized prospectively studied patients including a control group (ND) which permitted comparisons of HIV sequences associated with and without HIV dementia.

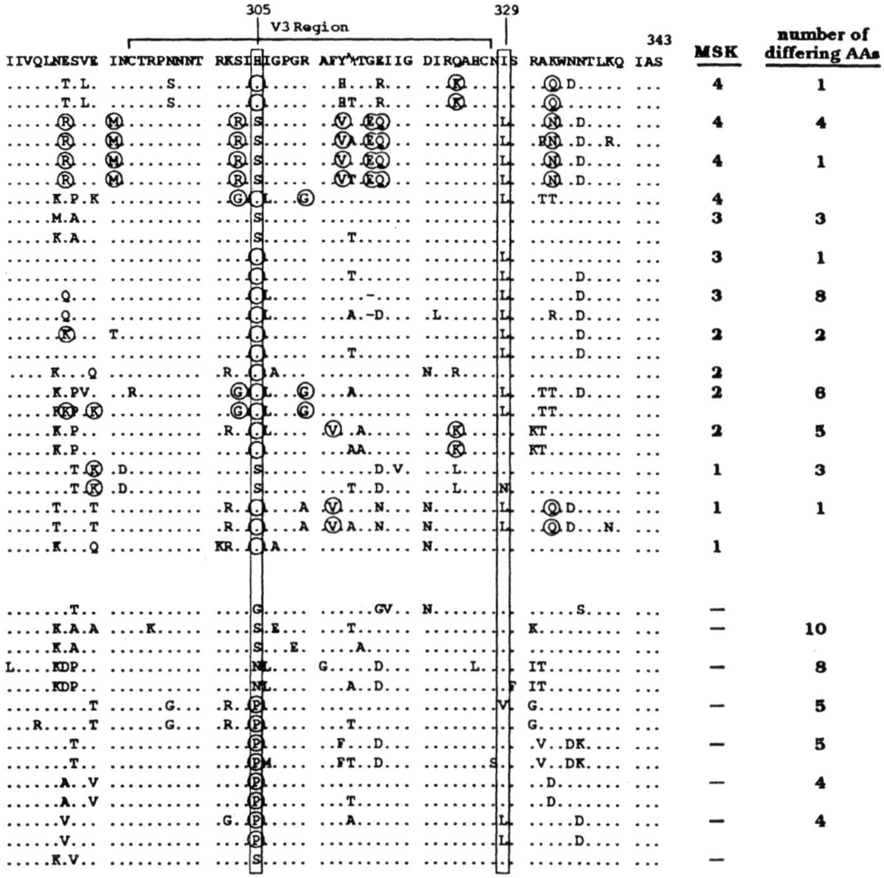

Fig. 3. Comparison of HIV envelope amino acid sequences from patients belonging to the HIV dementia group (HIVD) and the non-demented AIDS group (ND). Sequences are aligned with the macrophage-tropic consensus sequence (Mϕ) (CHESEBRO et al. 1992). HIVD sequences are presented in the order of severity of dementia according to the Memorial Sloan-Kettering (*MSK*) score. Two clones were sequenced from all but five patients, and the number of differing amino acids (*AAs*) between the two clones from the same patient is indicated in the *right hand column*. In agreement with previous reports (LI et al. 1991, 1992; PANG et al. 1991), clones from the same individual were closely related showing an average of four amino acids differing within the 135 amino acid fragment. In view of this homogeneity, we elected to sequence only one or two clones from each patient and instead study more individual patients. Positions 305 and 329, as indicated by *boxes*, show significant divergence in the residues between the HIVD and ND groups. Unique amino acids found at a particular position in two or more patients in one clinical group and absent in all patients of the other group are *circled*. In many cases only a single clone or single patient within a clinical group showed a unique amino acid at a particular position and in this situation it was impossible to know whether this was due to a PCR-induced mutation. Even when one considers only those changes in which unique amino acids were seen in two or more different patients, the HIVD group was found to have 21 unique amino acids and the ND group had two (Mann-Whitney, U = 99.5, p = 0.0033). Unique amino acids were significantly more frequent in clones from severely demented patients (MSK = 4) compared to the other HIVD-derived clones (Kruskal-Wallis, p = 0.027)

Table 4. Comparison of amino acids at positions 305 and 329 for brain-derived HIV envelope clones

Patient	Position 305				Total	Position 329		
group	His	Ser	Pro	Other		Leu	Ile	Other
HIVD	17*	8	0*	0	25	15*	9	1
ND	0	3	8	3	14	2	11	1
Database	20	6	6	27	59	6	41	12

HIVD, HIV dementia; ND, nondemented. Asterisk, significantly different compared to ND value. Database is from MYERS et al. (1993). The frequency of each amino acid at each position in the HIVD and ND groups was compared individually to the sum of all other amino acids frequencies in separate 2 × 2 contingency tables with Fisher's exact test. The following P values were observed: His-305 ($p < 0.0001$), Pro-305 ($p < 0.0001$), Leu-329 ($p < 0.0078$) and Ile –329 ($p < 0.0187$).

Furthermore, the group with HIVD demonstrated a spectrum in severity of dementia thus allowing for the diversity in the clinically observed phenotype. HIV genomes were detected in the brains of these clinically well-characterized individuals regardless of clinical status. It is clear from the results of this and earlier studies (POWER et al. 1994; SHARPLESS et al. 1992; PANG et al. 1991) that brain-derived HIV V3 sequences show marked homology to V3 sequences from macrophage-tropic strains and that recombinant viruses containing brain-derived V3 sequences demonstrate macrophage tropism and fail to infect 1022 cells and astrocytes. Thus, macrophage tropism appears to be required for infection of the CNS by HIV. Yet, the HIV sequences derived from the spleen also resembled the macrophage-tropic consensus sequence and recombinant viruses containing the spleen-derived *Stu* I-*Nhe* I fragment also infected macrophages at similar levels of infectivity to the brain-derived recombinant viruses (data not shown) despite multiple sequence differences between spleen and brain-derived fragments (Table 2). Thus neurotropism or the ability to infect the CNS may be dependent not only on macrophage tropism but also on additional sequence changes in *env* or other regions of the virus genome. For SIV also, macrophage tropism appears to be required for CNS infection, and distinct mutations in the SIV *env* gene may impart a capacity to cause disease in brain tissue (KODAMA et al. 1993).

3.2 HIV Envelope and Neurovirulence

The findings of highly significant differences in HIV *env* sequences between the HIVD and ND clinical groups in our study suggests that distinct *env* sequences might account for the presence of HIVD. Similarly, mouse retroviral *env* sequences have been shown to influence development and course of CNS disease (LYNCH et al. 1991; PORTIS et al. 1987). More recently, *env* sequences have also been shown to alter the location of CNS pathological lesions and the sites of viral antigen detection within the CNS (J. PORTIS personal communication). However, in these mouse models, *gag* and LTR sequences also influence clinical CNS disease (PORTIS et al. 1990), and this might suggest that other regions of the HIV genome may also be involved in the induction of dementia. Thus, future efforts should include analysis of other areas of the HIV genome derived from both of these clinical groups. Although our results show a strong association of certain HIV genotypes with two clinical groups, it is also possible that this correlation is related to other unknown factors differing in these two groups. For example, if the HIVD group had been infected for a much longer time before death than the ND group, there would have been more time to generate variable HIV genotypes. Similarly other factors such as host genetic background and source of infection might influence the associations we observed in this work. Thus, it will be important to extend this work to a larger group of patients from different geographic areas using larger regions of the HIV genome to confirm or refute these striking early observations.

Comparison of the infectivities of recombinant viruses in macrophage and MGC did not reveal significant biological differences between viral sequences derived from each of the clinical groups in this study. Thus, infectivity in macrophages and MGC is not an useful measure of the neuropathogenicity of HIV. However, it is unclear which type of assay will give the most valid correlation with the generation of HIVD in vivo. Resolution of this problem will require a better understanding of the complex mechanisms of pathogenesis of HIVD. Therefore, at present it may be more fruitful to use in vivo approaches such as the transgenic mouse gp120 expression models or the HIV-infected SCID mouse model (Jolicoeur et al. 1992b; Toggas et al. 1994; Tyor et al. 1993).

Acknowledgments. We thank Gary Hettrick and Bob Evans for assistance with graphics and Irene Cook Rodriguez for manuscript preparation. C.P. is an MRC Fellow (Canada). This study was supported by the following grants: NS 26643, AI 767234 and RR 00722.

References

Budka H (1991) Neuropathology of human immunodeficiency virus infection. Brain Pathol 1: 163–175

Budka H (1992) Cerebral pathology in AIDS: a new nomenclature and pathogenetic concepts. Curr Opin Neurol Neurosurg 5: 917–923

Budka H, Costanzi G, Cristina S et al. (1987) Brain pathology induced by infection with the human immunodeficiency virus (HIV). A histological, immunocytochemical, and electron microscopical study of 100 autopsy cases. Acta Neuropathol (Berl) 75: 185–198

Buller RS, Wehrly K, Portis J L, Chesebro B (1990) Host genes conferring resistance to a central nervous system disease induced by a polytropic recombinant Friend murine retrovirus. J Virol 64: 493–498

Chesebro B, Nishio J, Perryman S et al (1991) Identification of human immunodeficiency virus envelope gene sequences influencing viral entry into CD4-positive HeLa cells, T-leukemia cells, and macrophages. J Virol 65: 5782–5789

Chesebro B, Wehrly K, Nishio J, Perryman S (1992) Macrophage-tropic human immunodeficiency virus isolates from different patients exhibit unusual V3 envelope sequence homogeneity in comparison with T-cell-tropic isolates: definition of critical amino acids involved in cell tropism. J Virol 66: 6547–6554

Dawson VL, Dawson TM, Uhl GR, Snyder SH (1993) Human immunodeficiency virus type 1 coat protein neurotoxicity mediated by nitric oxide in primary cortical cultures. Proc Natl Acad Sci USA 90: 3256–3259

Dietzschold B, Wunner WH, Wiktor TJ et al (1983) Characterization of an antigenic determinant of the glycoprotein that correlates with pathogenicity of rabies virus. Proc Natl Acad Sci USA 80: 70–74
Dreyer EB, Kaiser PK, Offermann JT, Lipton SA (1990) HIV-1 coat protein neurotoxicity prevented by calcium channel antagonists. Science 248: 364–367
Epstein LG, Kuiken C, Blumberg BM et al (1991) HIV-1 V3 domain variation in brain and spleen of children with AIDS: tissue-specific evolution within host-determined quasispecies. Virology 180: 583–590
Fouchier RAM, Groenink M, Kootstra NA et al (1992) Phenotype-associated sequence variation in the third variable domain of the human immunodeficiency virus type 1 gp120 molecule. J Virol 66: 3183–3187
Gardner MB, Henderson BE, Officer, JE et al (1973) A spontaneous lower motor neuron disease apparently caused by indigenous type-C RNA virus in wild mice. J Natl Cancer Inst 51: 1243–1254
Georgsson G, Houwers DJ, Palsson PA, Pétursson G (1989) Expression of viral antigens in the central nervous system of visna-infected sheep: an immunohistochemical study on experimental visna induced by virus strains of increased neurovirulence. Acta Neuropathol (Berl) 77: 299–306
Glass JD, Wesselingh SL, Selnes OA, McArthur JC (1993) Clinical-neuropathologic correlation in HIV-associated dementia. Neurology 43: 2230–2237
Griffin DE, Wesselingh SL, McArthur JC (1994) Elevated central nervous system prostaglandins in human immunodeficiency virus-associated dementia. Ann Neurol 35: 592–597
Groenink M, Andeweg AC, Fouchier RAM et al (1992) Phenotype-associated *env* gene variation among eight related human immunodeficiency virus type 1 clones: evidence for in vivo recombination and determinants of cytotropism outside the V3 domain. J Virol 66: 6175–6180
Groenink M, Fouchier RAM, Broersen S et al (1993) Relation of phenotype evolution of HIV-1 to envelope V2 configuration. Science 260: 1513–1516
Heyes MP, Brew BJ, Martin A et al (1991) Quinolinic acid in cerebrospinal fluid and serum in HIV-1 infection: relationship to clinical and neurological status. Ann Neurol 29: 202–209
Hwang SS, Boyle TJ, Lyerly HK, Cullen BR (1991) Identification of the envelope V3 loop as the primary determinant of cell tropism in HIV-1. Science 253: 71–74
Janssen RS, Cornblath DR, Epstein LG, Foa RP, McArthur JC, Price RW (1991) Nomenclature and research case definitions for neurologic manifestations of human immunodeficiency virus-type 1 (HIV-1) infection. Neurology 41: 778–785
Johnson RT, McArthur JC, Narayan O (1988) The neurobiology of human immunodeficiency virus infections. Faseb J 2: 2970–2981
Jolicoeur P, Gravel C, Kay DG (1992a) Pathogenesis of murine spongiform myeloencephalopathy induced by a murine retrovirus. In: Roos RP (ed) Molecular neurovirology. Pathogenesis of viral CNS infections. Humana, Totowa, pp 199–224
Jolicoeur P, Laperriere A, Beaulieu N (1992b) Efficient production of human immunodeficiency virus proteins in transgenic mice. J Virol 66: 3904–3908
Kawano H, Rostapshov V, Rosen L, Lai C-J (1993) Genetic determinants of dengue type 4 virus neurovirulence for mice. J Virol 67: 6567–6575
Keys B, Karis J, Fadeel B et al (1993) V3 sequences of paired HIV-1 isolates from blood and cerebrospinal fluid cluster according to host and show variation related to the clinical stage of disease. Virology 196: 475–483
Kodama T, Mori K, Kawahara T, Ringler DJ, Desrosiers RC (1993) Analysis of simian immunodeficiency virus sequence variation in tissues of rhesus macaques with simian AIDS. J Virol 67: 6522–6534
Koenig S, Gendelman HE, Orenstein JM et al (1986) Detection of AIDS virus in macrophages in brain tissue from AIDS patients with encephalopathy. Science 233: 1089–1093
Koyanagi Y, Miles S, Mitsuyasu RT, Merrill JE, Vinters HV, Chen ISY (1987) Dual infection of the central nervous system by AIDS viruses with distinct cellular tropisms. Science 236: 819–822
Kure K, Weidenheim KM, Lyman WD, Dickson DW (1990) Morphology and distribution of HIV-1 gp41-positive microglia in subacute AIDS encephalitis. Pattern of involvement resembling a multisystem degeneration. Acta Neuropathol (Berl) 80: 393–400
Lackner AA, Dandekar S, Gardner MB (1991) Neurobiology of simian and feline immunodeficiency virus infections. Brain Pathol 1: 201–212
Li Y, Kappes JC, Conway JA, Price RW, Shaw GM, Hahn BH (1991) Molecular characterization of human immunodeficiency virus type 1 cloned directly from uncultured human brain tissue: identification of replication-competent and -defective viral genomes. J Virol 65: 3973–3985
Li Y, Hui H, Burgess CJ et al (1992) Complete nucleotide sequence, genome organization, and biological properties of human immunodeficiency virus type 1 in vivo: evidence for limited defectiveness and complementation. J Virol 66: 6587–6600

Lipton SA (1991) HIV-related neurotoxicity. Brain Pathol 1: 193–199
Lynch WP, Czub S, McAtee FJ, Hayes SF, Portis JL (1991) Murine retrovirus-induced spongiform encephalopathy: productive infection of microglia and cerebellar neurons in accelerated CNS disease. Neuron 7: 365–379
Masuda M, Remington MP, Hoffman PM, Ruscetti SK (1992) Molecular characterization of a neuropathogenic and nonerythroleukemogenic variant of Friend murine leukemia virus PVC-211. J Virol 66: 2798–2806
Masuda M, Hoffman PM, Ruscetti SK (1993) Viral determinants that control the neuropathogenicity of PVC-211 murine leukemia virus in vivo determine brain capillary endothelial cell tropism of the virus in vitro. J Virol 67: 4580–4587
McArthur JC (1987) Neurologic manifestations of AIDS. Medicine 66: 407–437
McArthur JC, Hoover DR, Bacellar H et al (1993) Dementia in AIDS patients: incidence and risk factors. Neurology 43: 2245–2252
Myers G, Wain-Hobson S, Korber B, Smith RF (1993) Human retorviruses and AIDS 1993. A compilation and analysis of nucleic acid and amino acid sequences, theoretical biology and biophysics group T-10, Los Alamos, NM
Narayan O, Clements JE (1989) Biology and pathogenesis of lentiviruses. J Gen Virol 70: 16 17–1639
Navia BA, Jordan BD, Price RW (1986) The AIDS dementia complex: I. Clinical features. Ann Neurol 19: 517–524
O'Brien WA, Koyanagi Y, Namazie A et al (1990) HIV-1 tropism for mononuclear phagocytes can be determined by regions of gp120 outside the CD4-binding domain. Nature 348: 69–73
Pang S, Vinters HV, Akashi T, O'Brien WA, Chen ISY (1991) HIV-1 *Env* sequence variation in brain tissue of patients with AIDS-related neurologic disease. J Acquir Immune Defic Syndr 4: 1082–1092
Park BH, Lavi E, Blank KJ, Gaulton GN (1993) Intracerebral hemorrhages and syncytium formation induced by endothelial cell infection with a murine leukemia virus. J Virol 67: 6015–6024
Petito CK, Cash KS (1992) Blood-brain barrier abnormalities in the acquired immunodeficiency syndrome: immunohistochemical localization of serum proteins in postmortem brain Ann Neurol 32: 658–666
Portis JL, McAtee FJ, Hayes SF (1987) Horizontal transmission of murine retroviruses. J Virol 61: 1037:1044
Portis JL, Czub S, Garon CF, McAtee FJ (1990) Neurodegenerative disease induced by the wild mouse ecotropic retrovirus is markedly accelerated by long terminal repeat and *gag-pol* sequences from nondefective Friend murine leukemia virus. J Virol 64: 1648–1656
Power C, Weinshenker BG, Dekaban GA, Kaufmann JC, Shandling M, Rice GP (1991) Pathological and molecular biological features of a myelopathy associated with HTLV-1 infection. Can J Neurol Sci 18: 352–355
Power C, Kong P-A, Crawford TO et al (1993) Cerebral white matter changes in acquired immunodeficiency syndrome dementia: alterations of the blood-brain barrier. Ann Neurol 34: 339–350
Power C, McArthur JC, Johnson RT et al (1994) Demented and non-demented patients with AIDS differ in brain-derived human immunodeficiency virus type 1 envelope sequences. J Virol 68: 4643–4649
Price RW, Brew BJ (1988) The AIDS dementia complex. J Infect Dis 158: 1079–1083
Roman GC (1988) The neuroepidemiology of tropical spastic paraparesis. Ann Neurol 23 [Suppl]: S113–S120
Royal W III, Selnes OA, Concha M, Nance-Sproson TE, McArthur JC (1994) Cerebrospinal fluid HIV-1 p24 antigen levels in HIV-1-related dementia. Ann Neurol 36: 32–39
Sharma DP, Zink MC, Anderson M et al (1992) Derivation of neurotropic simian immunodeficiency virus from exclusively lymphocytetropic parental virus: pathogenesis of infection in macaques. J Virol 66: 3550–3556
Sharpless NE, O'Brien WA, Verdin E, Kufta CV, Chen ISY, Dubois-DalcQ M (1992) Human immunodeficiency virus type 1 tropism for brain microglial cells is determined by a region of the *env* glycoprotein that also controls macrophage tropism. J Virol 66: 2588–2593
Shioda T, Levy JA, Cheng-Mayer C (1991) Macrophage and T cell-line tropisms of HIV-1 are determined by specific regions of the envelope gp120 gene. Nature 349: 167–169
Szurek PF, Floyd E, Yuen PH, Wong PKY (1990) Site-directed mutagenesis of the codon for Ile-25 gPr80 *env* alters the neurovirulence of *ts*1, a mutant of Moloney murine leukemia virus TB. J Virol 64: 5241–5429

Toggas SM, Masliah E, Rockenstein EM, Rall GF, Abraham CR, Mucke L (1994) Central nervous system damage produced by expression of the HIV-1 coat protein gp120 in transgenic mice. Nature 367: 188–193

Tucker PC, Strauss EG, Kuhn RJ, Strauss JH, Griffin DE (1993) Viral determinants of age-dependent virulence of Sindbis virus for mice. J Virol 67: 4605–4610

Tyor WR, Glass JD, Griffin JW et al (1992) Cytokine expression in the brain during the acquired immunodeficiency syndrome. Ann Neurol 31: 349–360

Tyor WR, Power C, Gendelman HE, Markham RB (1993) A model of human immunodeficiency virus encephalitis in *scid* mice. Proc Natl Acad Sci USA 90: 8658–8662

Watkins BA, Dorn HH, Kelly WB et al (1990) Specific tropism of HIV-1 for microglial cells in primary human brain cultures. Science 249: 549–553

Wesselingh SL, Power C, Glass JD et al (1993) Intracerebral cytokine messenger RNA expression in acquired immunodeficiency syndrome dementia. Ann Neurol 33: 576–582

Westervelt P, Gendelman HE, Ratner L (1991) Identification of a determinant within the human immunodeficiency virus 1 surface envelope glycoprotein critical for productive infection of primary monocytes. Proc Natl Acad Sci USA 88: 3097–3101

Westervelt P, Throwbridge DB, Epstein LG et al (1992) Macrophage tropisim determinants of human immunodeficiency virus type 1 in vivo. J Virol 66: 2577–2582

Wiley CA, Schrier RD, Nelson HA, Lampert PW, Oldstone MBA (1986) Cellular localization of human immunodeficiency virus infection within the brains of acquired immune deficiency syndrome patients. Proc Natl Acad Sci USA 83: 7089–7093

Role for Astrocytosis in HIV-1-Associated Dementia

L. Vitković[1] and A. da Cunha[2]

1 Introduction

HIV-1-associated dementia occurs in approximately one third of individuals infected with HIV-1, sometimes as the only presenting symptom (McArthur et al. 1994; for refinement of this number see Grant et al., this volume). It involves deficits in cognitive and higher motor functions. The dementia is linked to structural changes in neurons observed post mortem. The changes vary from dendritic dearborization to neuronal drop out (Masliah et al. 1992). Large pyramidal neurons especially vulnerable to as yet unknown pathogenic stimuli are located in the cerebral cortex. The cortex is the brain region known to control cognitive and motor functions. These functions are altered or deficient in AIDS patients. Thus, neuronal dysfunction is a likely direct cause of HIV-1-associated dementia. The precise cause and nature of neuronal dysfunction in AIDS are unknown. HIV-1-associated neuronal damage may be mediated by activation of one family of glutamate receptors (Lipton 1992). Glutamate is the major excitatory neurotransmitter in the brain. Glutamate receptor-mediated neuronal death—excitotoxicity—is implicated in the pathogenesis of several disorders affecting mental health such as Alzheimer's dementia (Choi 1994).

Excitotoxicity in AIDS is postulated to be initiated by HIV-1-infected macrophages and then amplified by astrocytes through a cell-cell interaction with the HIV-1-infected macrophages (Epstein and Gendelman 1993). This postulate

[1]Division of Neuroscience and Behavioral Science, National Institute of Mental Health, National Institutes of Health, 5600 Fishers Lane, Room 11C-06, Rockville, MD 20857, USA
[2]Laboratory of Cell Biology, Building 36, Room 3A17, National Institute of Mental Health, National Institutes of Health, Bethesda, MD 20892, USA

based on substantial data obtained in vitro and in tissue obtained post mortem, attempts to resolve a paradox between the small number of HIV-1-infected cells in the brain (particularly in the cortex) and the wide-spread astrocytosis. Astrocytosis is the most frequent pathologic change occurring in adult and developing brains of HIV-1-infected individuals (BUDKA et al. 1987; BUDKA 1990, CIARDI et al. 1990; KURE et al. 1990a,b, 1991; SCHMIDBAUER et al. 1992; FUNATA et al. 1991; SMITH et al. 1991) and SIV-infected juvenile macaques (WEIHE et al. 1993; RAUSCH et al. 1994) in every brain region thus far examined. We propose a hypothesis that *astrocytosis is: (a) the first cerebral response to HIV-1-infection and (b) alone sufficient to cause neuronal dysfunction underlying HIV-1-associated dementia.*

Astrocytosis is defined by increases in density (hyperplasia) and size (hypertrophy) of astrocytes (NORTON et al. 1992; EDDLESTON and MUCKE 1993). However, only the hypertrophy was correlated with a neuropathological diagnosis of astrocytosis in brain tissue of 22 individuals by three qualified observers (DA CUNHA et al. 1993b). Thus, neuropathological diagnosis of astrocytosis needs to be more rigorously defined. This is important because most, if not all, knowledge of astrocytosis in AIDS brain is based on the neuropathological diagnosis. Astrocytosis is poorly defined at the molecular level. A variety of biochemical changes have been attributed to astrocytes reacting to a host of insults (NORTON et al. 1992, EDDLESTON and MUCKE 1993). The most frequently reported molecular change is increase in intensity of staining with antibodies to glial fibrillary acidic protein (GFAP), a subunit of intermediate filaments unique to astrocytes. This change also is not correlated with the neuropathological diagnosis of astrocytosis (DA CUNHA et al. 1993b). The intensity of staining of GFAP depends on the extent of its polymerization and thus does not reflect the amount of protein present (LIPSKY and SILVERMAN 1987). This amount may decrease under some pathological conditions such as experimental autoimmune encephalomyelitis where GFAP mRNA and protein change with complicated kinetics (AQUINO et al. 1990). Regulation of GFAP gene expression is evidently very complex and needs to be better understood before GFAP staining will enlighten us about astrocytosis.

Astrocytic "response" to infection (and other insults), analogous to the immune cell response, is a protective and/or curative mechanism in the brain whose final result is restoration of homeostasis. However, as with immune cell response, astrocytosis can be harmful under some conditions. For example, astrocytosis in response to physical injury causes scars thought to prevent axonal regeneration, an ultimate healing in the brain. A dysregulated astrocytic response (we hypothesize) can cause neuronal dysfunction underlying HIV-1-associated dementia. If this is true, then astrocytosis must precede neuronal dysfunction, which cannot be directly tested in humans. Thus, a retrospective analysis of human brain tissue and experiments in animal models offer opportunities to test the hypothesis.

2 Staging of Astrocytosis

Astrocytosis occurs in asymptomatic HIV-1 seropositive individuals that is, in the earliest stage of the disease. This observation is documented in at least four independent studies, one involving 11 individuals (GRAY et al. 1992). Brain tissues of all 11 HIV-1 seropositive asymptomatic individuals (who died of unnatural causes such as heroin overdose [*n*=10] and gunshot injury [*n*=1]) exhibited astrocytosis which was more frequent and more severe than in controls. Controls were HIV-1 seronegative heroin addicts who died from heroin overdose and one from a gunshot wound. Astrocytosis was present throughout cerebral and cerebellar deep white matter of one individual in an early postinfection stage (LENHARDT et al. 1988). Astrocytosis was observed in the brain of another individual with an early HIV-1-associated dementia and early AIDS-related complex (MCARTHUR et al. 1989). A "mild prominence of glial cells" was detected as early as 15 days following infection with HIV-1 (DAVIS et al. 1992). Astrocytosis is consistently reported as the earliest neuropathological change in brains of HIV-1-infected individuals.

The earliest well-established time at which astrocytosis occurs is 1 week following infection of juvenile rhesus macaques with SIV (CHAKRABARTI et al. 1991; HURTREL et al. 1991). At this time, rapid and transient SIV replication (viremia) occurs in the periphery. Astrocytosis was observed in 11 SIV-infected rhesus macaques examined at various stages of the disease in another study, whereas it was absent in only one infected animal (WEIHE et al. 1993; RAUSCH et al. 1994).

A relationship between astrocytosis and the infection of the brain parenchyma with HIV-1 is unknown. HIV-1 expression in the brain is characterized by the presence of HIV-1-encoded mRNA and protein antigens, HIV-1 virions and multinucleated giant cells, the result of the productive infection of monocytes/macrophages. HIV-1 antigens were not detected by immunocytochemistry in any of the 11 brains from HIV-1 seropositive asymptomatic individuals displaying perivascular astrocytosis (GRAY et al. 1992). Astrocytosis of white matter in the absence of multinucleated giant cells was detected in an individual who died with early HIV-1-associated dementia (MCARTHUR et al. 1989). These observations are consistent with our data. We examined tissues from cerebral frontal cortex of eight seropositive individuals and one healthy seronegative control. Astrocytosis was present in all tissues from seropositive individuals, whereas multinucleated giant cells were only present in one tissue. p24 antigen was not detected in five tissues that were assayed. These observations suggest that astrocytosis may occur independently of HIV-1 expression (at least in the vicinity of astrocytosis) and as early as the time of seroconversion. This astrocytosis may be transient and does not cause dementia but could result in mild cognitive dysfunction (GRANT et al., this volume).

Astrocytosis may be triggered by circulating cytokines before the entry of HIV-1 into parenchyma. Interleukin-1 (IL-1) and tumor necrosis factor-α (TNF-α) concentrations in circulation increase upon systemic infection with HIV-1, secreted by HIV-1-infected monocytes (Locksley et al. 1988; Merrill et al. 1989). Circulating IL-1 stimulates directly and/or indirectly its own expression in the brain manifesting itself as fever during the initial HIV-1 viremia (see Vitkovic et al. 1994). [Elevation of IL-1 in the brain does not occur only in HIV-1 infection but also in a variety of other diseases (da Cunha et al. 1993a).] IL-1 induces transforming growth factor-β1 (TGF-β1) in astrocytes and other glial cells of the frontal cortex (da Cunha et al. 1993a; da Cunha and Vitkovic 1992). These and probably other cytokines initiate and control astrocytosis (da Cunha et al. 1993c). The notion that signals for astrocytosis stem from the circulation is consistent with the postulate in the following section, that perivascular astrocytosis precedes the diffuse astrocytosis.

IL-1-induced astrocytic "reaction" may lead to the entry of HIV-1-infected monocytes into the brain parenchyma (Vitković et al. 1994). Circulating monocytes including those that are infected with HIV-1 may be attracted by TGF-β1, a potent chemoattractant for monocytes (Wahl et al. 1987). TGF-β1 may be secreted from the astrocytic endfeet at the blood-brain barrier (BBB) where TGF-β1 immunoreactivity was visualized (da Cunha et al. 1993a). In addition to directly attracting monocytes, TGF-β1 can up-regulate astrocyte expression of monocyte chemoattractant protein-1 (W. D. Lyman, personal communication). The entry of monocytes is probably facilitated by IL-1-mediated stimulation of adhesion molecules on endothelial cells comprising the BBB (see Vitković et al. 1994). Thus, IL-1 may trigger in some regions of the brain a cascade of cytokine expression in glia and neurons that can spread throughout the brain (Tchleingerian et al. 1993). These observations indicate that astrocytosis may be the earliest neurocellular change in HIV-1 infection, setting the stage for the pathogenesis in the brain.

3 Heterogeneity of Astrocytes and Astrocytosis

Astrocytes are phenotypically extremely heterogeneous cells. Two types of astrocytes are morphologically (Privat and Ratboul 1986) and phenotypically (Raff 1989) distinguished. Early neuroanatomists distinguished fibrous astrocytes in the white matter from protoplasmic astrocytes in the gray matter. These morphologic distinctions took on a new meaning when Raff and collaborators discovered two astrocytic lineages, type 1 astrocytes and type 2 astrocytes (Raff 1989). Whether or not these two types of astrocytes exist in brain parenchyma and what their relationship to the fibrous and protoplasmic astrocytes may be are still unknown.

Astrocytosis occurs in the white and gray matter of HIV-1-infected individuals (Budka et al. 1987; Budka 1990; Ciardi et al. 1990; Kure et al. 1990a,b, 1991;

SCHMIDBAUER et al. 1992; FUNATA et al. 1991; SMITH et al. 1991). Whereas the significance of astrocytosis in white versus gray matter is a subject of some controversy in the literature, no quantitative data are available to resolve the controversy. We therefore, quantified astrocytosis in the cerebral frontal cortex of eight individuals with HIV-1 infection and one seronegative, healthy individual as a control. Seven HIV-1-infected individuals had HIV-1-associated dementia and one did not. Among seven subjects with HIV-1-associated dementia, all had encephalitis, one had mild cerebral atrophy, one had neuropathological signs of cytomegalovirus (CMV) infection and one had CNS lymphoma. One individual with no HIV-1-associated dementia had lymphadenopathy and sickle cell anemia prior to death. The mean age and standard error of the mean (SEM) of individuals with HIV-1-associated dementia was 32 ± 2 years (*n*=7). The nondemented individual with no neurological symptoms was 34 years old, and the healthy seronegative individual was 49 years old. The mean time and SEM between death and removal of tissue was 25 ± 4.5 h for individuals with dementia, 18 h for the individual without dementia and 24 h for the seronegative individual. The quantitative analysis of astrocytosis involved measurements of density and size of astrocytes and the intensity of staining with antibodies to GFAP, as previously described (DA CUNHA et al. 1993b,c). (Although some previously published studies were case reports or based on small numbers of individuals including *n*=1 for controls, such as here, we consider them of limited value for statistical reasons. The results presented here are therefore preliminary. A study of 23 brains from HIV-1-infected individuals, nine demented and 14 nondemented, and six from seronegative individuals, is in progress in collaboration with J. D. Glass, Johns Hopkins University, Baltimore, MD.)

All three astrocytic parameters were higher in the white matter than in the gray matter in all individuals (Table 1). This result is in contrast to the expectation that astrocytosis may be more apparent in gray matter than white matter, because of the increased intensity of GFAP staining (Table 1). However, it is

Table 1. Astrocytic changes attributed to astrocytosis in the cerebral frontal cortex of HIV-infected individuals[a]

Property of astrocytes	Area of frontal cortex	HIV–control (*n* = 1)	HIV⁺	
			Dementia⁻ (*n* =1)	Dementia⁺ (*n* =7)
Density[b]	White matter	6 +/– 1	9 +/– 2	20 +/– 6
(number/cm^2)	Gray matter	1 +/–2	3 +/– 2	11 +/– 3
Size	White matter	5 +/– 1	2 +/– 0	12 +/– 3
(μm^2)	Gray matter	2 +/– 2	7 +/– 5	11 +/– 7
Intensity	White matter	11 +/– 0	13 +/– 1	13 +/– 0
(OD/cell)	Gray matter	2 +/– 0	1 +/– 1	8 +/– 7

OD, Optical density.

[a] Astrocytic changes were measured as previously described (DA CUNHA et al. 1993b,c).

[b] The number were obtained at 100X magnification. Thus, to obtain the true number, multiply the number cited above, by 100. The true numbers may need to be corrected for tissue shrinkage, to get the absolute number of astrocytes.

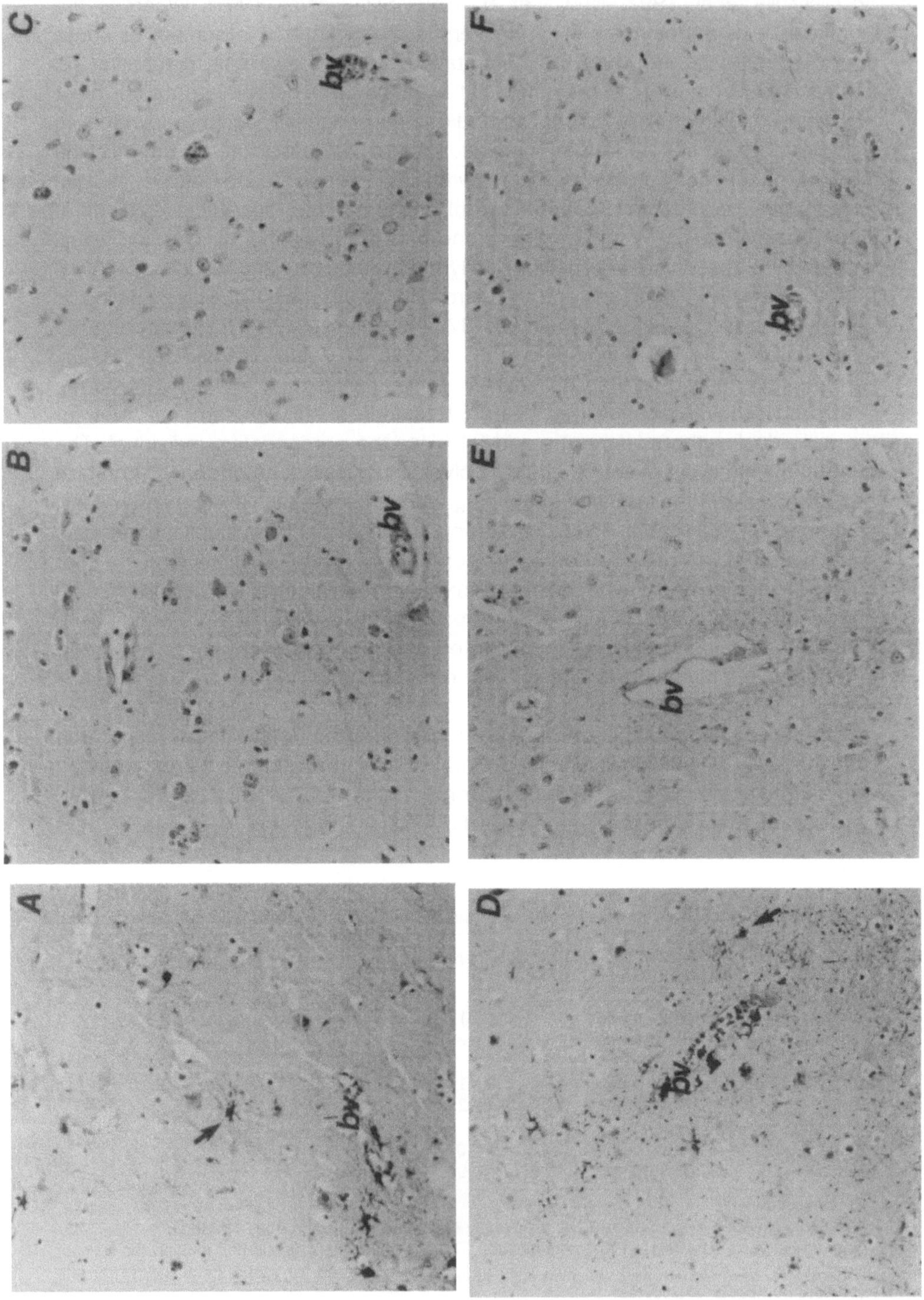
A
B
C
D
E
F
bv

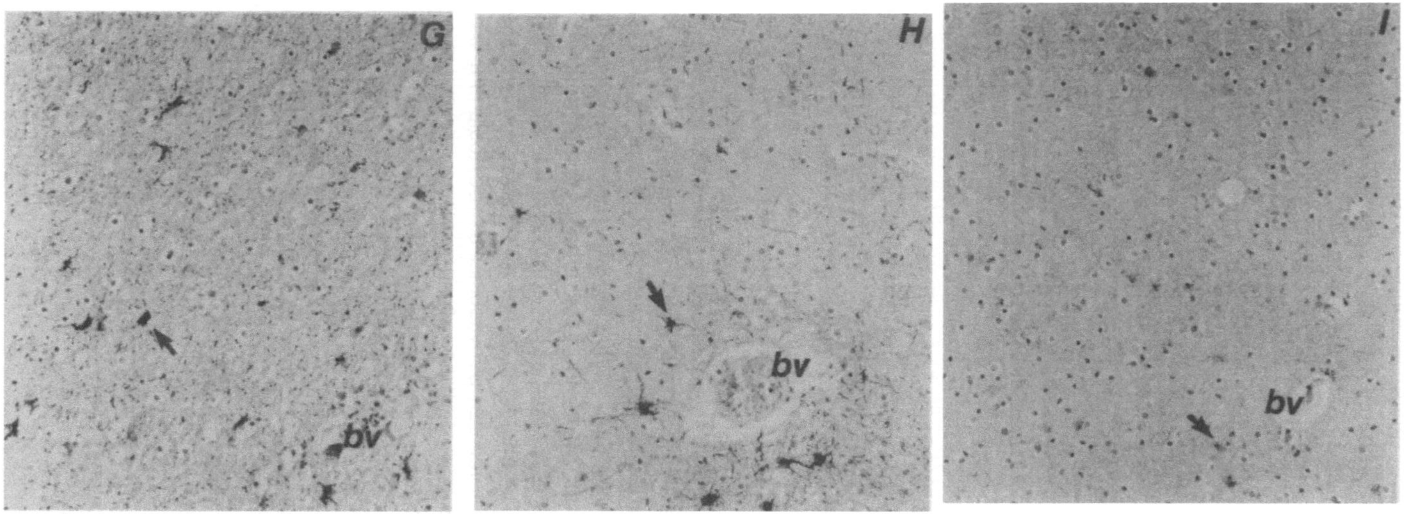

Fig. 1 A–I. Astrocytosis may be either diffuse and/or perivascular in the cerebral frontal cortex of HIV-infected individuals. Tissue sections stained with anti-glial fibrillary acidic protein (GFAP) antibody show GFAP⁺astrocytes (*arrows*) in cortical gray matter (**A–C**), gray/white matter interface (**D–F**) and subcortical white matter (**G–I**). The photomicrographs of tissues from an HIV-infected individual diagnosed with HIV-associated dementia (**A,D,G**) and without dementia (**B,E,H**) are shown. The density, size and staining of GFAP in these photomicrographs may be compared with those obtained from a seronegative, healthy individual who died in an accident (**C,F,I**). Data are representative of seven individuals with and one without HIV-associated dementia and one healthy seronegative individual. These results were obtained by previously described methods (DA CUNHA et al. 1993b,c). Magnification x 200; *bv*, blood vessels

consistent with the notion that differences among normal astrocytes are reflected in reactive astrocytes. All individuals with HIV-1 infection had increased density of astrocytes in both white and gray matter (Fig. 1A, D, G) relative to control (Fig. 1C,F,I). The size was consistently larger only in the gray matter of HIV-1-infected individuals. The intensity of staining was apparently unchanged in either region. Thus, HIV-1 infection results in white/gray matter-specific differences in astrocytosis.

The pattern of distribution of astrocytes was diffuse in the majority of tissues (Fig. 1A,D,G). Diffuse astrocytosis was strikingly absent in the gray and white matter of one AIDS patient who had no dementia (Fig. 1B,E,H). Whereas density of astrocytes in this tissue away from blood vessels, was low (Table 1), the density of astrocytes adjacent to blood vessels was high (Fig. 1H). The perivascular astrocytes were detected in the lowest layers of the gray matter (Fig. 1E) and in the white matter (Fig. 1H). We measured astrocytosis of the perivascular astrocytes and compared it to those distant from blood vessels. The density of perivascular astrocytes was 16 ± 3 cells/cm^2, comparable to the density of astrocytes in demented individuals with diffuse astrocytosis (Table 1). The size of astrocytes was larger (15 ± 7 μm^2) next to blood vessels than away from them (Table 1). Thus, perivascular astrocytosis was similar to the diffuse astrocytosis with regard to density and size of astrocytes. These data indicate that quantitation of astrocytosis may be necessary for an accurate diagnosis.

The diffuse astrocytosis was reported in five of six HIV-1-infected individuals with dementia (CIARDI et al. 1990) and three SIV-infected rhesus macaques with motor and cognitive impairments (WEIHE et al. 1993). However, perivascular rather than diffuse astrocytosis has been seen in three HIV-1-infected individuals with no behavioral symptoms (LENHARDT et al. 1988). Perivascular astrocytosis has also been described in 11 HIV-1 positive individuals who had no neurobehavioral or clinical signs of AIDS and were presumed asymptomatic (GRAY et al. 1992). Since the asymptomatic stage is the earliest stage of the disease and because dementia occurs late in AIDS, we infer that perivascular astrocytosis is a likely early event in the neuropathogenesis of HIV-1 infection. These observations support the hypothesis that astrocytosis may be induced by signals originating in the circulation and spreads from blood vessels to the parenchyma.

The morphologic distinction of astrocytes in white versus gray matter is less subtle than regionally determined heterogeneity. The brain region-determined heterogeneity appears to be conserved in culture. For example, only astrocytes from the cerebellum express somatostatin mRNA and peptides, astrocytes from the cerebellum, cortex and striatum express proenkephalin mRNA and enkephalin peptides whereas none express cholecystokinin (SHINODA et al. 1989). Also, brain region specificity of astrocytes may define the three dimensional shape of mouse dopaminergic neurons (DENIS-DONINI et al. 1984). Thus, astrocytic heterogeneity may be intimately related to neuronal structure and function. The heterogeneity of normal astrocytes is probably reflected in reactive astrocytes. These observations suggest that astrocytosis likely varies not only in white versus gray matter but also in different brain regions. This is presently being investigated.

As if the brain region-determined heterogeneity already is not complex enough, the heterogeneity exists among astrocytes which are near neighbors in tissue and in vitro. For example, glutamine synthetase is differentially expressed in individual astrocytes in the developing and adult rat brain (MEAROW et al. 1989) and in vitro (HALLERMAYER and HAMPRECHT 1984). Receptors for glutamate and other neurotransmitter receptors are heterogeneously expressed in cultured astrocytes (WILKIN et al. 1990; HOLZWORTH et al. 1994). IL-1 and TGF-β1 are detectable individually and together in some but not all, astrocytes of the human frontal cortex (DA CUNHA et al. 1993a). Whether this heterogeneity is random or regulated and how it is controlled is unknown. Nevertheless, because this heterogeneity is well documented and therefore real, we must account for it in analyzing and interpreting astrocytosis. The heterogeneity of astrocytosis may underlie the diversity of symptoms in HIV/CNS disease. This diversity probably reflects the numerous neuronal changes observed post mortem in different regions of the brain.

4 Astrocytosis Alone May Cause Neuronal Dysfunction Underlying HIV-1-Associated Dementia

Astrocyte functions are essential to neurons. The best understood such function in adult brain is the metabolism of glutamate and γ-amino butyric acid (GABA), the major inhibitory neurotransmitter. The glutamate-GABA cycle is catalyzed by several enzymes compartmentalized between astrocytes and neurons. One of these enzymes is glutamine synthetase. Glutamine synthetase is almost exclusively made in astrocytes (see MEAROW et al. 1989). It catabolizes glutamate to glutamine by consuming ammonia. This reaction detoxifies ammonia from the brain where, in contrast to other organs, the urea cycle does not operate. This reaction is also the sole means of removing excess glutamate which is excitotoxic at high concentrations. In addition, astrocytes take up synaptically released glutamate (MENNERICK and ZORUMSKI 1994). Thus, astrocytes do not only protect neurons from excitotoxicity but probably also control synaptic currents. These functions may be diminished during astrocytosis by TGF-β1, which downregulates glutamine synthetase in cultured astrocytes (TORU-DELBAUFFE et al. 1990). Thus, dysregulated astrocytosis may by itself disturb glutamatergic transmission and ultimately cause excitoxicity, the postulated mechanism for neuronal dysfunction in AIDS brain (LIPTON 1992). These observations suggest a potential relationship between astrocytosis *in appropriate brain regions* and HIV-1-associated dementia. To test this possibility, we measured astrocytosis in cerebral frontal cortex of seven individuals with and one without HIV-1 dementia. The density of astrocytes in both white and gray matter of individuals with HIV-1-dementia was higher by at least two standard deviations of the mean than the individual who either had no HIV-1-associated dementia or was seronegative (Table 1). The size of astrocytes in the white matter was also higher in HIV-1-demented individuals than in the nondemented and seronegative individuals. The increased size of astrocytes in the gray matter may not be statistically significant (Table 1). These results suggest, although they do not prove, that the extent of astrocytosis in the cerebral frontal cortex may be related to the severity of neurobehavioral deficits (as reflected in the diagnosis of dementia).

The relationship between astrocytosis and dementia, if confirmed, suggests that astrocytosis may cause dementia. Astrocytosis in the right place at the right time may cause changes in susceptible neurons that are more subtle than frank excitotoxicity or even the disturbance of glutamatergic transmission. Glutamate induces prompt and oscillatory elevation of cytoplasmic free Ca^{2+} in astrocytes. The increases in cytoplasmic free Ca^{2+} propagate as waves within the cytoplasm of individual astrocytes and between adjacent astrocytes in confluent culture (Cornell-Bell et al. 1990; Charles et al. 1991). Astrocytic Ca^{2+} waves are triggered by the firing of glutamatergic neuronal afferents (in organotypic culture of rat hippocampus resembling astrocyte-neuron relationships in vivo). Thus, astrocyte networks are able to respond to neuronal network activity (Dani et al. 1992). In addition, Ca^{2+} signals can propagate from astrocytes to neurons (Nedergaard 1994) suggesting that astrocytes can directly modulate neuronal intracellular Ca^{2+} concentration. These results indicate that astrocytes can bidirectionally interact with neurons via Ca^{2+}. "The astrocytic networks, like neuronal networks, may process information and participate in brain information processing to an extent comparable to and in cooperation with neuronal networks" (Dani et al. 1992). These observations together strongly suggest that dysregulated astrocytosis alone can lead to neuronal dysfunction underlying HIV-1-associated dementia. Thus, astrocytosis deserves a fresh new look in the context of HIV-1 infection and AIDS.

Acknowledgments. We thank W.D. Lyman for commenting on an earlier version of this manuscript. A. da Cunha is a Pediatric AIDS Foundation Scholar.

References

Aquino DA, Shafit-Zagardo B, Brosnan CF, Norton WT (1990) Expression of glial fibrillary acidic protein and neurofilament mRNA in gliosis induced by experimental autoimmune encephalomyelitis. J Neurochem 54: 1398–1404

Budka H (1990) Human immunodeficiency virus (HIV) envelope and core proteins in CNS tissues of patients with the acquired immune deficiency syndrome (AIDS). Acta Neuropathol (Berl) 79: 611–619

Budka H, Costanzi G, Cristina S, Lechi A, Parravicini C, Trabattoni R, Vago L (1987) Brain pathology induced by infection with the human immunodeficiency virus (HIV). Acta Neuropathol (Berl) 75: 185–198

Chakrabarti L, Hurtrel M, Maire M-A, Vazeux R, Dormont D, Montagnier L, Hurtrel B (1991) Early viral replication in the brain of SIV-infected rhesus monkeys. Am J Pathol 139: 1273–1280

Charles AC, Merrill JE, Dirksen ER, Sanderson MJ (1991) Intercellular signaling in glial cells: calcium ways and oscillations in response to mechanical stimulation and glutamate. Neuron 6: 983–992
Choi DW (1994) Glutamate receptor-mediated neuronal death In: Zalcman S, Scheler R, Tsien R (eds) Molecular neurobiology. US Department of Health and Human Services, Public Health Service, National Institutes of Health, Bethesda MD
Ciardi A, Sinclair E, Scaravilli F, Harcourt-Webster NJ, Lucas S (1990) The involvement of the cerebral cortex in human immunodeficiency virus encephalopathy: a morphological and immunohistochemical study. Acta Neuropathol (Berl) 81: 51–59
Cornell-Bell AH, Finkbeiner SM, Cooper MS, Smith SJ (1990) Glutamate induces calcium waves in cultured astrocytes: long-range glial signaling. Science 247: 470–473
da Cunha A, Vitković L (1992) Transforming growth factor beta-1 (TGF-β1) expression and regulation in rat cortical astrocytes. J Neuroimmunol 36: 157–169
da Cunha A, Jefferson JJ, Jackson RW, Vitković L (1993a) Glial cell-specific mechanisms of TGF-β1 induction by IL-1 iu cerebral cortex. J Neuroimmunol 42: 71–86
da Cunha A, Jefferson JJ, Tyor WR, Glass JD, Jannotta FS, Vitković L (1993b) Gliosis in human brain: relationship to size but not other properties of astrocytes. Brain Res 600: 161–165
da Cunha A, Jefferson JJ, Tyor WR, Glass JD, Jannotta FS, Vitković L (1993c) Control of astrocytosis by interleukin-1 and transforming growth factor-b1 in human brain. Brain Res 631: 39–45
Dani JW, Chernjavsky Y, Smith SJ (1992) Neuronal activity triggers calcium waves in hippocampal astrocyte networks. Neuron 8: 429–440
Davis LE, Hjelle BL, Miller VE, Palmer DL, Llewellyn AL, Merlin TL, Young SA, Mills RG, Wachsman W, Wiley CA (1992) Early viral brain invasion in iatrogenic human immunodeficiency virus infection. Neurology 42: 1737–1739
Denis-Donini S, Glowinski J, Prochiantz A (1984) Glial heterogeneity may define the three-dimensional shape of mouse mesencephalic dopaminergic neurones. Nature 307: 641–643
Eddleston M, Mucke L (1993) Molecular profile of reactive astrocytes-implications for their role in neurologic disease. Neuroscience 54: 15–36
Epstein LG, Gendelman HE (1993) Human immunodeficiency virus type 1 infection of the nervous system: pathogenetic mechanisms. Ann Neurol 33: 429–436
Funata N, Maeda Y, Koike M, Okeda R (1991) Neuropathology of the central nervous system in acquired immune deficiency syndrome (AIDS) in Japan. With special reference to human immunodeficiency virus-induced encephalomyelopathies. Acta Pathol Jpn 41: 206–211
Gray F, Lescs M-C, Keohane C, Paraire F, Marc B, Durigon M , Gherardi RJ (1992) Early brain changes in HIV infection: neuropathological study of 11 HIV seropositive, non-AIDS cases. J Neuropathol Exp Neurol 51: 177–185
Hallermayer K, Hamprecht B (1984) Cellular heterogeneity in primary cultures of brain cells revealed by immunocytochemical localization of glutamine synthetase. Brain Res 295: 1–11
Holzwarth JA, Gibbons SJ, Brorson JR, Philipson LH, Miller RJ (1994) Glutamate receptor agonists stimulate diverse calcium responses in different types of cultured rat cortical glial cells. J Neurosci 14: 1879–1891
Hurtrel B, Chakrabarti L, Hurtrel M, Maire MA, Dormont D, Montagnier LJ (1991) Early SIV encephalopathy. J Med Primatol 20: 159–166
Kure K, Wiedenheim KM, Lyman WD, Dickson DW (1990a) Morphology and distribution of HIV-1 gp41-positive microglia in subacute AIDS encephalitis. Acta Neuropathol (Berl) 80: 393–400
Kure K, Lyman WD, Weidenheim KM, Dickson DW (1990b) Cellular localization of an HIV-1 antigen in subacute AIDS encephalitis using an improved double-labelling immunohistochemical method. Am J Pathol 136: 1085–1092
Kure K, Llena JF, Lyman WD, Soeiro R, Weidenheim KM, Hirano A, Dickson DW (1991) Human immunodeficiency virus-1 infection of the nervous system: an autopsy study of 268 adult, pediatric, and fetal brains. Hum Pathol 22: 700–710
Lenhardt TM, Super MA, Wiley CA (1988) Neuropathological changes in an asymptomatic HIV seropositive man. Ann Neurol 23: 209–210
Lipsky RH, Silverman SJ (1987) Effects of mycophenolic acid on detection of alial filaments in human and rat astrocytoma cultures. Cancer Res 47: 4900–4904
Lipton SA (1992) Models of neuronal injury in AIDS: another role for the NMDA receptor? Trends Neurosci 15: 75–79
Locksley RM, Crowe S, Sadick MD, Heinzel FP, Gardner KD, McGrath MS, Mills J (1988) Release of interleukin 1 inhibitory activity (contra-IL-1) by human monocyte-derived macrophages infected with human immunodeficiency virus in vitro and in vivo. J Clin Invest 82: 2097–2104
Masliah E, Achim CL, Ge N, DeTeresa R, Terry RD, Wildy CA (1992) Spectrum of human immunodeficiency virus-associated neocortical damage. Ann Neurol 32: 321–323

McArthur JC, Becker PS, Parisi JE, Trapp B, Selnes OA, Cornblath DR, Balakrishnan J, Griffin JW, Price D (1989) Neuropathological changes in early HIV-1 dementia. Ann Neurol 26: 681–684
McArthur JC, Selnes OA, Glass JD, Hoover DR, Bacellar H (1994) HIV dementia. Incidence and risk factors. In: Price RW, Perry SW (eds) HIV, AIDS and the brain. Raven, New York, pp 251–272
Mearow KM, Mill JF, Vitković L (1989) The ontogeny and localization of glutamine synthetase gene expression in rat brain. Mol Brain Res 6: 223–232
Mennerick S, Zorumski CF (1994) Glial contributions of excitatory neurotransmission in cultured hippocampal cells. Nature 368: 59–62
Merrill JE, Koyangi Y, Chen ISY (1989) Interleukin-1 and tumor necrosis factor-α can be induced from mononuclear phagocytes by human immunodeficiency virus type 1 binding to the CD4 receptor. J Virol 63: 4404–4408
Nedergaard M (1994) Direct signaling from astrocytes to neurons in cultures of mammalian brain cells. Science 263: 1768–1769
Norton WT, Aquino DA, Hozumi I, Chiu C-F, Brosnan CF (1992) Quantitative aspects of reactive gliosis: a review. Neurochem Res 17: 877–885
Privat A, Rataboul P (1986) Fibrous and protoplasmic astrocytes. In: Federoff S, Vernadakis A (eds) Astrocytes, vol 1. Academic, New York, pp 105–129
Raff MC (1989) Glial cell diversification in the rat optic nerve. Science 243: 1450–1455
Rausch DM, Heyes MP, Murray EA, Lendvay L, Sharer LR, Ward JM, Rehm S, Nohr D, Weihe E, Eiden LE (1994) Cytopathological and neurochemical correlates of progression of motor/cognitive impairment in SIV-infected rhesus monkeys. J Neuropathol Exp Neurol 53: 165–175
Schmidbauer M, Huemer M, Cristina S, Trabattonic GR, Budka H (1992) Morphological spectrum, distribution and clinical correlation of white matter lesions in AIDS brains. Neuropathol Appl Neurobiol 18: 489–501
Shinoda H, Marini AM, Cosi C, Schwartz JP (1989) Brain region and gene specificity of neuropeptide gene expression in cultured astrocytes. Science 245: 415–417
Smith TW, DeGirolami U, Henin D, Bolgert F, Hauw JJ (1991) Human immunodeficiency virus (HIV) leukoencephalopathy and the microcirculation. J Neuropathol Exp Neurol 49: 357–370
Tchelingerian J-L, Quinonero J, Booss J, Jacque C (1993) Localization of TNF-α and IL-1α immunoreactivities in striatal neurons after surgical injury to the hippocampus. Neuron 10: 213–224
Toru-Delbauffe D, Bagdassarian-Chalaye D, Gavaret JM, Courtin F, Pomerance M, Pierre M (1990) Effects of transforming growth factor β1 on astroglial cells in culture. J Neurochem 54: 1056–1061
Vitković L, Tyor RW, da Cunha A (1994) Cytokine expression and regulation of pathogenesis in AIDS brain. In: Price RW, Perry SW (eds) HIV, AIDS and the brain. Raven, New York, pp 203–222
Wahl SM, Hunt DA, Wakefield LM, McCartney-Francis N, Wahl LM, Roberts AB, Sporn MB (1987) Transforming growth factor type β induces monocyte chemotaxis and growth factor production. Proc Natl Acad Sci USA 84: 5788–5792
Weihe E, Nohr D, Sharer L, Murray E, Rausch D, Eiden L (1993) Cortical astrocytosis in juvenile rhesus monkeys infected with simian immunodeficiency virus. Neuroreport 4: 263–266
Wilkin GP, Marriott DR, Cholewinski AJ (1990) Astrocyte heterogeneity. Trends Neurosci 13: 43–46

Therapeutic Approaches to HIV Infection Based on Virus Structure and the Host Pathogen Interaction

C.D. PAUZA and D.N. STREBLOW

1 Introduction

Molecular dissection of the primate immunodeficiency viruses has led to identification of their principal structural components and defined many of their required activities. The resulting picture is one of highly complex viruses that execute subtle yet catastrophic destruction of host immune capacity and thus rate among the most insidious and lethal of the infectious viral agents. The identification and functional analysis of viral genes, gene products, and regulatory elements proceeded more quickly than studies on viral pathogenesis. This situation reflects the intense and early interest on full molecular characterization of these viral agents. However, our lack of understanding of viral pathogenesis in AIDS and its related disorders also reflects the genuine difficulty that is associated with developing experimental approaches to identify key mechanisms in disease progression and point the way to effective therapeutic intervention. In the absence of definitive views of viral pathogenesis, we must rely on basic knowledge about how virus structure and unique features of the viral life cycle can be exploited for the development of effective therapeutic agents.

Department of Pathology and Laboratory Medicine, University of Wisconsin, 1300 University Avenue, Madison, WI 53706, USA

2 Synthetic Antiviral Compounds That Inhibit Function of Viral Proteins

The initial and most direct approach to antiviral therapy in AIDS was based on the early finding that human immunodeficiency virus type 1 (HIV-1) was a retrovirus that used a magnesium-dependent reverse transcriptase for conversion of virion genomic RNA into viral DNA (Popovic et al. 1984). With few exceptions, the act of copying RNA into complementary DNA is unique to retroviruses (Swanstrom and Varmus 1982), and the subsequent chromosomal integration of viral DNA to form the provirus establishes the productive infection in susceptible cells. Direct inhibition of viral reverse transcriptase provided a simplistic and straightforward target for therapeutic intervention. The reverse transcriptase reaction was easily modeled in vitro and conventional screening for inhibitory molecules was initiated. Soon thereafter, nucleoside analogues (Mitsuya and Broder 1986; Mitsuya et al. 1985) were identified that inhibited viral DNA synthesis. These inhibitors are incorporated into growing DNA chains which cannot be extended past the point of analogue insertion (Furman et al. 1986). The nucleoside analogues are partly selective for the RNA-dependent DNA polymerase (reverse transcriptase) found in retroviruses and also show a limited inhibition of cellular DNA-dependent DNA polymerases that are required for cellular replication (Mitsuya et al. 1985). Given their favorable therapeutic indices, reasonable oral bioavailability, and ease of chemical synthesis, these agents were advanced quickly through clinical trials and were approved for therapeutic use in 1987. The most widely used nucleoside analogues include zidovudine (3'azido-3'-deoxythymidine, AZT, zidovudine), dideoxyadenosine (ddA, didanosine), and dideoxycytosine (ddC, zalcitabine). They have widespread use and are generally recommended for patients with declining CD4 cell counts and collateral evidence of disease progression. In general, these agents are characterized as having short-term effects on constitutional parameters although their ability to prolong survival is questionable and they exhibit modest effects on virus burden (Fischl et al. 1987, 1990; Ho et al. 1989; Parks et al. 1988; Yarchoan et al. 1986). In vitro studies revealed that zidovudine could actually increase virus loads by inhibiting HIV-1 cytopathic effects to a greater degree than they inhibit primary infection (Pauza et al. 1990). Such a mechanism may reflect the requirement for sequential cycles of reinfection in already infected cells to induce apoptosis and cell death (Terai et al. 1991).

Nonnucleoside inhibitors of reverse transcriptase were introduced in 1989 and have not yet been approved for widespread clinical application. These agents include TIBO, HEPT, nevirapine, and many other analogues which bind a modulatory site that is distinct from the polymerase and RNase active sites of reverse transcriptase. These molecules are effective at inhibiting HIV-1 reverse transcriptase (including AZT resistant isolates) but not HIV-2 reverse transcriptase or other DNA polymerases (Grob et al. 1992). Some of these agents have good oral bioavailability and are being tested in clinical trials. The eventual value of non-nucleoside reverse transcriptase inhibitors is not yet established.

A second viral function with unique properties is the retroviral protease. The viral protease is a component of the pr160$^{gag\text{-}pol}$ precursor protein that arises by translation of the 55 kDal *gag* gene followed by frame-shifting of the ribosome (~10% frequency) and read-through translation of the *pol* gene. A presumed dimerization of the pr160$^{gag\text{-}pol}$ precursor protein permits functional association of two copies of the protease protein. The dimerized protein becomes active and initiates processing of the viral pr160$^{gag\text{-}pol}$ and p55^{gag} precursor proteins at unique cleavage sites. The viral protease dimer is composed of two 10 kDa monomers and is classified as an aspartyl protease by virtue of two conserved DTG domains that project into the crevice of the homodimer and are critical for activity (KEMPF et al. 1993; MILLER et al. 1989). The small size and low complexity of the protease facilitated structural analysis that provided the complete structure at 2.8 Å (MILLER et al. 1989). Numerous studies documented that viral protease is required for the production of infectious virions from already infected cells (DREYER et al. 1989) and some studies suggest that protease activity is also required for an initial step in the pathway of virus penetration (BABOONIAN et al. 1991).

Numerous inhibitors of viral protease have been synthesized and tested, including active site inhibitors (substrate analogs), C-2 symmetric inhibitors and inhibitors of protease dimerization. Peptidomimetics that bind in the active site often contain the unique amino acid statine, also found in fungal protease inhibitors such as peptstatin (RICH et al. 1992). The statine residue acts as a transition state analogue and the protease-peptidomimetic complex is inactive. Others have used small molecule approaches to develop inhibitors like dihydroxyethylene isosteres that also block the active site. In many of these cases, crystal structures were obtained for the enzyme-inhibitor complex and the contact points were identified at atomic resolution (JACOBSEN et al. 1992). C-2 symmetric inhibitors are generally less peptide-like and bind tightly to the active site (ERICKSON et al. 1990). Prevention of protease homodimerization also seems to be an effective way to specifically inhibit viral maturation; this may be accomplished by introducing altered protease monomers or by drugs that bind to interaction domains on the monomers. Protease inhibitors are often effective in the nanomolar to micromolar concentration ranges and most are specific for the viral protease. Delivery has proved a difficult problem for protease inhibitors. The peptidomimetics must be modified to increase permeability across cell membranes and the oral bioavailability for all of these compounds is poor. Once the compounds are absorbed across the gastrointestinal mucosa, loss through the biliary circulation has also proved a major problem. Although initial clinical trials are underway, progress has been slow in identifying clinically useful inhibitors of the retroviral protease.

Aromatic polycyclic diones (hypericin and pseudohypericin) also prevent the maturation of immature viral particles and/or cause abnormal assembly that inactivates the virus. Significantly, they inactivate directly mature virus particles.

These compounds associate with the cell membrane and may block HIV protease activity and interfere with gag maturation and assembly. The exact mechanism of action is unclear. Hypericin, an antidepressant, can cross the blood-brain barrier (LAVIE et al. 1989; MERUELO et al. 1988) and may provide useful for preventing HIV infection in the brain and CNS.

3 Virus Variation Leads to Rapid Evolution of Drug-Resistant Variants

The design and application of synthetic antiviral compounds based on the structure and activity of HIV-1 proteins assumes that the target enzyme has a stable sequence and structure. Clinical and laboratory experience with the nucleoside analogues revealed a significant weakness in this approach to developing antiviral agents.

Within 1 year of initiating zidovudine therapy, a significant proportion of patients produced a drug-resistant HIV-1 variant that was easily isolated from peripheral blood samples (RICHMAN 1991; ROOKE et al. 1989). Resistance was variable, but many isolates were discovered to be resistant to more than 100 times the therapeutic dose of zidovudine (LARDER et al. 1989; LARDER and KEMP 1989). The evolution of resistance seemed directly related to selective pressure exerted by drug therapy, inasmuch as resistant variants were not recovered from untreated patients who were infected prior to widespread use of zidovudine (LARDER et al. 1991). Sequences were determined from the *pol* gene of resistant variants and specific amino acid changes were associated with growth in the presence of zidovudine in vitro (LARDER et al. 1991). The amino acid changes each contributed to the level of resistance in a nearly additive manner (LARDER and KEMP 1989), and there are preliminary indications that the course of infection in vivo was related to the development of drug-resistant variants. Viruses with drug resistance phenotypes grow well in vitro and retain their capacity for inducing cytopathic effect in lymphoblastoid cell cultures (LARDER and KEMP 1989). The development of zidovudine-resistant variants in patients can be monitored reliably by direct DNA sequence analysis for specific amino acid substitutions (LARDER et al. 1991), and it is technically possible to use these criteria for discontinuation of therapy in long-term treated individuals. Importantly, the development of zidovudine resistant HIV-1 variants limits the utility of this drug for chronic therapy and served to highlight the problems associated with developing drugs to inhibit viral proteins.

Although extensive clinical data are not yet available, we anticipate similar problems will arise during therapy with viral protease inhibitors. Indeed, recent publications report the identification of protease inhibitor-resistant variants of HIV-1 that evolved after long-term in vitro passage of virus in the presence of these agents (Ho et al. 1994). It is not yet possible to compare the rates of developing resistant variants for protease inhibitors as compared to reverse transcriptase inhibitors. However, the value of these two major categories of antiviral compounds is reduced by the rapid rates of virus variation in vivo that permit outgrowth of inhibitor-resistant variants.

4 Synthetic Viral Components That Inhibit HIV-1 Replication

In response to problems of evolving resistance to synthetic inhibitors of virus replication, new approaches were developed that target virus regulation or assembly events. These targets include components of self-assembling oligomeric complexes and crucial functions in the pathway of virus transcription and translation. The guiding principle is that activities that require the interaction of two or more virus structural elements (protein or nucleic acid) will be less likely to evolve drug-resistant variants than elements that act alone (such as reverse transcriptase).

The most apparent oligomeric complex in HIV-1 is the capsid structure. The matrix protein gag p17 is associated with the viral envelope and, until maturation, it interacts physically with the capsid protein p24. Viral p24 forms the shell of a ribonucleoprotein complex which also consists of the viral RNA, gag p6 and p7, integrase, reverse transcriptase, and the tRNA lys primer of reverse transcription. In a fraction of virions, premature initiation of reverse transcription produces a fragment of viral DNA that is also encapsidated (Lori et al. 1992; Trono 1992). The gag pr55 proteins can self-assemble into oligomeric structures when expressed by recombinant baculoviruses (Gheysen et al. 1989; Hughes et al. 1993).

Complex oligomers are particulary prone to "poison" subunits that can be incorporated into the complex even though they interfere with functions of the oligomer. The "poison" subunit can thus confer a dominant negative virus replication phenotype and can accomplish this effect when present as only a minor component in the pool of assembling monomers. Alterations in HIV-1 *gag* genes were developed and tested for their ability to inhibit infecting virus. Some variants were completely effective in blocking virus replication in vitro and many independent mutants exerted partial negative effects (Trono et al. 1989). Mutations in the major homology region of p24 interfere with viral capsid asembly (Ebbets-Reed et al. 1994). Thus, the goal of generating dominant negative phenotypes for virus replication was achieved for in vitro models of infection.

Similar approaches were used to target activities of the viral tat and rev proteins that are required for virus expression. The tat protein relieves a block to elongation of virus transcription by transactivating a complex RNA structure termed *TAR* (Berkhout et al. 1990; Feng and Holland 1988). The rev protein is required for expression of viral structural genes. In the absence of rev protein, inhibitory sequences present in HIV-1 structural genes prevent efficient transport of unspliced, singly spliced and genomic viral RNA out of the cell nucleus (Malim et al. 1990). The rev protein binds to a RNA structure within the envelope gene sequence, the *RRE* and facilitates RNA transport out of the nucleus. Searches were initiated for dominant negative mutants that could block the normal function of these viral regulatory genes. *Trans*-dominant mutants of both tat and rev were discovered and both inhibited virus replication in vitro.

Synthetic oligonucleotides were also explored for their therapeutic potential against HIV-1 infection. Oligonucleotides complementary to translation initiation or splicing sites were used to block RNA expression or processing (Agrawal et al. 1988). The value of modified bases was evaluated (Zaia et al. 1988) with consideration for efforts to increase penetration and half-life of the agents. Overall, synthetic oligonucleotides appear selective and effective against HIV-1 infection in vitro. Problems of bioavailability and pharmacodynamics remain largely unexplored.

5 Compounds That Attack HIV-1 Structural Components

One of the first compounds that targeted virus structural components was the synthetic liposome AL721 (Sarin et al. 1985). Previous studies showed that this cholesterol-deficient liposome could bind to virions and thus render them noninfectious. The HIV-1 lipid envelope is rich in cholesterol and this composition seems required for efficient virus entry (Aloia et al. 1987). The synthetic liposome is thought to work by depleting the virus envelope of cholesterol and thus reducing the infectivity of individual particles. The early reports on AL721 activity in vitro prompted an avalanche of popular press that provided home recipes for generating liposomes from egg whites and kitchen oils. Despite these large scale, uncontrolled trials little information was produced to encourage further development of AL721 as a clinical agent. Aurothiolate compounds also manifest antiviral activity in vitro. Aurothioglucose and aurothiomalate react in molar equivalency with thiol groups on exposed proteins. In the case of HIV-1 virions, these compounds react with envelope glycoprotein resulting in a permanent modification of free sulfhydryl groups with an attendant loss of infectivity (Okada et al. 1993). Approximately 50% of the mass of env is in the form of N-linked oligosaccharides. These are highly structured compounds that play a critical role in the processing of gp160, budding, syncytium formation, binding, and entry. Processing of the carbohydrates is crucial for these activities. Inhibitors of the ER-associated α-glucosidase activity, e.g., castanospermine or N-butyl deoxynojirimycin (Ratner et al. 1991), block viral replication. The compounds prevent processing of the terminal glucose residues which in turn antagonizes proper folding leading to increased degradation. Compounds such as deoxymannojirimycin and swainsonine inhibit mannosidase activity in the cis and medial Golgi apparatus, respectively. Both show some antiviral activity by inactivating later processing steps of envelope glycoprotein (GRUTERS et al. 1987).

Physical blocking or masking of the viral attachment proteins is another way to block virus entry and prevent infection. The polyanionic compounds dextran sulfate, heparin, and suramin all block virus binding, fusion and syncytium formation. Polyanions bind and coat env to prevent gp120/CD4 interaction. Dextran sulfate at concentrations below levels that inhibit blood coagulation blocks virus replication (BABA et al. 1988; MITSUYA et al. 1988). The activities of polyanions suggest they could be used in conjucntion with other antiviral compounds.

The stress-induced mannose binding protein (MBP) binds to the high mannose glycans of the envelope of HIV. This is an innate host defense mechanism that resists infections by gram-negative bacteria, yeast, mycobacteria, fungi, and enveloped viruses. MBP binding to env prevents env/CD4 interactions and thus inhibits infection (EZEKOWITZ et al. 1989). MBP may also be a useful ligand for binding toxins to the surface of infected cells, although the specificity has not been established.

A popular approach to designing antivirals to attack virus structural components relies on molecular biology techniques. Ribozymes, for example, are catalytic RNA molecules that recognize and cleave target RNA molecules by a complex endonucleolytic mechanism (SULLENGER and CECH 1993). Considerable selectivity is afforded by sequence-specific interactions and their short half-life renders them of little danger to the host. Ribozymes are generally produced after DNA transfer into target cells. Expression can be linked to inducible viral promoters, such as the HIV-1 LTR, to coordinate expression of the ribozyme with expression of the viral *Tat* gene product (DROPULIC et al. 1992; WEERASINGHE et al. 1991). Cech showed that it is possible to deliver ribozymes by tethering them to retroviral packaging sequences. This may facilitate specific delivery to defined cell types such as CD4^{+} T cells (CECH 1993).

Ribozymes are proven to be effective against persistent RNA viruses, such as arenaviruses. Ribozymes specific for the lymphocytic choriomeningitis virus were developed. High level expression of these ribozymes rendered cells resistant to LCMV infection (XING and WHITTON 1992). Ribozymes against HIV-1 and HIV-2 have been developed and tested in vitro (DROPULIC et al. 1992; WEERASINGHE et al. 1991). The agents are effective at blocking virus replication (ROSSI et al. 1992; SARVAR et al. 1990), with the important consideration that the cells remain infected by virtue of the integrated provirus. In the example of neuropathologic manifestations of HIV-1 infection in the brain, the selective ability to block expression of viral gene products, particularly envelope glycoprotein, may favorably affect the clinical situation (LEVI et al. 1993). Stable maintenance of the integrated provirus without virus expression and without particle release may be an acceptable endpoint for therapy. The problems of targeted gene delivery especially to protected sites such as the central nervous system, selection of the best cell population for modification, and controlled ribozyme expression stand as important barriers before these inventions are reduced to practice. Situations such as modification of bone marrow before immune reconstitution or direct delivery to cellular targets in the brain provide attractive and compelling opportunities for practical appplication of ribozyme and other gene delivery-based therapeutic approaches. Mechanisms active in *cis* or *trans* can be combined with little loss of activity. The strategy of attacking multiple steps in the virus lifecycle and affecting host functions, serves to reduce the liklihood of evolving resistant virus strains.

Synthetic oligonucleotides are being developed for direct blocking of virus gene expression. Modified oligonucleotides specific for the *tat* gene (Poly TAR and TAR decoys) block virus replication in vitro (Agrawal et al. 1988; Zaia et al. 1988). Phosphorothioate modified oligonucleotides complementary to HIV-1 *gag*, *rev*, or *RRE* sequences inhibited virus replication in tissue culture systems for as much as 80 days (Lisziewicz et al. 1992, 1993). These oligonucleotide inhibitors of virus replication were effective at micromolar concentrations; comparisons among several agents suggested that appropriate sequences could be selected to provide the most favorable targets for inhibition.

Synthetic oligonucleotide approaches are also envisioned as a means of ridding the cell of integrated provirus and thus purging the infection. Triple helix formation involving the chromosomal provirus locus could target a modified oligonucleotide that would be capable of destroying the integrated viral DNA without extensive damage to surrounding host DNA. The concept was reduced to practice in the SV40 system. Psoralen-linked oligonucleotides were targeted to SV40 sequences and then activated by light. Specific mutagenesis was accomplished with such specificity for the targeted gene that 56% of all identified mutations were at the same base pair (Havre and Glazer 1993; Havre et al. 1993. It may be possible to utilize this approach, perhaps in combination with other antiviral methods, to eliminate HIV-1 DNA sequences from the genome of infected individuals.

6 Therapeutic Approaches Based on Modifying the Host

The extensive variation among HIV-1 isolates stands as an important consideration for any therapeutic approaches based on direct attacks on virus structural components. In response to this situation, many investigators are seeking to better understand the host components required for virus infection and replication.

Many studies addressed the contributions of cellular differentiation state to HIV-1 susceptibility. The requirement for T cell activation in virus replication is well established; however, less is understood of the relationship between macrophage differentiation and HIV-1 replication and this has immediate relevance for infection of macrophages in the CNS. Early studies on HIV-1 infection of monoblastoid cell lines described the phenomenon of restricted virus replication (PAUZA et al. 1988). In these examples, soluble products released by infected cells inhibited virus replication. Interferon-α IFN-α was proposed as the mediator of restriction (MACÉ et al. 1989; SZEBNI et al. 1991) and could work by altering the pathway of virus assembly (MELTZER et al. 1990). The IFN-α mediated block to HIV-1 replication may be early in the virus replication cycle (SHIRAZI and PITHA 1992,1993) in addition to the restriction phenomenon described above. Infected monoblastoid cells, primary monocyte derived macrophages differentiated in vitro by phorbol ester or 1,25- dihydroxyvitamin D_3 (PAUZA et al. 1992), and primary monocyte derived macrophages that differentiate in response to colony stimulating factors failed to produce cell-free virus and were shown (KORNBLUTH et al. 1989; MELTZER and GENDELMAN 1988) to contain abundant virions contained within intracytoplasmic vesicles. IFN-α genes linked to the HIV-1 long terminal repeat were expressed specifically in infected cells and they effectively modulated virus replication (BEDNARIK et al.1989). IFN-γ also has specific effects on virus replication and was reported to specifically inhibit translation of p24 gag antigen in infected cells (YAHI et al. 1992a). The interferons have many effects on HIV-1 replication depending on the individual cytokine and the type of target cell.

Manipulations of the normal pathways for viral protein processing and assembly provided one approach to antiviral therapy. ROSE and colleagues modified the CD4 virus receptor molecule to include the endoplasmic reticulum retention signal KDEL in its cytoplasmic tail. The modified CD4 engaged HIV-1 gp60 within the endoplasmic reticulum and prevented export to the cell surface (BUONCORE and ROSE 1990). This form of inhibition prevented virus release, syncytium formation, and secretion of gp120 that might be toxic to neighboring cells. Although this approach was not proposed as a practical method for inhibiting virus replication, it does demonstrate a key step in the virus life cycle that can be targeted. These investigations may well have inspired others to examine more facile host factors that might provide a means for blocking virus replication. The complex pathway for envelope glycoprotein synthesis requires proteolytic cleavage to separate the gp120 and gp41 subunits before they are assembled into oligomers and transported to the cell surface. This cleavage is accomplished by the host protein furin. Inhibitors of furin were previously known to inhibit avian influenza viruses and were recently verified as effective inhibitors of HIV-1 envelope glycoprotein processing (HALLENBERGER et al. 1992).

A very exciting new approach was suggested by the work of Goff and colleagues, relating properties of the host protein cyclophilin to HIV-1 replication. Cyclophilin refers to a small family of proteins typified by CycA and CycB in mammalian cells. The cyclophilins are crucial to lymphocyte activation pathways in their capacity as prolyl isomerases and as molecular chaperones that interact with the regulatory phosphatase calcineurin. The interaction of cyclophilin and calcineurin probably serves to control phosphorylation of transcription factors important to lymphocyte activation and early gene expression. The roles of cyclophilin A and B were highlighted because these proteins are important binding sites for the cyclic undecapeptide cyclosporine A, which has potent immunosuppressive properties. Using a yeast system that allows selection of proteins which interact physically, host proteins were identified that bound to the HIV-1 p55^{gag} protein of HIV-1; the predominant species were cyclophilin A and cyclophilin B, and cyclosporine A disrupted the *gag*/cyclophilin complexes (LUBAN et al. 1993). These data suggested a possible role for cyclosporine A in blocking virus assembly. Although the mechanisms are not yet apparent, these investigations suggest that cyclosporine A, a drug that binds to the same family of host proteins that also bind p55^{gag}, may be useful for therapy of HIV-1 infection. Clinical trials of Cyclosporine A prior to or concomitant with HIV-1 infection provided a conflicting picture for the activity of this agent (ALLEN 1992; BRINER et al. 1989; JACOBSON et al. 1991; KRAUSE et al. 1988; LAPORTE et al. 1987; PHILLIPS et al. 1989). Selected clinical reports (LAPORTE et al.1987; TORLONTANO et al. 1992) and many in vitro studies (ANDRIEU et al.1988; KARPAS et al. 1992; SANDSTROM and KAPLAN 1987; SCHMIDT et al. 1990) encourage continued study of cyclopsorine A as a clinically useful therapy in HIV infection. The specific effects of this drug on microglial cells (SAWADA et al. 1992) suggest that therapy of central nervous system HIV-1 infections might also be possible. Perhaps modified derivatives of the drug or more selective administration may still define an application for cyclosporine A in treatment of HIV-1 infection.

7 New Directions for Antiviral Development

Our search for effective antivirals to block HIV-1 infection assumes our capacity to identify unique aspects of HIV-1 that set this agent apart from other microbial pathogens.

Moreover, high priority targets have been identified owing to their absolute requirement in the virus life cycle and whether or not these viral or host functions contribute to pathogenesis and disease progression. Most importantly, therapeutic objectives must be matched to the situation of immediate interest including considerations of predominant infected cells types, principal pathogenetic sequelae, and likely mechanisms of pathogenesis.

Many reports describe HIV-1 infection of CD4-negative cell lines. One recent study showed that HIV-1 could infect brain-derived primary endothelial cells in culture, even though these cells lacked both CD4 and the galactosyl cerebroside that was proposed as an alternate receptor (MOSES et al. 1993). The infection of CD4-negative cells clearly plays a role in pathogenesis of the central nervous system and may also contribute to enteropathogenic complications in CD4-negative gut epithelial cells (NELSON et al. 1988; PHILLIPS and BOURINBAIAR 1991; YAHI et al. 1992b). The search for alternative or accessory receptors may still yield a novel target for future efforts to prevent HIV-1 infection and virus dissemination.

Another potential target was revealed in recent stuides of HIV-1 infection in unstimulated primary T lymphocytes. Virus attaches and penetrates these unstimulated cells with good efficiency. However once inside, synthesis of viral DNA proceeds only to the first step that produces the "negative strand strong stop" DNA. The replication complex contains HIV-1 RNA, strong stop DNA, reverse transcriptase, integrase, and the p17 matrix protein, along with unspecified host components (BUKRINSKY et al. 1993). This complex is reasonably stable in unstimulated lymphocytes and may remain for days to weeks until the lymphocytes are activated, viral DNA synthesis is completed, the complex is transported to the nucleus and proviral integration occurs (BUKRINSKY et al. 1991). We do not know the host factors that change after lymphocyte activation to permit completion of the viral DNA synthesis pathway. It is possible that low deoxynucleotide triphosphate levels in unstimulated lymphocytes prevent viral DNA synthesis or that a specific host protein must be induced during activation. This step represents an important point at which HIV-1 replication is blocked for prolonged intervals and this could constitute a target for antiviral intervention. Studies by LORI and colleagues (GAO et al. 1993) showed that hydroxyurea was an effective inhibitor of HIV-1 infection in vitro, presumably acting through its effect on the host ribonucleotide reductase that would have the consequence of lowering nucleoside triphosphate levels. Hydroxyurea or related compounds might be useful for medium-term therapy of HIV-1 infection and could provide an important adjunct when other agents are contraindicated because of associated toxicity or the development of resistant variants.

Once viral DNA synthesis is complete, subsequent steps in the replication pathway may also provide targets for therapeutic intervention. The linear double-stranded unintegrated viral DNA entering the nucleus can proceed either along the pathway toward proviral integration or become circularized (PAUZA 1990). The distribution of circular and linear unintegrated DNA forms is consistent among cell lines and for in vitro infected primary T lymphocytes (PAUZA et al. 1990). The circularization reactions appear to be mediated by host components, in as much as no viral proteins are known to have these activities, and circular DNA are present at reasonable levels in infected patients (JURRIAANS et al. 1992; PAUZA et al. 1994). Compounds that affect the cost determinant of circularization might increase the probability that viral DNA would enter this pathway that is not productive for infection (PAUZA et al. 1990). Failing this approach, decoy molecules or direct inhibitors that block integration could be considered.

8 Summary

The HIV-1 infection of central nervous system, with attendant neuropathy and dementia, poses a unique challenge for antiviral therapy. For practical considerations, it is important to define carefully the precise therapeutic objectives. (1) Is it necessary to inhibit spreading HIV-1 infection in the central nervous system? (2) What is the role of inflammatory responses in central nervous system disease during HIV-1 infection? (3) Is there a correlation between pathology and dementia? (4) Are virions or virus gene products toxic in the central nervous system? (5) Is there a role for immune suppression and opportunistic pathogens in AIDS dementia?

The development of therapeutic agents for HIV-1 infection is guided by our knowledge of virus structure, the function of viral proteins, the interactions with host components, and detailed features of the virus life cycle. In each case, unique features of the virus can be identified and established as targets for unique antiviral compounds. Drugs acting as inhibitors of virus enzymatic functions are plagued by the rapid development in vivo of drug-resistant virus variants, although combination or alternating chemotherapeutic regimens may obviate some of these concerns. Novel approaches to inhibiting virus are flourishing. In vitro studies show the value of agents as diverse as molecular decoys for tat activity to efforts to mutagenize integrated proviruses by modified oligonucleotides that form triple helices with chromosomal genes. As each particular clinical situation is better defined, the design and application of these agents can be refined to inhibit HIV-1 replication and reduce the associated morbidity.

References

Agrawal S, Goodchild J, Civeira M, Thornton A, Sarin P, Zamecnik P (1988) Oligodeoxynucleoside phosphoramidates and phosphorothioates as inhibitors of human immunodeficiency virus. Proc Natl Acad Sci USA 85: 7079–7083

Allen BR (1992) Use of cyclosporin for psoriasis in HIV-positive patient [letter] (see comments). Lancet 339: 686

Aloia R, Jensen F, Curtain C, Mobley P, Gordon LM (1988) Lipid composition and fluidity of the human immunodeficiency Virus (HIV). Proc Natl Acad Sci USA 85: 900–904

Andrieu JM, Even P, Venet A, Tourani JM, Stern M, Lowenstein W, Audroin C, Eme D, Masson D, Sors H et al. (1988) Effects of cyclosporin on T-cell subsets in human immunodeficiency virus disease. Clin Immunol Immunopathol 47: 181–198

Baba M, Pauwels R, Balzarini J, Arnout J, Desmyter J, DeClercq E (1988) Mechanism of inhibitory effect of dextran sulfate and heparin on replication of human immunodeficiency virus in vitro. Proc Natl Acad Sci USA 85: 6132–6136

Baboonian C, Dalgleish A, Bountiff L, Gross J, Oroszlan S, Rickett G, Smith-Burchnell C, Troke P, Merson J (1991) HIV-1 proteinase is required for synthesis of pro-viral DNA. Biochem Biophys Res Commun 179: 17–24

Bednarik DP, Mosca JD, Raj NB, Pitha PM (1989) Inhibition of human immunodeficiency virus (HIV) replication by HIV-trans-activated alpha 2-interferon. Proc Natl Acad Sci USA 86: 4958–4962

Berkhout B, Gatignol A, Rabson A, Jeang K (1990) TAR-independent activation of the HIV-1 LTR: evidence that Tat requires specific regions of the promoter. Cell 62: 757–767

Briner V, Zimmerli W, Cathomas G, Landmann J, Thiel G (1989) HIV infection caused by kidney transplant: case report and review of 18 published cases. Schweiz Med Wochenschr 119: 1046–1052

Bukrinsky MI, Stanwick TL, Dempsey MP, Stevenson M (1991) Quiescent T lymphocytes as an inducible virus reservoir in HIV-1 infection. Science 254: 423–427

Bukrinsky MI, Sharova N, McDonald TL, Pushkarskaya T, Tarpley WG, Stevenson M (1993) Association of integrase, matrix, and reverse transcriptase antigens of human immunodeficiency virus type 1 with viral nucleic acids following acute infection. Proc Natl Acad Sci USA 90: 6125–6129

Buoncore L, Rose J (1990) Prevention of HIV-1 glycoprotein transport by soluble CD4 retained in the endoplasmic reticulum. Nature 245: 625–628

Dreyer G, Metcalf B, Tomaszek T, Carr T, Chandler A, Hyland L, Fakahoury S, Magaard V, Moore M, Strickler J, Debouck C, Meek T (1989) Inhibition of human immunodeficiency virus 1 protease in vitro: rational design of substrate analogue inhibitors. Proc Natl Acad Sci USA 86: 9752–9756

Dropulic B, Lin NH, Martin MA, Leang K-T (1992) Functional characterization of a U5 ribozyme: intracellular suppression of human immunodeficiency virus type 1 expression. J Virol 66: 1432–1441
Ebbets-Reed D, Zybarth G, Ehlrich L, Carter C (1994) HIV-1 mutant encoding MHR-deleted gag precursors interferes with assembly of wild-type particles. J Cell Biochem (Keystone symposia) 18 B: 161
Erickson J, Neidhart D, VanDrie J, Kempf D, Wang X, Norbeck D, Plattner J, Rittenhouse J, Turon M, Wideburg N, Kohlbrenner W, Simmer R, Helfrich R, Paul D, Knigge M (1990) Design, activity, and 2.8 Å crystal structure of a C_2 symmetric inhibitor complexed to HIV-1 protease. Science 249: 527–533
Ezekowitz A, Kuhlman M, Groopman J, Byrn R (1989) A human serum mannose-binding protein inhibits in vitro infection by the human immunodeficiency virus. J Exp Med 169: 185–196
Feng S, Holland E (1988) HIV-1 tat trans-activation requires the loop sequence within tar. Nature 334: 165–167
Fischl M, Richman D, Grieco M, Gottlieb M, Volberding P, Laskin O, Leedom J, Groopman J, Mildvan D, Schooley R, Jackson G, Durack D, King D (1987) The efficacy of azidothymidine (AZT) in the treatment of patients with AIDS and AIDS-related complex. N Engl J Med 317: 185–191
Fischl M, Parker C, Pettinelli C, Wulfsohn M, Rasheed S (1990) The efficacy and safety of a lower dose of zidovudine in the treatment of patients with AIDS-associated PCP. Sixth international conference on AIDS (abstract)
Furman PA, Fyfe JA, St Clair MH, Weinhold K, Rideout JL, Freeman GA, Lehrman NS, Bolognesi DP, Broder S, Mitsuya H, Barry DW (1986) Phosphorylation of 3'-azido-3'deoxythymidine and selective interaction with human immunodeficiency virus reverse transcriptase. Proc Natl Acad Sci 83: 8333–8337
Gao WY, Cara A, Gallo RC, Lori F (1993) Low levels of deoxynucleotides in peripheral blood lymphocytes: a strategy to inhibit human immunodeficiency virus type 1 replication. Proc Natl Acad Sci USA 90: 8925–8928
Gheysen D, Jacobs E, de Foresta F, Thiriart C, Francotte M, Thines D, De Wilde M (1989) Assembly and release of HIV-1 precursor $Pr55^{gag}$ virus- like particles from recombinant baculovirus-infected insect cells. Cell 59: 103–112
Grob P, Wu J, Cohen K, Ingrahm R, Shih C, Hargrave K, McTague T, Merluzzi V (1992) Nonnucleoside inhibitors of HIV-1 Reverse Transcriptase: Nevirapine as a prototype drug. AIDS Res Hum Retroviruses 8: 145–152
Gruters R, Neefjes J, Tersmette M, de Goede R, Tulp A, Huisman H, Miedema F, Plodegh H (1987) Interference with HIV-induced syncytium formation and viral infectivity by inhibitors of trimming glucosidase. Nature 330: 74–77
Hallenberger S, Bosch V, Angliker H, Shaw E, Klenk H-D, Garten W (1992) Inhibition of furin-mediated cleavage activation of HIV-1 glycoprotein gp160. Nature 360: 358–361
Havre PA, Glazer PM (1993) Targeted mutagenesis of simian virus 40 DNA mediated by a triple helix-forming oligonucleotide. J Virol 67: 7324–7331
Havre PA, Gunther EJ, Gasparro FP, Glazer PM (1993) Targeted mutagenesis of DNA using triple helix-forming oligonucleotides linked to psoralen. Proc Natl Acad Sci USA 90: 7879–7883
Ho D, Moudgil T, Alam M (1989) Quantitation of human immunodeficiency virus type 1 in the blood of infected persons. N Engl J Med 321: 1621–1625
Ho DD, Toyoshima T, Mo T, Kempf DJ, Norbeck D, Chen CM, Wideburg NE, Burt SK, Erickson JW, Singh MK (1994) Characterization of human immunodeficiency virus type-1 variants with increased resistance to a C2-symmetric protease inhibitor. J Virol 68: 2016–2020
Hughes BP, Booth TF, Elyaev AS, McIlroy D, Jowett J, Roy P (1993) Morphogenic capabilities of human immunodeficiency virus type 1 gag and gag-pol proteins in insect cells. Virology 193: 242–255
Jacobsen H, Ahlborn-Laake L, Gugel R, Mous J (1992) Progression of early steps of human immunodeficiency virus type 1 replication in the presence of an inhibitor of viral protease. J Virol 66: 5087–5091
Jacobson SK, Calne RY, Wreghitt TG (1991) Outcome of HIV infection in transplant patient on cyclosporin (letter) (see comments). Lancet 337:
Jurriaans S, de Ronde A, Goudsmit J, Cornelissen M (1992) Analysis of human immunodeficiency virus type 1 LTR-LTR junctions in peripheral blood mononuclear cells of infected individuals. J Gen Virol 73: 1537–1541
Karpas A, Lowdell M, Jacobson SK, Hill F (1992) Inhibition of human immunodeficiency virus and growth of infected T cells by the immunosuppressive drugs cyclosporin A and FK 506. Proc Natl Acad Sci USA 89: 8351–8355
Kempf DJ, Codacovi L, Wang XC, Kohlbrenner WE, Wideburg NE, Saldivar A, Vasavanonda S, Marsh KC, Bryant P, Sham HL et al. (1993) Symmetry-based inhibitors of HIV protease. Structure-activity studies

of acylated 2,4-diamino-1,5-dipheny1-3- hydroxypentane and 2,5-diamino-1,6-diphenylhexane-3,4-diol. J Med Chem 36: 320–330

Kornbluth R, Oh P, Munis J, Cleveland P, Richman D (1989) Interferons and bacterial lipopolysaccharides protect macrophages from productive infection by human immunodeficiency virus in vitro. J Exp Med 169: 1137–1151

Krause M, Frei D, Rudlinger R, Kewusch G, Joller H, Binswanger U (1988) Kaposi's sarcoma following kidney transplantation: remission following reduction of immunosuppression and consequent HIV infection (in German). Schweiz Med Wochenschr 118: 100–105

Laporte JP, Meyohas MC, Eliaszewicz M, Najman A, Frottier J, Woehrle R (1987) Addition of HPA 23 to cyclosporin in the treatment of thrombopenic purpura in HIV-positive subjects (in French). Pathol Biol 35: 870–872

Larder B, Kemp S (1989) Multiple mutations in HIV-1 reverse transcriptase confer high-level resistance to Zidovudine (AZT). Science 246: 1155–1158

Larder B, Darby G, Richman D (1989) HIV with reduced sensitivity to Zidovudine (AZT) isolated during prolonged therapy. Science 243: 1731–1734

Larder BA, Kellam P, Kemp SD (1991) Zidovudine resistance predicted by direct detection of mutations in DNA from HIV-infected lymphocytes. AIDS 5: 137–144

Lavie G, Valentine F, Levin B, Mazur Y, Gallo G, Lavie D, Weiner D, Meruelo D (1989) Studies of the mechanisms of action of the antiretroviral agents hypericin and pseudohypericin. Proc Natl Acad Sci USA 86: 5963–5967

Levi G, Patrizio M, Bernardo A, Petrucci TC, Agresti C (1993) Human immunodeficiency virus coat protein gp20 inhibits the beta-adernergic regulation of astroglial and microglial functions. Proc Natl Acad Sci USA 90: 1541–1545

Lisziewicz J, Sun D, Klotman M, Agrawal S, Zamecnik P, Gallo R (1992) Specific inhibition of human immunodeficiency virus type 1 replication by antisense oligonucleotides: an in vitro model for treatment. Proc Natl Acad Sci USA 89: 11209–11213

Lisziewicz J, Sun D, Metelev V, Zamecnik P, Gallo RC, Agrawal S (1993) Long-term treatment of human immunodeficiency virus-infected cells with antisense oligonucleotide phosphorothioates. Proc Natl Acad Sci USA 90: 3860–3864

Lori F, di Marzo Veronese F, De Vico AL, Lusso P, Reitz MS Jr, Gallo RC (1992) Viral DNA carried by human immunodeficiency virus type 1 virions. J Virol 66: 5067–5074

Luban J, Bossolt KL, Franke EK, Kalpana GV, Goff SP (1993) Human immunodeficiency virus type 1 gag protein binds to cyclophilins A and B. Cell 73: 1067–1078

Macé K, Dodon M, Gazzolo L (1989) Restriction of HIV-1 replication in promonocytic cells: A role for IFN-alpha. Virology 168: 399–405

Malim M, Tiley L, McCarn D, Rusche J, Hauber J, Cullen B (1990) HIV-1 structural gene expression requires binding of the Rev trans-activator to its RNA target sequence. Cell 60: 675–683

Meltzer M, Gendelman H (1988) Effects of colony stimulating factors on the interaction of monocytes and human immunodeficiency virus. Immunol Lett 19: 193–198

Meltzer M, Skillman D, Hoover D, Hanson B, Turpin J, Kalter D, Gendelman H (1990) Macrophages and the human immunodeficiency virus. Immunol Today 11: 217–223

Meruelo D, Lavie G, Lavie D (1988) Therapeutic agents with dramatic antiretroviral activity and little toxicity at effective doses: aromatic polycyclic diones and pseudohypericin. Proc Natl Acad Sci USA 85: 5230–5234

Miller M, Schneider J, Sathyanarayana B, Toth M, Marshall G, Clawson L, Selk L, Kent S, Wlodawer A (1989) Structure of complex of synthetic HIV-1 protease with a substrate-based inhibitor at 2.3 A resolution. Science 246: 1149–1152

Mitsuya H, Broder S (1986) Inhibition of the in vitro infectivity and cytopathic effect of human T-lymphotropic virus type III/lymphadenopathy-associated virus (HTLV-III/LAV) by 2',3'-dideoxynucleosides. Proc Natl Acad Sci USA 83: 1911–1915

Mitsuya H, Weinhold K, Furman P, St. Clair M, Lehrman S, Gallo R, Bolognesi D, Barry D, Broder S (1985) 3'-Azido-3'-deoxythymidine (BW A509U): An antiviral agent that inhibits the infectivity and cytopathic effect of human T-lymphotropic virus type III/lymphadenopathy-associated virus in vitro. Proc Natl Acad Sci USA 82: 7096–7100

Mitsuya H, Looney D, Kuno S, Ueno R, Wong-Staal F, Broder S (1988) Dextran sulfate suppression of viruses in the HIV family: Inhibition of virion binding to CD4+ cells. Science 240: 649-649

Moses AV, Bloom FE, Pauza CD, Nelson JA (1993) Human immunodeficiency virus infection of human brain capillary endothelial cells occurs via a CD4/galactosylceramide-independent mechanism. Proc Natl Acad Sci USA 90: 10474–10478

Nelson JA, Wiley CA, Reynolds-Kohler C, Reese CE, Margaretten W, Levy JA (1988) Human immunodeficiency virus detected in bowel epithelium from patients with gastrointestinal symptoms. Lancet 1: 259–262

Okada T, Patterson BK, Ye S-Q, Gurney ME (1993) Aurothiolates inhibit HIV-1 infectivity by Gold (I) ligand exchange with a component of the virion surface. Virology 192: 631–642

Parks W, Parks E, Fischl E, Leuther M, Allain J, Nusinoff-Lehrman S, Barry D, Makuch R (1988) HIV-1 inhibition by azidothymidine in a concurrently randomized placebo-controlled trial. J AIDS 1: 125–130

Pauza C, Galindo J, Richman D (1988) Human immunodeficiency virus infection of monoblastoid cells: cellular differentiation determines the pattern of virus replication. J Virol 62: 3558–3564

Pauza CD (1990) Two bases are deleted from the termini of HIV-1 linear DNA during integrative recombination. Virology 179: 886–889

Pauza CD, Galindo J, Richman DD (1990) Reinfection results in accumulation of unintegrated viral DNA in cytopathic and persistent human immunodeficiency virus type 1 infection of CEM cells. J Exp Med 172: 1035–1042

Pauza CD, Kornbluth R, Emau P, Richman DD, Deftos LJ (1992) Vitamin D_3 compounds regulate human immunodeficiency virus type 1 replication in U937 monoblastoid cells and in monocyte-derived macrophages. J Leukoc Biol 53: 157–164

Pauza CD, Trivedi P, McKechnie TM, Richman DD, Graziano FM (1994) 2-LTR circular viral DNA as a marker for human immunodeficiency virus type 1 infection in vivo. Virology 205: 470–478

Phillips A, Wainberg MA, Coates R, Klein M, Rachlis A, Read S, Shepherd F, Vellend H, Walmsley S, Halloran P et al. (1989) Cyclosporine-induced deterioration in patients with AIDS. Can Med Assoc J 140: 1456–1460

Phillips DM, Bourinbaiar AS (1991) Mechanism of HIV spread from lymphocytes to epithelia. Virology 186: 261–273

Popovic M, Sarngadharan M, Read E, Gallo R (1984) Detection, isolation, and continuous production of cytopathic retroviruses (HTLV-III) from patients with AIDS and pre-AIDS. Science 224: 497–500

Ratner L, Vander Heyden N, Dedera D (1991) Inhibition of HIV adn SIV infectivity by blockade of alpha-glucosidase activity. Virology 181: 180–182

Rich D,Prasad V, Sun C, Green J, Mueller R, Houseman K, MacKenzie D, Malkovsky M (1992) New Hydroxyethylamine HIV protease inhibitors that suppress viral replication. J Med Chem 35: 3803–3812

Richman DD (1991) Selection of AZT-resistant variants of HIV by therapy. J NIH Res 3: 83–87

Rooke R, Tremblay M, Soudeyns H, DeStephano L, Ya X-J, Fanning M, Montaner JSG, O'Shaughnessy M, Gelmon K, Tsoukas C, Gill J, Ruedy J, Wainberg MA (1989) Isolation of drug-resistant variants of HIV-1 from patients on long-term zidovudine therapy. AIDS 3: 411–415

Rossi J, Elkins D, Zaia J, Sullivan S (1992) Ribozymes as anti-HIV-1 Therapeutic agents: Principles, applications and problems. AIDS Res Hum Retroviruses 8: 183–189

Sandstrom EG, Kaplan JC (1987) Antiviral therapy in AIDS. Clinical pharmacological properties and therapeutic experience to date (Review). Drugs 34: 372–390

Sarin P, Gallo R, Scheer D, Crews F, Lippa A (1985) Effects of a novel compound (AL721) on HTLV-III infectivity in vitro. N Engl J Med 313: 1289–1290

Sarvar N, Cantin E, Chang P, Zaia J, Ladne P, Stephens D, Rossi J (1990) Ribozymes as potential anti-HIV-1 therapeutic agents. Science 2247: 1222–1225

Sawada M, Suzumura A, Marunouchi T (1992) Down regulation of CD4 expression in cultured microglia by immunosuppressants and lipopolysaccharide. Biochem Biophys Res Commun 189: 869–876

Schmidt A, Hennighausen L, Siebenlist U (1990) Inducible nuclear factor binding to the kappa B elements of the human immunodeficiency virus enhancer in T cells can be blocked by cyclosporin A in a signal-dependent manner. J Virol 64: 4037–4041

Shirazi Y, Pitha PM (1992) Alpha interferon inhibits early stages of the human immunodeficiency virus type 1 replication cycle. J Virol 66: 1321–1328

Shirazi Y, Pitha PM (1993) Interferon α-mediated inhibition of human immunodeficiency virus type 1 provirus synthesis in T-cells. Virology 193: 303–312

Sullenger BA, Cech TR (1993) Tethering ribozymes to a retroviral packaging signal for destruction of viral DNA. Science 262: 1566–1569

Swanstrom R, Varmus H (1982) Replication of RNA tumor viruses. In: Weiss R, Teich N, Varmus H, Coffin J (eds) RNA tumor viruses. Cold spring Harbor Press, Cold Spring Harbor, pp 396–512

Szebni J, Dieffenbach C, Wahl SM, Venkateshan CN, Yeh A, Popovic M, Gartner S, WL, Peterfy M, Freidman RM, Weinstein JN (1991) Induction of alpha interferon by human immunodeficiency virus type 1 in human monocyte-macrophage cultures. J Virol 65: 6362–6364

Terai C, Kornbluth RS, Pauza CD, Richman DD, Carson DA (1991) Apoptosis as a mechanism of cell death in cultured T lymphoblasts acutely infected with HIV-1. J Clin Invest. 87: 1710–1715

Torlontano G, Di BP, Di GG, Angrilli F, Verani P, Maggiorella MT, Dragani A, Iacone A, Papalinetti G, Olioso P et al. (1992) AIDS-related complex treated by antiviral drugs and allogeneic bone marrow transplantation following conditioning protocol with busulphan, cyclophosphamide and cyclosporin. Haematologica 77: 287–290

Trono D (1992) Partial reverse transcripts in virions from human immunodeficiency and murine leukemia viruses. J Virol 66: 4893–4900

Trono D, Feinberg M, Baltimore D (1989) HIV-1 gag mutants can dominantly interfere with the replication of the wild-type virus. Cell 59: 113–120

Weerasinghe M, Liem SE, Asad S, Read SE, Joshi S (1991) Resistance to human immunodeficiency virus type 1 (HIV-1) infection in human CD4⁺ lymphocyte-derived cell lines conferred by using retroviral vectors expressing an HIV-1 RNA-specific ribozyme. J Virol 65: 5531–5534

Xing Z, Whitton JL (1992) An anti-lymphocytic choriomeningitis virus ribozyme expressed in tissue culture diminishes viral RNA levels and leads to reduction in infectiour virus yields. J Virol 67: 1840–1847

Yahi N, Baghdiguian S, Bolmont C, Fantini J (1992a) Inhibition of human immunodeficiency virus infection in human colon epithelial cells by recombinant interferon-gamma. Eur J Immunol 22: 2495–2499

Yahi N, Baghdiguian S, Bolmont C, Fantini J (1992b) Replication and apical budding of HIV-1 mucous-secreting colonic epithelial cells. J AIDS 5: 993–1000

Yarchoan R, Weinhold K, Lyerly H, Gelmann E, Blum R, Shearer G, Mitsuya H, Klecker R, Markham P, Durack D, Lehrmann S, Barry D, Fischl M, Gallo R, Collins J, Bolognesi D, Myers C, Broder S (1986) Administration of 2'-azido-3'-deoxythymidine, an inhibitor of HTLV-III/LAV replication, to patients with AIDS or AIDS, related complex. Lancet i: 575–580

Zaia J, Rossi J, Murakawa G, Spallone P, Stephens P, Kaplan B, Eritja R, Cantin E (1988) Inhibition of human immunodeficiency virus by using an oligonucleoside methylphosphonate targeted to the tat-3 gene. J Virol 62: 3914–3917

B Lentivirus Animal Models of HIV Central Nervous System Disease

Neurobiology of Simian and Feline Immunodeficiency Virus Infections

M.B. Gardner[1] and S. Dandekar[2]

1 Introduction

The simian and feline immunodeficiency lentivirus infections of captive macaques and domestic cats (SIV and FIV, respectively) are excellent models for HIV infection of humans (Gardner and Luciw 1989). Among the animal lentiviruses, SIV is the most closely related to HIV and both SIV and FIV models manifest a disease pattern and viral encephalopathy with many similarities to that observed in HIV infection of humans (Lackner et al. 1991a; Johnson et al. 1988). As in HIV infection of humans, infection of the CNS with SIV and FIV occurs in virtually all infected individuals, many of whom develop clinical and pathologic evidence of the associated lentivirus encephalopathy.

Infection of the CNS occurs early and apparently persists throughout the lifetime of the infected human or animal. In both animals and humans, the target cells infected in the CNS are primarily blood-borne monocyte/macrophages and resident microglial cells. As in all animal lentivirus infections, the virus expression is restricted in circulating monocytes but virus activation occurs with maturation of

[1]Department of Pathology, School of Medicine, University of California, Davis, CA 95616, USA
[2]Department of Internal Medicine, University of California, Davis, CA 95616, USA

the monocytes to macrophages in the tissues, including the CNS (HAASE 1986). The genomic structures of SIV, FIV and HIV are similar (although not identical), as are the molecular mechanisms governing viral latency, activation and mutation over the time course of infection (CLEMENTS and PAYNE 1994). The pattern of systemic virus expression in relation to host immune response and the gradual decline in CD4+ lymphocytes and end-stage opportunistic infections are similar in the SIV- and FIV-infected animals and HIV-infected humans (Fig. 1). Although the asymptomatic incubation period (2 months to 2 years) is considerably shorter and the course of disease thus more accelerated in the SIV-infected macaques than in FIV-infected cats (> 4 years) or most HIV-infected humans (average 8 years), the biologic similarities are so remarkable that it is quite safe to say that whatever basic tenets of pathogenesis, vaccine protection or therapy are learned in one model will apply equally to all. The slower disease course in spontaneous or experimentally FIV-infected cats may be attributable to this being a naturally occurring infection that has been present in domestic and wild felidae species over eons, thus allowing time for a stable host virus relationship to develop. By contrast, SIV infection of captive macaques is entirely an experimental disease where virus from a naturally infected African monkey host (e.g., sooty mangabey), in whom the virus is nonpathogenic, has been introduced into Asian macaques that are not the natural host and have has no opportunity to adapt to the virus. In this sense, SIV infection of macaques represents a virulent, "worse case scenario" whereas FIV infection of domestic cats and HIV infection of humans are naturally occurring and therefore somewhat less virulent. The more indolent course of FIV infection in domestic cats does ultimately terminate in feline AIDS which often includes encephalopathy. Coinfection with feline leukemia virus

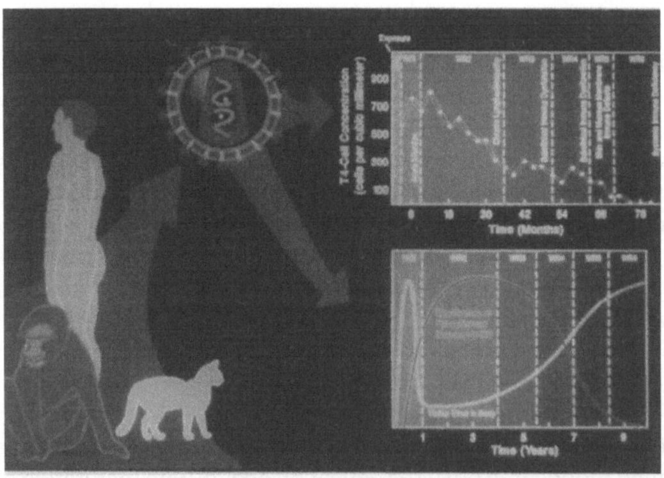

Fig. 1. The identical reciprocal relationship between viral load (*Yellow line, lower right*) and antiviral immune response (*red line, lower right*) in relationship to declining CD4 T cell count (*upper right*) over time in HIV-infected humans, SIV-infected macaques and FIV-infected cats.

(FeLV) greatly potentiates the severity of the primary and secondary stages of FIV infection and increases the level of FIV proviral DNA in lymphoid and nonlymphoid tissues including the brain (PEDERSEN et al. 1990). The mechanism underlying this FIV potentiation is attributed to the nonspecific immunosuppressive effect of FeLV (BEEBE et al. 1994). For vaccine and antiviral research, both SIV and FIV models are quite suitable, but for prevention of AIDS, the monkey model has the advantage of a more rapid disease course.

Comparative studies on viral pathogenesis, vaccines and antiviral therapy have already used both of these animal lentivirus models to advantage (GARDNER 1991). We review here the neurobiology and neuropathology associated with SIV and FIV infection in comparison with HIV encephalitis (Table 1) and emphasize the usefulness of these animal models for better understanding of the possible pathogenic mechanisms of the related CNS disease.

2 Simian Immunodeficiency Virus

2.1 Clinical Pathologic Correlates

The neurologic disease manifestations and the clinical pathologic correlates are remarkably similar between SIV and HIV infection (BASKIN et al. 1988; RINGLER et al.

Table 1. Comparison of human, simian and feline immunodeficiency virus infection of the nervous system. (From LACKNER et al. 1991a)

	HIV	SIV	FIV
Encephalitis	Yes	Yes	Yes
Giant cells	Yes	Yes	No
Vascular orientation	Yes	Yes	Yes
Vascular mineralization	In pediatric cases	In juveniles and adults	?
Brain atrophy	Yes	Yes	?
Diffuse white matter pallor	Yes	No	No
Most commonly affected areas of the brain	Cerebral white and deep grey matter	Cerebral white matter	Midbrain and thalamus
Myelitis with giant cells	Yes, rare	Yes	?
Vascular myelopathy	Yes	No	?
Peripheral neuropathy	Yes	No	?
Primary cell type infected in the nervous system	Macrophages	Macrophages	Macrophages and astrocytes
Neurobehavioral abnormalities	Yes	Yes	Yes
Motor deficits	Yes	Yes	No
Virus isolation from CSF and the brain	Yes	Yes	Yes
Virus-specific intrathecal immune response	Yes	Yes	Yes
Increased concentration of quinolinic acid in the CSF	Yes	Yes	?

1988; SHARER et al. 1988; LACKNER et al. 1991b). About 50% of rhesus macaques inoculated with the prototype SIV strains develop, within a few months to several years, an encephalitis characterized by accumulations of virus infected macrophages and giant cells. The encephalitis occurs in juvenile and adult macaques of both sexes and has been seen as early as 90 days and as late as 2 years postinoculation, depending mainly on the effectiveness of the immune response in controlling the viral infection. Whether or not newborn or infant macaques will be more susceptible to SIV encephalitis is now being studied. Animals with giant cell encephalitis often have SIV-containing giant cells in multiple organs, particularly the lymph nodes and lung, along with other evidence indicating a high level of replicating SIV throughout the body and a weakening or deficient anti-SIV immune response. During the primary infection, as early as 1–2 weeks after intravenous (IV) inoculation, SIV containing monocytes or macrophages can be seen in the meninges and cell-free virus can be isolated from the cerebrospinal fluid (CSF) (HURTREL et al. 1991; BOTTIGER et al. 1991). During the asymptomatic interval, infectious virus remains in the CNS and CSF but at a lower level. At necropsy, SIV can be isolated from the CSF and detected by in situ hybridization or immunocytochemistry in the brain and spinal cord from nearly all SIV-infected rhesus monkeys, whether or not they have detectable lesions or overt neurologic disease.

Despite the high prevalence (> 50%) of neurologic lesions at necropsy, clinical evidence of neurologic disease in SIV-infected animals is rather infrequent and generally observed only near death. Overt neurologic signs have included loss of equilibrium, blindness, diminished agility, and posterior paresis. However, SIV-infected macaques exhibit some impairment in visual recognition memory and rapid response tasks early in the course of disease prior to detection of overt clinical signs (MURRAY et al. 1992). Neuroimaging studies (JORDAN et al. 1990) have revealed various degrees of cortical atrophy in about 40% of SIV-infected macaques, and at necropsy many animals (~75%) have cortical atrophy. In addition, magnetic resonance imaging of terminally infected macaques revealed ventricular dilatation and focal areas of demyelination. Neuropsychological and immunological testing is currently being monitored in SIV-infected macaques to characterize the progression of cognitive changes over time in relation to social stress, immunologic changes and SIV disease, and to permit testing of potential psychotherapeutic agents (GOLD et al. 1994; CAPITANIO 1994). Thus, SIV infects the CNS during primary infection, and with neurobehavioral testing and neuroimaging, associated neurologic deficits can often be detected. However, the number of animals with detectable neurologic abnormalities, at least using standard techniques, is never as high as the number of infected macaques from which SIV can be detected in the CNS.

SIV specific Ig is present in the CSF of about 50% infected macaques (SMITH et al. 1994). In approximately half of these animals the SIV specific IgG appears to be derived from serum by leakage across a disrupted blood-brain barrier, whereas in the remaining animals this is evidence of intrathecal IgG synthesis. The IgG index [(CSF IgG/serum IgG)/CSF albumin/serum albumin)], a means of evaluating the integrity of the blood-brain barrier and the occurrence of intrathecal immunoglobulin synthesis, is elevated in all animals which also have evidence of SIV antibodies in the CSF by immunoblot. An elevated IgG index and SIV-specific immune response in the CSF is occasionally seen independent of the SIVimmune response in the peripheral blood suggesting intrathecal antibody synthesis. Usually, however, the intrathecal immune response parallels the systemic immune response and animals which fail to develop a systemic humoral immune response usually have no intrathecal immune response and often die with SIV encephalitis in just a few months. In individual infected monkeys, therefore, the presence or absence of an intrathecal SIV-specific IgG response in CSF may have a modulating effect on the development of the neurologic disease. It remains to be determined if SIV-infected macaques like HIV-1-infected humans (SINGER et al. 1994a, b) show an increased severity of cognitive impairment associated with higher levels of IgG synthesis.

3 Neuropathologic Features in the CNS of Simian Immunodeficiency Virus-Infected Macaques

3.1 Gross Findings

At necropsy, the brains and spinal cords of SIV-infected macaques usually appear grossly normal (BASKIN et al. 1988; SHARER et al. 1988; LACKNER et al. 1991b). Gross changes which have been observed include congestion and granularity of the meninges, cerebral edema, and multifocal areas of malacia involving the white matter of the cerebral cortex, spinal cord or optic tracts and lymphomas (FEICHTINGER et al. 1990).

3.2 Microscopic Findings

Perivascular infiltrates of macrophages and multinucleated giant cells, the "microglial nodules," are present in the meninges and throughout the gray and white matter of the brain and spinal cord, but are most common in the cerebral white matter. Similar giant cells may be noted in the lung, lymph nodes, intestine, spleen, and other organs of SIV infected macaques. Most of these giant cells contain detectable SIV antigen (Fig. 2a), nucleic acid (Fig. 2b) and lentiviral particles. Identical giant cell-containing lesions are a major feature of the neuropathology of HIV encephalitis, but in contrast to the infected macaques, such giant cells are not seen outside the CNS in infected humans. Microglial nodules may be surrounded by a zone of spongiosis. The discrete macrophage and giant cell infiltrates sometimes coalesce into large, grossly visible masses ("syncytiomas"). Individual multinucleated giant cells can also be found alone

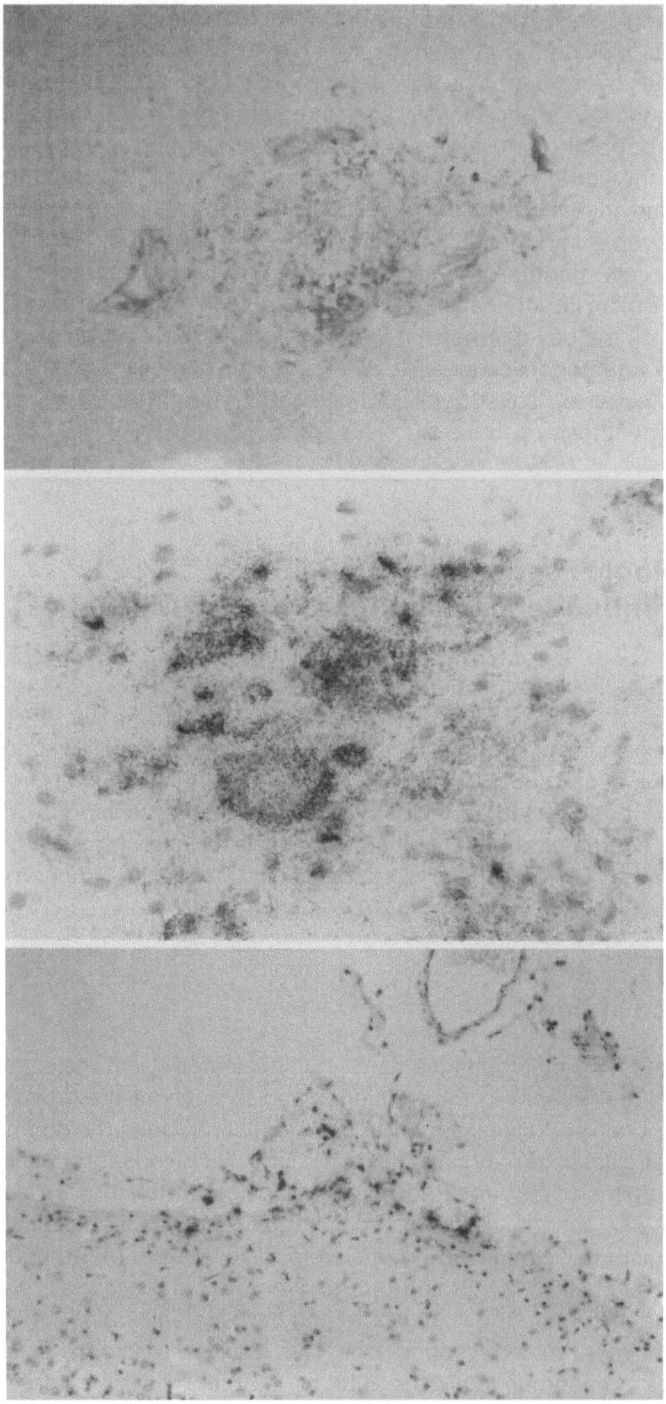

a

b

c

Fig. 2a–c (legend see p.142)

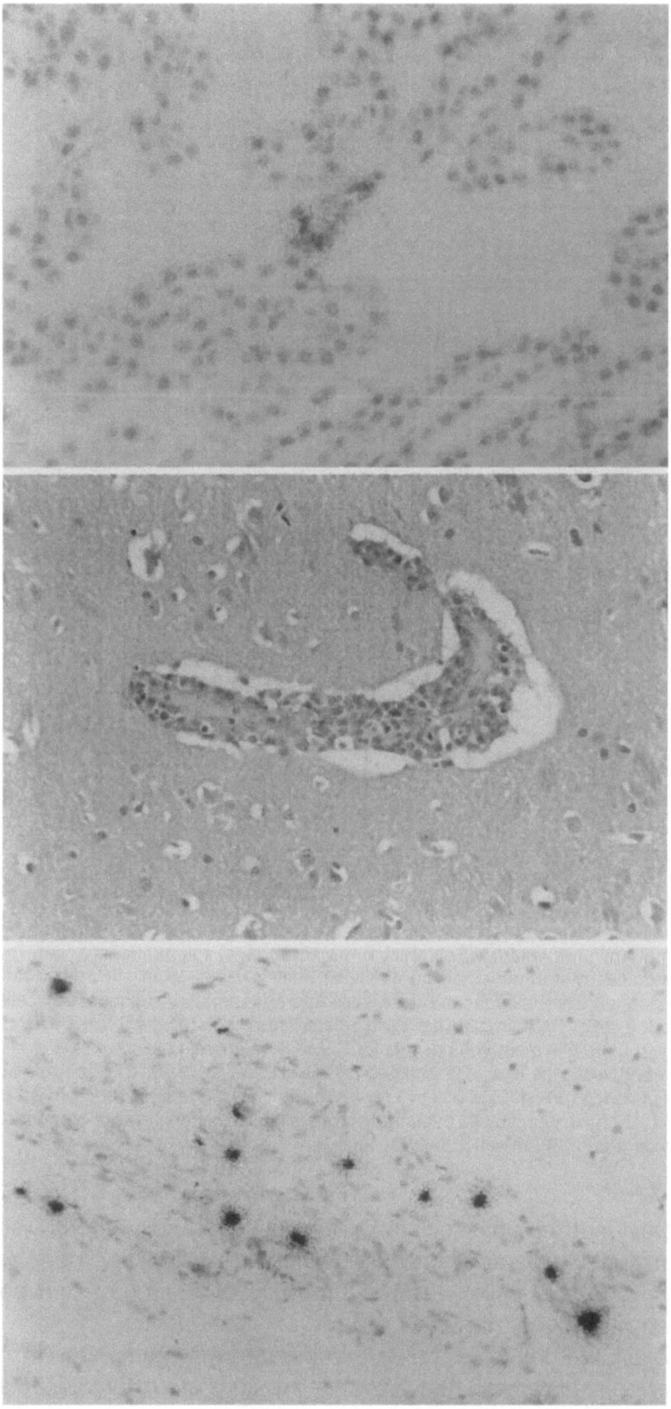

d

e

f

Fig. 2d–f (legend see p.142)

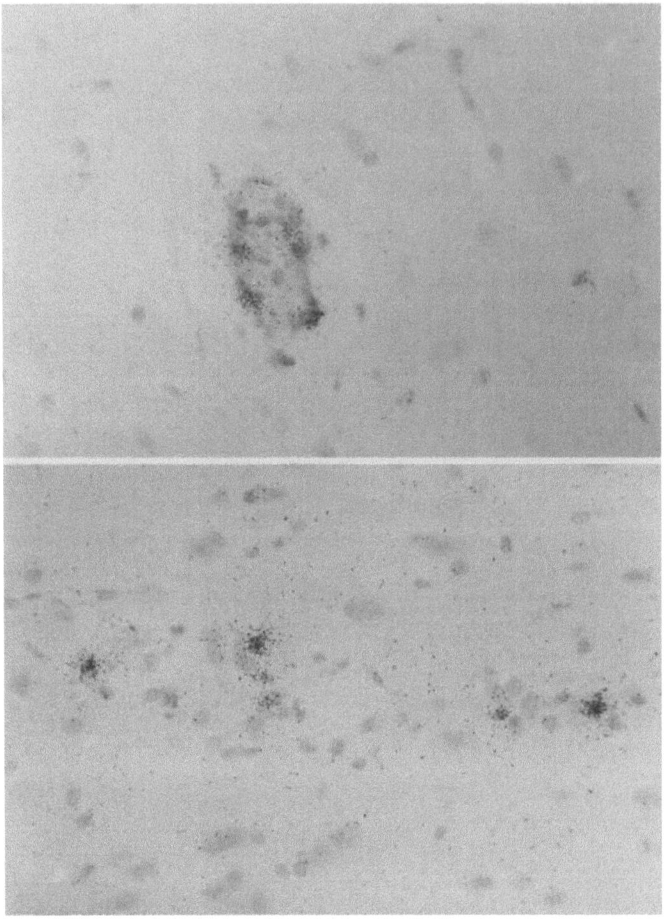

g

h

Fig. 2a–h. Examination of brain tissues from SIV-infected macaques (**a–d**) and FIV-infected cats (**e–h**). Viral antigens were detected by immounostaining and viral nucleic acids by in situ hybridization. **a** Presence of SIV p27 antigen in multinucleated giant cells and inflammatory mononuclear cells in cerebrum, x 80. **b** Presence of SIV RNA in giant cells and inflammatory cells in cerebrum, x 170. **c** SIV RNA present in meninges, x 65. **d** A giant cell positive for SIV RNA in choroid plexus, x 200. **e** Histopathologic exam of cerebral vasculitis in an FIV-infected cat, H & E stain, x 150. **f** FIV nucleic acids positive cells in choroid plexus, x150. **g** Cells positive for FIV nucleic acid in brain parenchyma, x 250. **h** Combined immunohistochemistry and in situ hybridization of brain tissue showing cells of monocyte/macrophage lineage as targets of FIV injection, x 250

within white or gray matter, unassociated with vessels, and without visible changes in the surrounding parenchyma. The vacuolar myelopathy, peripheral neuropathy and white matter pallor described in AIDS patients have not been seen in SIV-infected macaques.

The macrophage/giant cell infiltrates in SIV-infected macaques are virtually always perivascular. Vessels associated with these lesions are often dilated and lined by swollen endothelial cells, occasionally penetrated by lymphocytes and mononuclear cells. Mineralization associated with vessels in the basal ganglia may be seen similar to that described in pediatric AIDS patients (SHARER et al. 1986). This vascular mineralization is generally not associated with a cellular infiltrate.

SIV infected macaques resemble HIV infected children in that they are much more likely to have evidence of primary lentiviral infection within the CNS and less likely to show secondary infections or lymphoma (SHARER et al. 1986). Cytomegalovirus (CMV) is the most common opportunistic pathogen in the CNS of infected macaques. Activated CMV usually presents as a poly-radiculoneuritis and meningitis without involvement of the CNS parenchyma. Infrequent cases of progressive multifocal leukoencephalopathy (SV40) and toxoplasmosis have been noted (BASKIN et al. 1988; BASKIN 1987; GRIBBLE et al. 1975). Interestingly, reactivated Herpes simiae (B virus) infections have not been described despite the fact that greater than 95% of adult rhesus macaques are seropositive for this virus (WEIGLER et al. 1990).

3.3 Localization of Virus

Within the CNS of SIV-infected macaques, infiltrates of macrophages and multinucleated giant cells, located primarily in the white matter of the cerebral hemispheres and cerebellum, are consistently positive for SIV-p27 antigen by immunohistochemistry (LACKNER et al. 1991b; WARD et al. 1987). SIV antigen has not been identified in neurons, astrocytes or oligodendrogliocytes. Capillaries within the macrophage/giant cell infiltrates generally show antigen adjacent to or within vessel walls. In addition, vessels not associated with lesions are also occasionally positive for viral antigen. In animals without giant cell encephalitis viral antigen is limited to individual cells or clusters of cells associated with capillaries and small vessels in the parenchyma (primarily white matter) and meninges. It is not possible to distinguish microscopically if the viral antigen is present within endothelial cells, mononuclear cells transgressing the vessel wall or within perivascular microglia.

The wide distribution of SIV nucleic acid positive cells detected in the CNS by in situ hybridization correlates well with detection of SIV antigen by immunohistochemistry (LACKNER et al. 1991b). The macrophage/giant cell infiltrates are consistently positive for viral RNA. In addition, SIV RNA is also commonly associated with small parenchymal vessels, meninges (Fig. 2c) and choroid plexus (Fig. 2d). Hybridization is evident over endothelial cells, mononuclear cells in the vessel wall and in the perivascular space. In the choroid plexus and meninges, SIV RNA is detected in individual mononuclear cells and multinucleated giant cells. No labeling of neurons, astrocytes, or oligodendrogliocytes has been detected.

By electron microscopy (EM), typical lentivirus particles (MUNN et al. 1985) have been identified only within or near multinucleated giant cells in which the viral particles lie within vesicles of smooth endoplasmic reticulum and small cytoplasmic vacuoles. The giant cells also contain immature and occasional budding particles.

4 Immunophenotype of Cellular Infiltrates

The cellular infiltrates in the CNS of SIV infected animals are primarily monocyte/macrophage that express CD68 and rhesus MHC class II antigens (LACKNER et al. 1991b). Capillary endothelial cells and microglia also express MHC class II antigens. Lesser numbers of T cells expressing CD2 and CD8 antigens are also present. No cells expressing CD4 or CD20 have been identified within the cellular infiltrates. Numerous CD68-positive macrophages and lesser numbers of CD2 and CD8-positive T lymphocytes are intimately associated with capillaries, and similar infiltrates are present in the meninges and stroma of the choroid plexus.

5 Feline Immunodeficiency Virus

5.1 Clinical Pathologic Correlates

Although FIV is clearly neurotropic and associated with neurologic disease (DOW et al. 1990), neurologic signs have not been as prominent a clinical feature as seen with SIV or HIV infection (PEDERSEN 1992; YAMAMOTO et al. 1989). FIV commonly infects the CNS but the neurologic lesions are less florid than in either SIV or HIV infection (DOW et al. 1990; HURTREL et al. 1992). In naturally and experimentally FIV-infected cats, about 20% show clinical or pathologic evidence of lentivirus encephalitis at necropsy. FIV can be recovered from the CSF and brain in the majority of naturally and experimentally infected cats and FIV-specific intrathecal antibody production is present in most infected cats (DOW et al. 1990). FIV reaches the CNS within 1 week after peripheral inoculation, FIV specific antibodies are detected in the CSF by 8 weeks postinoculation and CSF pleocytosis is seen 4–8 weeks before that.

Neurologic abnormalities reported in FIV-infected cats generally include altered behavior and attitude without any evidence of motor deficits, suggesting involvement of subcortical structures (YAMAMOTO et al. 1989; PODELL et al. 1993; PHILLIPS et al. 1994). Pupillary and sleep abnormalities and altered auditory and visual evoked potentials have been consistently observed in acutely infected cats. This correlates with the localization of histologic lesions primarily to the midbrain and thalamus. Virus can be isolated from most regions of the CNS, regardless of the site of detectable pathology, probably because of contamination of brain tissues by FIV-infected peripheral blood cells.

5.2 Neuropathologic Features

5.2.1 Gross and Microscopic Findings

Gross brain abnormalities have not been described in FIV-infected cats. The microscopic lesions consist of perivascular accumulations of mononuclear cells and occasional discrete glial nodules confined to the meaninges, midbrain and thalamus (Fig. 2e) (Dow et al. 1990; Hurtrel et al. 1992; Podell et al. 1993; Phillips et al. 1994; Beebe et al. 1994a, b). Such lesions have not yet been observed in the cerebral or cerebellar cortex (Dow et al. 1990). The histopathology is similar to those observed in HIV-infected humans and SIV-infected macaques except for the striking rarity of multinucleated giant cells.

5.2.2 Localization of Virus

Within 4 weeks postinoculation and at necropsy, in situ hybridization and combined in situ hybridization-immunohistochemistry have localized FIV to mononuclear cells in and around vessels in the meninges, choroid plexus (Fig. 2f) and parenchyma (Fig. 2g) (Beebe et al. 1994). The majority of infected cells are of monocyte macrophage lineage (Fig. 2h). In vitro studies have shown replicative differences in early passage feline brain cells among three different FIV isolates (Kawaguchi et al. 1992). Feline brain macrophages and astrocytes are permissive to infection whereas neurons, oligodendrogliocytes and brain endothelial cells are resistant to infection. Astrocyte infection resulted in syncytia formation and cell death while microglial cells remained persistently and productively infected, without cytopathic effect (Dow et al. 1992). However, infection of astrocytes in vivo has not yet been described. Neurons and oligodendrocytes appear free of virus in vivo.

6 Discussion

The similarity of the CNS disease caused by SIV, FIV and HIV infection indicates that these animal models will continue to be very useful for better understanding of the pathogenesis of AIDS encephalopathy. However, the macaque model has the advantage of a shorter incubation period (2–24 months), a more fulminant encephalitis and the ability to induce fatal immunodeficiency with molecular clones of SIV. Unlike HIV and SIV, FIV apparently induces a lesser extent of CNS pathology without multinucleated giant cells. Infectious molecular clones of FIV have been derived but, as yet, have not proven pathogenic (Phillips et al. 1990).

As in HIV infection, the evolution of CNS infection in both simian and feline models probably depends principally on the vigor of the host antiviral immune response and, possibly, to mutations in the infecting virus. A weak immune response during primary infection or a waning immune response during chronic infection allows the virus to escape from immune control, thus increasing the invasiveness and replication of virus within the brain. Increased virus replication may allow for the development of neurotropic variants that more readily infect brain macrophages and microglia (Simon et al. 1992). Neurotropic viruses are macrophage-tropic in vitro, but not all macrophage-tropic viruses are neurotropic. No single SIV molecular clone or biologic isolate reliably induces encephalitis in all infected monkeys. The virus is therefore essential but not sufficient in the absence of other unknown cofactors. However, the ability to induce simian AIDS in many recipients (~50%) with molecular clones of SIV (Kestler et al. 1990) and to construct recombinants between neuropathogenic and nonneuropathogenic molecular clones should facilitate an answer to the question of possible viral determinants of neurovirulence (Luciw et al. 1992). In experimental transmission studies, macrophage-tropic clones of SIV tend to be neurotropic without being immunosuppressive (Desrosiers et al. 1991; Sharma et al. 1992b) whereas lymphocyte-tropic clones are immunosuppressive but not neurotropic (Sharma et al. 1992a). This difference in cell tropism is primarily a function of the envelope (Mori et al. 1992). However, the immunosuppression caused by the lymphocyte-tropic variants may well be required for induction of the encephalitis by the macrophage-tropic variants (Joag et al. 1994), Biologic isolates of SIV, e.g., $SIV_{mac\ 251}$, are, of course, a mixture of lymphotropic and macrophage tropic variants. Although no biologic or molecular clones of SIV reliably produce encephalopathy in all recipient monkeys, CNS disease appears most likely to develop in individual monkeys with a high SIV load in the blood and lymphoid tissues and the presence of macrophage tropic variants. Infectious molecular clones of FIV have been derived but, as yet, have not been shown to be either immunosuppressive or neuropathogenic for domestic cats. Therefore, we must rely primarily on the SIV model for the time being to understand the molecular mechanisms of neuropathogenesis.

In SIV and FIV infection of monkeys and cats, respectively as in HIV infection of humans, the brain apparently becomes infected during the primary infection, principally by blood-borne mononuclear cells and macrophages, and to a lesser extent by circulating cell-free virus. Infected mononuclear cells in the blood carry the virus to the leptomeninges, choroid plexus and brain parenchyma, like "trojan horses" (Haase 1986), and allow secondary spread of the virus to other CD4 receptor-bearing permissive cells such as microglia. Whether or not cell-free virus in blood infects the brain by way of endothelial cells remains uncertain. The association of lentiviruses with vessels in quite common, however, having been described for HIV, SIV and FIV as well as caprine arthritis encephalitis virus (Lackner et al. 1991b; Wiley et al. 1986; Moses et al. 1993; Zink et al. 1990). This suggests that vessels play a key role in the pathogenesis of lentiviral-induced neurologic disease. Although brain capillary endothelial cells can be infected in vitro with SIV (Zink et al. 1994), it remains uncertain whether or not endothelial cells are infected in vivo. Interesting new studies indicate that recruitment of monocytes to the CNS during the development of AIDS encephalitis induced by HIV or SIV involves monocyte adhesion to endothelium, mediated by the vascular adhesion molecule–1 (VCAM-1)–variable lymphocyte 4 (VLA-4) pathway (Sasseville et al. 1992,1994).

The mechanisms accounting for the neurologic dysfunction of SIV, FIV and HIV associated AIDS dementia remain uncertain (for reviews see Johnson et al. 1988; Price et al. 1988). The foci of infected macrophages, microglia and derivative multinucleated giant cells seem too sparsely distributed to account for the diffuse cerebral atrophy observed in end stage disease in the human and monkey.

Furthermore, in the animals and humans, there is a striking absence of productive in vivo infection or lysis of neurons, astrocytes or oligodendrogliocytes. In addition, opportunistic infections of the CNS, particularly in the monkey and cat, are too infrequent to account for the changes seen. Therefore, indirect mechanisms of tissue damage caused by the release of toxic agents from infected cells (GUILIAN et al. 1993) or immune activated endothelial or glial cells appear likely to be involved. Included here are cytokines such as tumor necrosis factor, interleukin-1 or neurotoxins, such as quinolinic acid (BROUWERS et al. 1993), or viral gene products such as viral envelope glycoprotein (DAWSON et al. 1993; BENOS et al. 1994). tat or nef proteins, all of which may act by way of NMDA receptors and enhanced Ca^{2+} entry into neurons (DREYER et al. 1990). Altered functions of astrocytes, either uninfected or latently infected, may play a central role in amplifying these neurotropic effects (BENOS et al 1994; GENIS et al. 1992; BLUMBERG et al. 1994). Also, lentiviral infection of endothelial cells in the CNS might alter the vascular permeability and contribute to the encephalopathy. The apparent absence of diffuse white matter pallor and vacuolar myelopathy, as described with HIV-1 infection of humans, in the macaque or cat models suggests that these lesions may not be directly related to lentiviral infection of the CNS or may have a different etiology.

A deeper understanding of the potential mechanisms that underlie the neuropathogenesis of AIDS can now best be approached by systematic time-course study of the CNS in the SIV macaque and FIV cat models.

Acknowledgments. We thank Tobie Faith for preparation of the figures and Kate Maney for typing of the manuscript.

References

Baskin GB (1987) Disseminated cytomegalovirus infection in immunodeficient rhesus monkeys. Am J Pathol 129: 345–352

Baskin GB, Murphey-Corb M, Watson EA, Martin LN (1988) Necropsy findings in rhesus monkeys experimentally infected with cultured simian immunodeficiency virus (SIV)/Delta. Vet Pathol 25: 456–467

Beebe AM, Dua N, Faith TG, Moore PF, Pedersen NC, Dandekar S (1994a) Primary stage of feline immunodeficiency virus infection: viral dissemination and cellular targets. J Virol 68: 3080–3991

Beebe AM, Faith TG, Moore PF, Torten M, Pedersen NC, Dandekar S (1994b) Macrophages are the predominant cellular targets of feline immunodeficiency virus in tissues of terminally ill experimentally infected cats (submitted)

Beebe AM, Faith TG, Sparger EE, Torten M, Pedersen NC, Dandekar S (1994) Evaluation of in vivo and in vitro interactions of feline immunodeficiency virus and feline leukemia virus. AIDS 8: 873–878

Benos DJ, Hahn BH, Bubien JK, Ghosh SK, Mashburn NA, Chaikin MA, Shaw GM, Benveniste EN (1994) Envelope glycoprotein gp120 of human immunodeficiency virus type 1 alters ion transport in astrocytes: implications for AIDS dementia complex. Proc Natl Acad Sci USA 91: 494–498

Blumberg BM, Gelbard HA, Epstein LG HIV-1 infection of the developing nervous system: Central role of astrocytes in pathogenesis. Virus Res (in press)

Bottiger D, Keys B, Putkonen P, Walther L, Fenyo EM, Norrby E, Oberg B, Chiodi F (1991) Simian immunodeficiency virus (SIVsm) isolation from blood and brain of experimentally infected macaques. AIDS 5: 445–449

Brouwers P, Heyes MP, Moss HA, Wolters PL, Poplack DG, Markey SP, Pizzo PA (1993) Quinolinic acid in the cerebrospinal fluid of children with symptomatic human immunodeficiency virus type 1 disease: relationships to clinical status and therapeutic response. J Infect Dis 168: 1380–1386
Capitanio JP (1994) Variability in HIV disease: toward an animal model of psychosocial influences. National Institute of Mental Health Neuroscience Findings in AIDS Research, (Rockville) 12 (abstract)
Clements JE, Payne SL. Molecular basis of the pathobiology of lentiviruses. Virus Res (in press)
Dawson VL, Dawson TM, Uhl GR, Snyder SH (1993) Human immunodeficiency virus type 1 coat protein neurotoxicity mediated by nitric oxide in primary cortical cultures. Proc Natl Acad Sci USA 90: 3256–3259
Desrosiers RC, Hansen-Moosa A, Mori K, Bouvier DP, King NW, Daniel MD, Ringler DJ (1991) Macrophage-tropic variants of SIV are associated with specific AIDS-related lesions but are not essential for the development of AIDS. Am J Pathol 139: 29–35
Dow SW, Poss ML, Hoover EA (1990) Feline immunodeficiency virus: a neurotropic lentivirus. J Acquir Immune Syndr 3: 658–668
Dow SW, Dreitz MJ, Hoover EA (1992) Feline immunodeficiency virus neurotropism: evidence that astrocytes and microglia are the primary targets cells. Vet Immunol Immunopathol 35: 23–35
Dreyer EB, Kaiser PK, Offermann JT, Lipton SA (1990) HIV-1 coat protein neurotoxicity prevented by calcium channel antagonists (see comments) Science 2: 364–367
Feichtinger H, Putkonen P, Parravicini C, Li SL, Kaaya EE, Bottiger D, Biberfeld G, Biberfeld P (1990) Malignant lymphomas in cynomolgus monkeys infected with simian immunodeficiency virus. Am J Pathol 137: 1311–1315
Gardner MB (1991) Simian and feline immunodeficiency viruses: animal lentivirus models for evaluation of AIDS vaccines and antiviral agents. Antiviral Res 15: 267–286
Gardner MB, Luciw PA (1989) Animal models of AIDS. FASEB J 3: 2593–2606
Genis P, Jett M, Bernton EW, Boyle T, Gelbard HA, Dzenko K, Keane RW, Resnick L, Mizrachi Y, Volsky DJ (1992) Cytokines and arachidonic metabolites produced during human immunodeficiency virus (HIV)-infected macrophage-astroglia interactions: implications for the neuropathogenesis of HIV disease. J Exp Med 176: 1703–1718
Gribble DH, Haden CC, Schwartz LW, Henrickson RV (1975) Spontaneous progressive multifocal leukoencephalopathy (PML) in macaques. Nature 254: 602–604
Giulian D, Wendt E, Vaca K, Noonan CA (1993) The envelope glycoprotein of human immunodeficiency virus type 1 stimulates release of neurotoxins from monocytes. Proc Natl Acad Sci USA 90: 2769–2773
Gold L, Sahakian B, Elliott R, Pozniak A, Low N, Roberts A, Robbins T, Bloom F, Koob G (1994) Neuropsychological performance following immunodeficiency virus infection in humans and rhesus monkeys. In: Pathogenesis of HIV Infection of the Brain: Impact on Function and Behavior, Chantilly, VA, p (abstract)
Haase AT (1986) Pathogenesis of lentivirus infections. Nature 322: 130–136
Hurtrel B, Chakrabarti L, Hurtrel M, Maire MA, Dormont D, Montagnier L (1991) Early SIV encephalopathy. J Med Primatol 20: 159–166
Hurtrel M, Ganiere JP, Guelfi JF, Chakrabarti L, Maire MA, Gray F, Montagnier L, Hurtrel B (1992) Comparison of early and late feline immunodeficiency virus encephalopathies. AIDS 6: 399–406
Joag SV, Stephens EB, Foresman L, Pinson DM, Narayan O (1994) Neurovirulence of SIVmac is predicated on prior immunosuppression of the animals with a T lymphocyte tropic virus. Chantilly VA, 4–7 April: pathogenesis of HIV infection of the brain: impact on function and behavior 42 (abstract)
Johnson RT, McArthur JC, Narayan O (1988) The neurobiology of human immunodeficiency virus infections. FASEB J 2: 2970–2981
Jordan EK, Gravell M, Hamilton R, Nusbaum W, Frank JA, Snoy P, Beilke K, Pomeroy K, Gibbs CJ (1990) Neuroimaging and immunohistochemical evaluation of neurological AIDS in rhesus monkeys. Neurological and neuropsychological complications of HIV infection, abstract BS-1: 33 (abstract)
Kawaguchi Y, Maeda K, Tohya Y, Furuya T, Miyazawa T, Horimoto T, Norimine J, Kai C, Mikami T (1992) Replicative difference in early-passage feline brain cells among feline immunodeficiency virus isolates. Arch Virol 125: 347–354
Kestler H, Kodama T, Ringler D, Marthas M, Pedersen N, Lackner A, Regier D, Sehgal P, Daniel M, King N, Desrosiers R (1990) Induction of AIDS in rhesus monkeys by molecularly cloned simian immunodeficiency virus. Science 248: 1109–1112
Lackner AA, Dandekar S, Gardner MB (1991a) Neurobiology of simian and feline immunodeficiency virus infections. Brain Pathol 1: 12–24

Lackner AA, Smith MO, Munn RJ, Martfeld DJ, Gardner MB, Marx PA, Dandekar S (1991b) Localization of simian immunodeficiency virus in the central nervous system of rhesus monkeys. Am J Pathol 139: 609–621

Luciw PA, Shaw KE, Unger RE, Planelles V, Stout MW, Lackner JE, Pratt-Lowe E, Leung NJ, Banapour B, Marthas ML (1992) Genetic and biological comparisons of pathogenic and nonpathogenic molecular clones of simian immunodeficiency virus (SIVmac). AIDS Res Hum Retroviruses 8: 395–402

Mori K, Ringler DJ, Kodama T, Desrosiers RC (1992) Complex determinants of macrophage tropism in env of simian immunodeficiency virus. J Virol 66: 2067–2075

Moses AV, Bloom FE, Pauza CD, Nelson JA (1993) Human immunodeficiency virus infection of human brain capillary endothelial cells occurs via a CD4/galactosylceramide-independent mechanism. Proc Natl Acad Sci USA 90: 10474–10478

Munn RJ, Marx PA, Yamamoto JK, Gardner MB (1985) Ultrastructural comparison of the retroviruses associated with human and simian acquired immunodeficiency syndrome. Lab Invest 53: 194–199

Murray EA, Rausch DM, Lendvay J, Sharer LR, Eiden LE (1992) Cognitive and motor impairments associated with SIV infection in rhesus monkeys. Science 255: 1246–1249

Pedersen NC (1992) Feline immunodeficiency virus infection, In: Levy JA (ed) The retroviridae, vol 2. Plenum, New York

Pedersen NC, Torten M, Rideout B, Sparger E, Tonachini T, Luciw PA, Ackley C, Levy N, Yamamoto J (1990) Feline leukemia virus infection as a potentiating cofactor for the primary and secondary stages of experimentally induced feline immunodeficiency virus infection. J Virol 64: 598–606

Phillips TR, Talbott RL, Lamont C, Muir S, Lovelace K, Elder JH (1990) Comparison of two host cell range variants of feline immunodeficiency virus. J Virol 64: 4605–4613

Phillips TR, Prospero-Garcia O, Puaoi DL, Lerner DL, Fox HS, Olmsted RA, Bloom FE, Henriksen SJ, Elder JH (1995) Neurological abnormalities associated with FIV infection. J Gen Virol (in press)

Podell M, Oglesbee M, Mathes L, Krakowka S, Olmstead R, Lafrado L (1993) AIDS-associated encephalopathy with experimental feline immunodeficiency virus infection. J Acquir Immune Defic Syndr 6: 758–771

Price RW, Brew B, Sidtis J, Rosenblum M, Scheck AC, Cleary P (1988) The brain in AIDS: central nervous system HIV-1 infection and AIDS dementia complex. Science 239: 586–592

Ringler DJ, Hunt RD, Desrosiers RC, Daniel MD, Chalifoux LV, King NW (1988) Simian immunodeficiency virus-induced meningoencephalitis: natural history and retrospective study. Ann Neurol 23 [Suppl]: S101–S107

Sasseville VG, Newman WA, Lackner AA, Smith MO, Lausen NCG, Deall D, Ringler DJ, (1992) Elevated vascular cell adhesion molecule-1 in AIDS encephalitis induced by simian immunodeficiency virus (short communication), Am J Pathol 141: 1021–1030

Sasseville VG, Newman W, Brodie SJ, Hesterberg P, Pauley D, Ringler DJ (1994) Monocyte adhesion to endothelium in simian immunodeficiency virus-induced AIDS encephalitis is mediated by vascular cell adhesion molecule-1/alpha 4 beta 1 integrin interactions. Am J Pathol 144: 27–40

Sharer LR, Epstein LG, Cho E-S, Joshi VV, Meyenhofer MF, Rankin LF, Petito CK (1986) Pathologic features of AIDS encephalopathy in children: evidence for LAV/HTLV-III infection of the brain. Hum Pathol 17: 271–284

Sharer LR, Baskin GB, Cho E-S, Murphey-Corb M, Blumberg BM, Epstein LG (1988) Comparison of simian immunodeficiency virus and human immunodeficiency virus encephalitides in the immature host. Ann Neurol 23 [Suppl]: S108–S112

Sharma DP, Anderson M, Zink MC, Adams RJ, Donnenberg AD, Clements JE, Narayan O (1992a) Pathogenesis of acute infection in rhesus macaques with a lymphocyte-tropic strain of simian immunodeficiency virus. J Infect Dis 166: 738–746

Sharma DP, Zink MC, Anderson M, Adams R, Clements JE, Joag SV, Narayan O (1992b) Derivation of neurotropic simian immunodeficiency virus from exclusively lymphocytotropic parental virus: pathogenesis of infection in macaques. J Virol 66: 3550–3556

Simon MA, Chalifoux LV, Ringler DJ (1992) Pathologic features of SIV-induced disease and the association of macrophage infection with disease evolution. AIDS Res Hum Retroviruses 8: 327–337

Singer EJ, Syndulko K, Fahy-Chandon B, Shapshak, P, Resnick L, Schmid P, Conrad A, Tourtellotte WW (1994a) Cerebrospinal fluid p24 antigen levels and intrathecal immunoglobulin G synthesis are associated with cognitive disease severity in HIV-1. AIDS 8: 197–204

Singer EJ, Syndulko K, Fahy-Chandon B, Schmind P, Conrad A, Tourtellotte WW (1994b) Intrathecal IgG synthesis and albumin leakage are increased in subjects with HIV-1 neurologic disease. J Acquir Immune Defic Syndr 7: 265–271

Smith MO, Sutjipto S, Lackner AA (1994) Intrathecal synthesis of IgG in simian immunodeficiency virus (SIV)-infected rhesus macaques (Macaca mulatta). AIDS Res Hum Retroviruses 10: 81–89

Ward JM, O'Leary TJ, Baskin GB, Benveniste R, Harris CA, NaraPL, Rhodes RH (1987) Immunohistochemical localization of human and simian immunodeficiency viral antigens in fixed tissue sections. Am J Pathol 18: 217–226

Weigler BJ, Roberts JA, Hird DW, Lerche NW, Hilliard JK (1990) A cross sectional survey for B virus antibody in a colony of group housed rhesus macaques. Lab Anim Sci 440: 257–261

Wiley CA, Schrier RD, Nelson JA, Lampert PW, Oldstone MBA (1986) Cellular localization of human immunodeficiency virus infection within the brains of acquired immune deficiency syndrome patients. Proc Natl Acad Sci USA 83: 7089–7093

Yamamoto JK, Hansen H, Ho EW, Morishita TY, Okuda T, Sawa TR, Nakamura RM, Pedersen NC (1989) Epidemiologic and clinical aspects of feline immunodeficiency virus infection in cats from the continental United States and Canada and possible mode of transmission. JAVMA 194: 213–220

Zink MC, Yager JA, Myers JD (1990) Pathogenesis of caprine arthritis encephalitis virus. Cellular localization of viral transcripts in tissues of infected goats. Am J Pathol 136: 843–854

Zink MC, Mankowski JL, Spelman JP, Miller SE, Clements JE (1994) Neurovirulent SIVmac replicates in CNS endothelial cells in vivo and in vitro. Workshop on HIV/SIV pathogenesis and mucosal transmission and AIDS postdoctoral fellows meeting 4, Bethesda (abstract)

Selected Models of HIV-Induced Neurological Disease

O. NARAYAN, S.V. JOAG, and E.B. STEPHENS

The encephalopathy associated with HIV infection consists of a constellation of histopathological abnormalities including gliosis, multinucleated giant cell encephalitis, vacuolar myelopathy and white matter pallor, each of which may be evident singly or in combination in different individuals (KURE et al. 1990; BUDKA 1986). In infants, the lesions are more severe and include degeneration and mineralization in the basal ganglia, necrotizing encephalitis in cortical and subcortical structures and acquired microcephaly (SHARER et al. 1986), reflecting the massive loss of brain substance that occurs during infection in the brain. These changes are accompanied by production of β2-microglobulin, neopterin and quinolinic acid which are markers of immune activation, especially of macrophages (BREW et al. 1990, 1992; HEYES et al. 1991). Histological changes in such brains are marked by nodular accumulations of microglial cells, activation of perivascular macrophages and microglia, many of which express MHC II antigens, and activation and proliferation of astrocytes. Paradoxically, these signs of immune activation in the brain occur on a background of severe immunosuppression in the affected persons (NAVIA et al. 1986; PRICE 1994).

Whereas the immunosuppression of the host is associated with HIV replication in $CD4^+$ T lymphocytes, pathological changes in the brain are associated principally if not exclusively with productive HIV replication in macrophages in the brain (KURE et al. 1990; GENDELMAN et al. 1989). This finding drew immediate parallels to infection in sheep with visna virus and goats with caprine arthritis encephalitis (CAE) virus, two of the prototypes of the lentivirus family of which HIV is a member. Infection in macrophages is a generic biological property of the lentiviruses (JOAG et al. 1994d). In both sheep and goats, productive virus replication in macrophages in brain is accompanied by severe meningoencephalitis, involving gray and white matter in the brain with frequent extension of lesions to the spinal cord where the myelitic lesions are usually located in the lateral and dorsal columns of the spinal cord (CORK et al. 1974b). Interestingly, these regions in the spinal cord are the same that are involved in HIV-associated lesions in spinal cord. A further parallel between the human and animal lentiviral disease is the increased severity of lesions in the newborn. This is seen in newborn goats that

Department of Microbiology, Marion Merrell Dow Laboratory of Viral Pathogenesis, University of Kansas Medical Center, Kansas City, KS 66160-7424, USA develop progressive paralysis which appears within 2–4 weeks of birth of the animals (CORK et al. 1974a).

A brief review of current knowledge on the pathogenesis of disease in sheep and goats shows that the virus in these animals replicates productively only in macrophages and not lymphocytes (GORRELL et al. 1992). Thus, affected animals do not become immunosuppressed, unlike HIV-infected humans. Virus in these animals is transmitted mainly in milk and colostrum containing infected macrophages (HOUWERS et al. 1988; KENNEDY-STOSKOPF et al. 1985) and this results in infection in cells of macrophage lineage in the bone marrow and lymphoid tissues of the young animals. Infected animals remain in a subclinical state for months to years, during which the virus is largely unexpressed in the infected cells. In many cases, during this period only viral DNA can be detected by the polymerase chain reaction (PCR) technique (RIMSTAD et al. 1993; CHEBLOUNE and NARAYAN, submitted). Lack of production of virus proteins in these infected animals resembles true viral latency. However, under conditions not completely understood, virus expression (i.e., viral gene transcription, translation, and assembly) does occur and this induces immune responses to the virus and, later, onset of clinical disease. This ends the prolonged incubation period for which these lentiviruses are well-known (SIGURDSSON 1954). However, unique to animals infected with maedi-visna and CAE viruses, viral gene expression develops in only selected populations of macrophages and these are located in special tissues such as the lungs, the synovia, the mammary glands and the central nervous system (NARAYAN and CORK 1985). Inflammation and degenerative lesions peculiar to each of these particular tissues then ensues. A number of poorly understood factors surround the onset of virus replication, the selectivity of tissues for virus replication and the pathogenesis of the lesions, all of which have an impact on the encephalopathy.

Several environmental factors are associated with the "switch on" of virus gene expression in the mature macrophages in the "tissues-at-risk" as mentioned above. Among these are poor husbandry practices, including overcrowded housing, poor nutrition, poor ventilation, and concomitant infections (NARAYAN and CORK 1985). Genetic factors may also play a role in disease since encephalitis occurred more frequently in naturally infected Icelandic sheep than American sheep (PALSSON 1976), in newborn goats than adult goats (CORK et al. 1974a), and in milk-producing sheep (seen mainly in Europe) than American sheep which are bred mainly for wool and meat (VITU et al. 1993).

The pathogenesis of infection during the period of CNS disease involves unrestricted replication of virus in macrophages in the brain of these animals. In contrast, precursor monocytes in bone marrow, monocytes in blood and mature macrophages in other tissues such as spleen contain viral DNA, but this largely is unexpressed. However, in comparison to infected animals with subclinical infection, animals with clinical disease have much higher numbers of infected cells in blood and lymphoid tissues. Thus, pneumonic sheep, arthritic animals or animals paralyzed with visna disease all have high levels of infected cells in blood, lymph node, spleen and bone marrow (KENNEDY-STOSKOPF et al. 1989).

The brain lesions in sheep with visna or newborn goats with CAE virus-induced encephalitis, are characterized by massive meningo-encephalitis and can be malacic in some regions, especially in young animals. Microglia are activated and express MHC II antigens; glial nodules consisting of activated macrophages are common; astrocytosis is intense. These responses also occur in HIV-infected human brain. However, unique to sheep and goats there are the high concentrations of lymphocytes and macrophages in perivascular spaces in the brain. Inflamed lesions are most severe at junctions of gray and white matter, but with no obvious regional selection, although lesions are more intense in the brain stem in animals with paralytic disease (SHEFFIELD et al. 1980). The lesions are remarkably similar to those of experimental allergic encephalomyelitis in sheep and are therefore suggestive of an autoimmune process triggered by the virus (PANITCH et al. 1976).

Studies in cell culture systems showed that one of the factors associated with enhanced virus replication in macrophages was the maturation of monocytes to adherent macrophages (NARAYAN et al. 1982, 1983). Thus, cultivation of peripheral blood mononuclear cells (PBMC), splenocytes, and bone marrow cells (that had viral DNA but no virus proteins) in medium that potentiated development of adherent macrophages, resulted in production of cell-free virus in the culture medium. Mechanisms of this induction of virus replication involved binding of c-*jun* and c-*fos* transcription factors in the incipient macrophages to AP-1 sites in the long terminal repeat (LTR) of the integrated viral DNA (GABUZDA et al. 1989; SHIH et al. 1992). Another observation was that addition of fresh PBMC to mature, virus-producing macrophages resulted in a burst of cytokine release. The cytokines included a unique type of ovine/caprine interferon, as defined by the ability of the supernatant fluids to inhibit cytopathicity of vesicular stomatitis virus (NARAYAN et al. 1985). This cytokine may be relevant to the disease process because it is present in synovial fluid of CAE virus-induced arthritic joints (KENNEDY-STOSKOPF et al. 1989) and in supernatant fluids of mononuclear cells cultivated from lungs with maedi-visna-induced interstitial pneumonia (LAIRMORE et al. 1988). Addition of fluids containing this LV-IFN to macrophage cultures caused enhanced expression of MHC II antigens resembling macrophages in affected tissue (KENNEDY et al. 1985). Thus, this cytokine may be important for the massive inflammatory lesions seen in affected tissues. Repeat of this protocol in the SIV/macaque system described below, yielded interleukin (IL)-6 and tumor necrosis factor (TNF)-a but no interferons (MCENTEE et al. 1992). This may account for the lesser inflammatory nature of the brain lesions seen in humans and macaques infected with HIV and SIV, respectively, compared to sheep and goats infected with visna and CAE virus, respectively.

Another important lesson from the ovine/caprine infection was that CNS disease was not easily reproducible. Inoculations of animals by any route resulted in the persistent subclinical infection typically seen in naturally infected animals. Interstitial pneumonia was more easily reproduced by intratracheal inoculation of specific strains of virus (DEMARTINI et al. 1993), but only transient focal inflammatory lesions resulted at the site of inoculation in brain when virus was given intracerebrally into American sheep (NARAYAN et al. 1984). Interestingly, much more severe lesions developed in intracerebrally inoculated Icelandic sheep (KENNEDY et al. 1988; GEORGSSON et al. 1989). Thus Icelandic sheep were more susceptible to natural as well as experimentally induced disease than American sheep. "Neuroadaptation" of the virus by sequential intracerebral passage of infectious material through brains, together with intracerebral administration of tissue culture fluid containing the "interferon" material caused more severe lesions than inoculation of virus alone, but this protocol did not result in a predictable development of neurological disease (ZINK et al. 1991). The failure to obtain virus of predictable neurovirulence similar to other neurovirulent RNA viruses, (such as rabies and neuroadapted flaviviruses) was probably due to the strict tropism of the lentiviruses for macrophages, whereas the other RNA viruses are tropic for neuroectodermal cells, mainly neurons (JOHNSON 1980; MONATH 1990).

The failure of lentivirus infected sheep and goats to become immunosuppressed and develop AIDS, together with the finding that CNS lesions in sheep and goats were much more inflammatory than those seen in humans, suggested that this animal model was not appropriate for studying pathogenesis of HIV-associated encephalopathy. In contrast, macaques infected with SIV_{mac} developed immunosuppression, AIDS and CNS disease in a manner very similar to HIV-infected people (SHARER et al. 1988; LACKNER et al. 1991; CHAKRABARTI et al. 1991), with histopathological changes in the CNS similar to those seen in HIV-associated CNS disease (Fig.1). Furthermore, an infectious molecular clone of SIV, SIV_{mac}239, developed in the Desrosiers lab and shown by his group to be pathogenic, caused all three aspects of the disease (KESTLER et al. 1988, 1990). This clone provided a critical tool for beginning investigation of the viral genetic basis of neurological disease in macaques. Studies on infection in macaque-activated $CD4^+$ T lymphocytes, in mitogen-treated PBMC and in monocyte-derived macrophages with SIV_{mac}239 have shown that the virus replicated permissively in the T cells, with accompanying development of multinucleated giant cell formation, but replicted inefficiently in macrophage cultures (MORI et al. 1993; SHARMA et al. 1992; ANDERSON et al. 1993; STEPHENS et al. 1995).

Investigations on about 20 macaques inoculated with SIV_{mac}239 showed that the infection could be separated into three phases (JOAG et al. 1994a, c). Phase 1 was the period characterized by productive virus replication in activated T cells in vivo. Within days following inoculation in the animal, infected, activated T lymphocytes appeared in peripheral blood and in lymphoid tissues. Cells bearing IL-2 receptor (IL-2R) appeared in PBMC and virus could be recovered from these cells following cultivation in medium containing only IL-2. In this phase, treatment of PBMCs with mitogens was not required for virus production since cells were already activated in vivo. Plasma also had infectious virus, probably indicating the result of production of infectious virus in vivo by activated T cells in the lymphoid tissues. Another interesting aspect of phase 1 was that infected cells appeared in the cerebrospinal fluid, a finding in keeping with the known biological properties

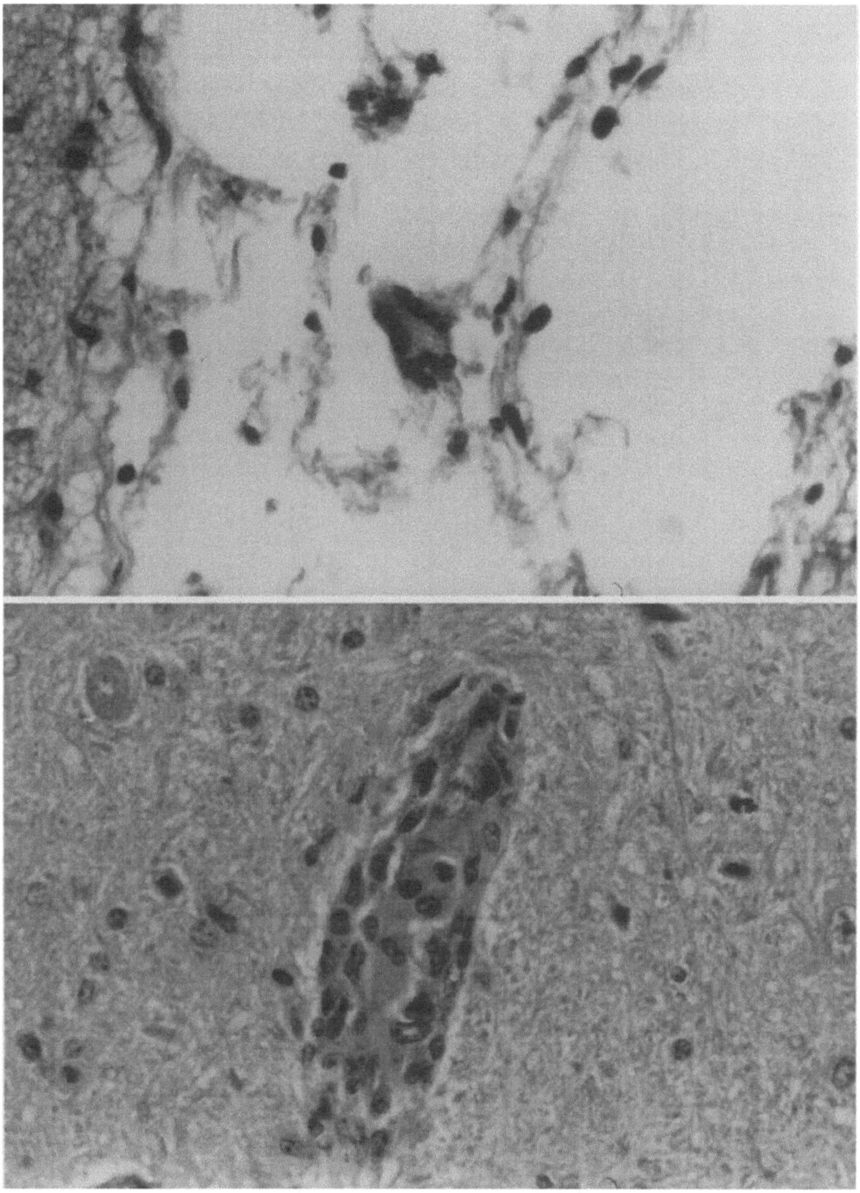

A

B

Fig. 1A–E. Neuropathological changes associated with SIV infection in macaques. **A** Pigtail macaque inoculated with SIV_{mac} Low grade meningeal inflammation is characterized by the presence of a few inflammatory cells and multinucleated giant cells; H & E, ×100. **B** Pigtail macaque inoculated with SIV_{mac} Lesions of encephalitis include perivascular cuffing by inflammatory mononuclear cells resembling macrophages; H & E, ×100

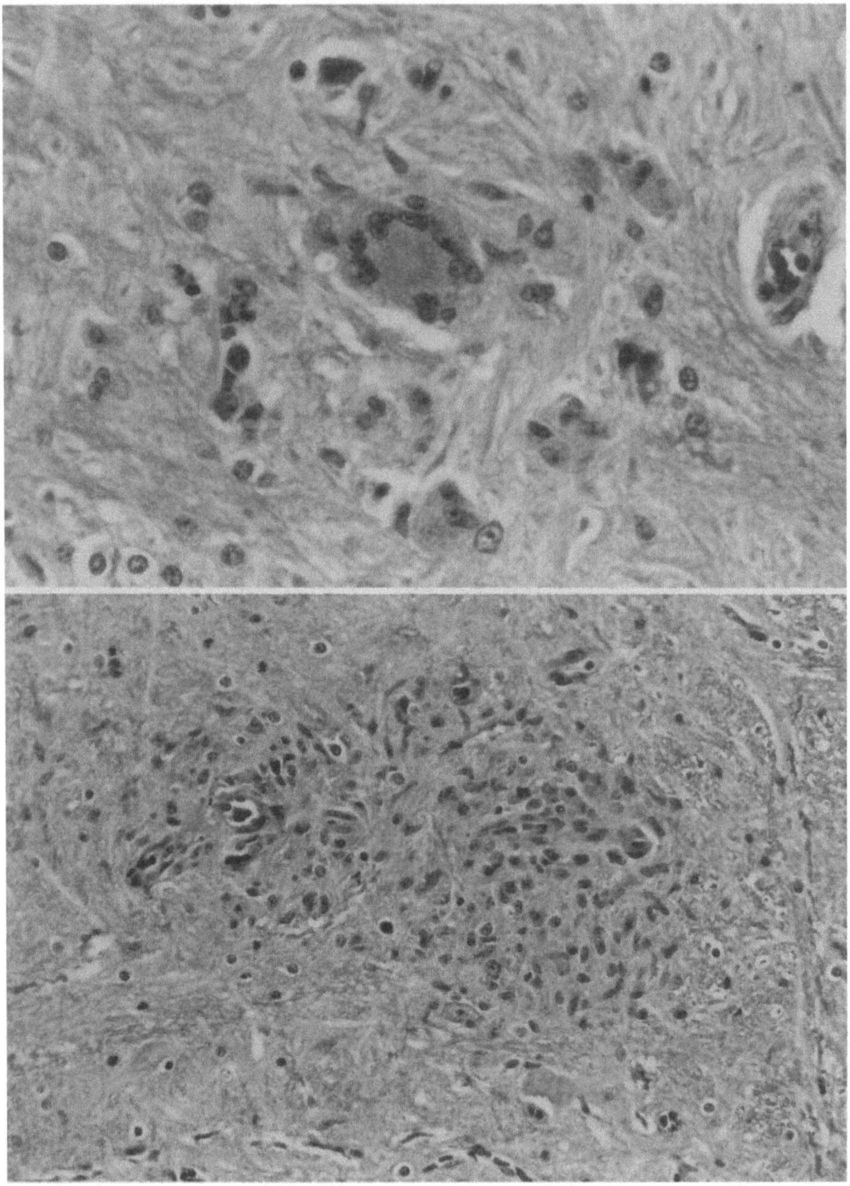

C

D

Fig. 1. C Pigtail macaque inoculated with SIV_{mac}. Multinucleate giant cells are associated with foci of gliosis in the brain; H & E, × 100. **D** Pigtail macaque inoculated with SIV_{mac} Inflammatory lesions extend into the spinal cord. Gray matter in this section contains a large glial nodule; H & E, × 50. **E** Rhesus macaque inoculated with SIV_{mac} Glial nodules are also found in white matter tracts of the spinal cord; H & E, × 100

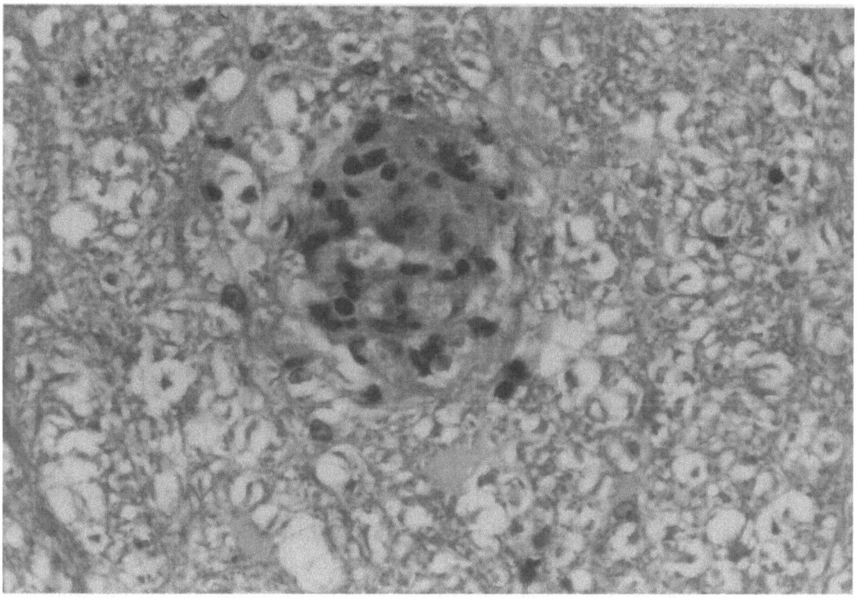
E

of activated T cells to disseminate throughout the body and cross physiological barriers such as the blood-brain barrier (HICKEY et al. 1991). Studies on the pathogenesis of this infection in macaques in other labs had shown that animals developed meningitis early during infection (SHARER et al. 1988) and that many of the mononuclear inflammatory cells in these exudates expressed viral RNA when examined by in situ hybridization (LACKNER et al. 1991). Such cells were not present in animals examined several months later suggesting that lesions in the CNS during the early phase of infection were only transient. Interestingly, people in the early phase of infection with HIV-1 also experience mild symptoms of meningeal infection (PRICE et al. 1988). Infected cells were found only rarely in the neuropil. These findings are in keeping with viral infection in activated T cells as seen in other viral meningitides such as mumps and vaccinia virus infections (JOHNSON 1982; GINSBERG and JOHNSON 1976). In our studies, virus could not be rescued from blood or spleen by cultivation of macrophages from these two tissues, despite the high concentration of infectious virus in the spleen (SHARMA et al. 1992).

Phase 2 was characterized by the appearance of antiviral $CD8^+$ T cells in blood. Circulation of infected activated $CD4^+$ T cells in the peripheral blood ceased, presumably the result of killing by antiviral $CD8^+$ T cells. Infected T cells were no longer found in the CSF and the detection of infected T cells in the meninges was rare (LACKNER et al. 1994; JOAG et al. 1994a). Some animals progressed to phase 3, a period characterized by effective control and equivalent to infected nonprogressor state of HIV infection (SHEPPARD et al. 1993). In animals in this stage of infection, $CD8^+$ T cell-mediated inhibition of virus replication became more effective in the blood and the lymphoid tissues. Virus recovery required depletion of $CD8^+$ T cells prior to treatment of the cells with mitogen, followed by cultivation in medium containing IL-2. Animals reaching phase 3 remained clinically well for 1.5–2 years before developing AIDS.

The level of immunosuppression observed in macaques inoculated with $SIV_{mac}239$ was variable. Immunosuppression was characterized by loss of $CD4^+$ T cells or $CD4^+$ T cell function and was frequently accompanied by development of high susceptibility to infection with opportunistic agents and occurrence of neoplasia. The most severe type of immunosuppression was seen in animals that failed to mount any immune responses to the virus (as determined by the failure to detect antibodies by immunoprecipitation). These animals usually developed acquired infections and died within 6 months. A less severe type of immunosuppression was characterized by the ability of animals to produce antiviral antibodies but whose $CD8^+$ T cell responses were inadequate to control virus replication in lymphoid tissues. This type of immune response was also characteristic of animals in phase 2 and the animals lasted for about 1 year before succumbing to AIDS. The third type of immunosuppression was the type that heralded end-stage disease and was characterized by the inability of lymphocytes to respond to mitogens. This indicated global anergy and inability of the animal to mount cell mediated immune responses.

Animals succumbed to infection with $SIV_{mac}239$ at various periods following inoculation, depending on the degree of immunosuppression and the type of AIDS that they developed. Most animals developed only acquired infections and tumors and only rarely neurological disease. In one study encephalopathy developed about 2 years after inoculation of the animal with virus 239 and virus isolated from the brain was typically macrophage-tropic, unlike virus 239 (Kodama et al. 1993; Desrosiers et al. 1991). A probable explanation for this phenomenon was that virus 239 mutated to a macrophage-tropic phenotype and this agent invaded the brain and subsequently caused encephalitis.

Attempts to cause encephalitis directly have proved as elusive in the case of SIV as in studies on visna in sheep. Inoculation of $SIV_{mac}239$ intracerebrally into macaques resulted in the typical $SIV_{mac}239$-like infection in blood and lymphoid tissues, without infection or development of lesions in the CNS during a 3 month test period (Sharma et al. 1992). We therefore changed the strategy and used infectious bone marrow cells (Sharma et al. 1992) for intracerebral (IC) inoculation as shown in Table 1. The rationale for this approach was threefold: (1) It is well known that lentiviruses exist as a swarm of genotypes. (2) The bone marrow contains the greatest variety of stem cells and mature and immature mononuclear cells compared to other tissues in the body. We speculated that different subpopulations of these cells may select different members of the virus swarm. (3) Inoculation of infectious bone marrow cells intracerebrally would expedite conditions for selection of a macrophage-tropic virus from the potential swarm by the microglia in the brain. The third passage of infectious bone marrow cells into a new macaque resulted in a disease associated with macrophage-tropic virus; the animal developed interstitial pneumonia similar to lentiviral pneumonia

Table 1. Neuroadaptation of SIV in rhesus macaques

Passage	Inoculum	Route	Animal	Status	Disease	Virus in brain		
						DNA	Explants	Homogenates
1	239	IC	5C	Killed 11 weeks	No	ND	Neg	Neg
2	5C BR+239	IC	2C	Killed 10 weeks	No	ND	Neg	Neg
3	2C BR+BM	IC, SC	10C	Died 12 weeks	Encephalitis Pneumonia	ND	ND	ND
4	10C BM	IC, BM	9D	Died 24 weeks	Encephalitis Pneumonia	ND	Pos	Pos
	10C BM	IC, BM	R71	Died 23 weeks	Encephalitis Pneumonia	ND	Pos	Pos
5	R71 BR	IC	17E	Died 16 weeks	Encephalitis	Pos	Pos	Pos
	R71 BR	IC	24E	Died 28 weeks	Renal failure	Pos	Pos	Neg
	R71 BR	BM	12E	Died 18 weeks	Encephalitis	Pos	ND	Neg
	R71 BR	BM	14E	Killed 43 weeks	No	Neg	Neg	Neg
	9D BR	BM	16D	Killed 26 weeks	No	ND	Neg	Neg
	9D BR	BM	22D	Killed 26 weeks	No	ND	Neg	Neg
	R71 BR	BM	6A	Alive 26 weeks	No	ND	ND	ND
	R71 BR	BM	6C	Alive 26 weeks	No	ND	ND	ND
	R71 BR	BM	6E	Died 8 weeks	Encephalitis	Pos	Pos	Pos
6	17E BR	IC, SC	P34R	Killed 5 weeks	No	Pos	Pos	Neg
	17E BR	IC, SC	O109	Killed 5 weeks	No	Pos	Pos	Neg
	17E BR	SC	O159	Killed 5 weeks	No	Pos	Pos	Pos
	17E BR	BM	6B	Alive 26 weeks	No	ND	ND	ND
	17E BR	BM	6F	Alive 26 weeks	No	ND	ND	ND
	17E BR	BM	6D	Died 4 weeks	AIDS	Pos	ND	Neg

ND, not done; Pos, positive; Neg, negative; BR, brain homogenate; IC, intracerebral; SC, subcutaneous; BM, bone marrow.

(maedi) observed in sheep infected with visna-maedi virus. A further passage of infectious bone marrow cells into two macaques (9D and R71) resulted in development of encephalitis and pneumonia in both animals with typical productive virus replication in macrophages in brain and lung, but not in macrophage populations in other tissues. Again, this pathogenesis was remarkably similar to maedi-visna of sheep. The major difference between the disease in macaques and the disease complex in sheep and goats was that in macaques virus was isolated from macrophages cultivated from brain and lung, but not from the spleen or blood although the agent was isolated from lymphocytes from the latter two tissues (SHARMA et al. 1992). In contrast, the virus infection of the ruminant animals was confined to the macrophage lineage cells in all tissues but virus replication was greatly enhanced in the two target organs (brain and lung). Thus, in macaques, two types of infections, one in lymphocytes and one in macrophages, coincided in the pathogenesis of neurological disease. Moreover, both animals with neurological disease had become severly immunosuppressed as evidenced by failure to develop antibodies against virus and the development of other acquired infections.

Homogenates of brain from the two animals that developed encephalitis contained infectious virus and the virus in the R71 homogenate was tested for its tropism for lymphocytes and macrophages. The brain homogenate virus replicated efficiently in both activated lymphocytes in phytohemagglutinin (PHA)-treated rhesus PBMCs as well as rhesus blood-derived macrophages (ANDERSON et al. 1993). The agent thus had macrophage and lymphocyte-tropic properties. The R71 homogenate was inoculated into nine macaques in three cohorts to evaluate potential neurovirulence of viruses in this homogenate (Table 1). The first cohort comprised four 1-year-old animals. Two were inoculated in the bone marrow and two intracerebrally. All four animals developed the type of infection seen during phase 1 of virus 239-induced infection (JOAG et al. 1994 a, b). Cell-free virus and activated infected T cells appeared in blood 1 week postinoculation and virus infected T cells were found in the CSF. Macaque 17E remained in phase 2 and was euthanized after developing wasting, tremors and ataxia at 4 months postinoculation. This animal was severely immunosuppressed, never having produced antibodies against the virus. The remaining three animals progressed into phase 2 of infection by 6 weeks postinoculation and macaques 12E and 24E remained in this status until they were killed. Macaque 12E was euthanized with severe anemia at 5 months. This animal had encephalitis and also failed to produce antiviral antibodies. Macaque 24E was clinically healthy at 5 months. This animal produced binding antibodies to the virus at 3 months, but no antibodies were detectable at 6 months postinoculation by which time it had become severly ill. While still clinically healthy, this animal was subjected to a brain biopsy, which upon examination had normal histological appearance. Cell-free virus was not detected in the brain homogenate but macrophages cultivated from the brain produced virus. This animal became moribund from severe glomerulonephritis at 6 months postinoculation and was euthanized. Necropsy studies showed that except for severe inflammation at the biopsy site in the brain, the rest of the neuropil was normal histologically. Interestingly, macrophages cultivated from the biopsy site, which had become infiltrated with inflammatory macrophages, failed to produce virus, although macrophages from other regions of the brain produced virus (JOAG et al. 1994b). This suggested that microglial macrophages, rather than inflammatory macrophages, probably contained the viral genome which was not expressed in vivo. Cultivation of macrophages from this brain resulted in viral gene expression and virus production, a phenomenon observed repeatedly in sheep infected with visna virus or goats with CAE virus (NARAYAN et al. 1982, 1983).

Macaque 14E progressed through phases 1, 2, and 3 by 20 weeks postinoculation. The animal developed splenomegaly and lymphoid hyperplasia similar to SIV_{mac}239-infected animals. This animal produced binding antibodies against all antigens of SIV_{mac}239, although these antibodies did not neutralize infectivity of the virus. Examination of a brain biopsy of this animal showed no lesions or infection in macrophages cultivated from the biopsy. This was remarkable since virus-infected cells had been found in the CSF during phase 1 of infection yet, even latent infection was not detectable in brain macrophages when examined during phase 3 (JOAG et al. 1994b).

The data thus far showed that the four animals that developed severe encephalitis (9D, R71, 12E, and 17E) had become immunosuppressed prior to onset of encephalitis. Two other animals (16D and 22D) were inoculated with macaque 9D brain homogenate (Table 1) and killed 6 months later while they were still in good health. Macaque 22D produced antibodies to the virus although macaque 16D did not. Examination of brain showed normal histological architecture. No infectious virus or viral antigen was detected in brain homogenate or brain sections and brain macrophage explant cultures did not yield virus. Three other animals (6A, 6C, 6E) were inoculated in the bone marrow with R71 brain homogenate. All three animals went through phases 1 and 2. Macaque 6E developed progressive weakness, quadriparesis, choreoathetoid movements, and signs of bulbar paralysis at 8 weeks and was euthanized. At necropsy, virus was recovered from brain homogenate and from macrophage explant cultures. This animal did not develop antibodies to SIV. The other two animals (6A and 6C) were still healthy after 6 months of infection. Both have produced antiviral antibodies.

Brain homogenate from 17E was also examined for neurovirulence using two cohorts of animals. The first consisted of three adult animals (more than 10 years old), 0109, 0159, and P34R (JOAG et al. 1994b). They were inoculated with virus and given continuous intracerebral infusion of TNF-α and IL-6. In an earlier study (MCENTEE et al. 1992) we found that addition of $CD4^+$ T lymphocytes to cultures of macaque macrophages infected for 5 days earlier with SIV_{mac} resulted in a burst of cytokine release by the T cells during the next 2 days. The supernatant fluid contained 16 U/ml TNF-α and 100 U/ml IL-6. This fluid was loaded into Alzet pumps which were installed subcutaneously and an attached catheter delivered the material into the left cerebral ventricle at the rate of 70 μl/day for 4 weeks. All three animals developed viremia typical of phase 1 of the infection and were killed between weeks 4 and 6 to examine the brain for evidence of virus replication and possible early encephalitis. Histological and virological analysis of the three brains showed that there was no evidence of virus replication, except latent infection in brain macrophages, and only equivocal histological evidence of encephalitis. Macrophages in brain explant cultures produced virus. Thus, infusion of TNF-α and IL-6 did not induce viral replication in the brain (JOAG et al. 1994b). The lack of lesions in these brains at the time suggested that the lesions do not develop gradually during the course of subclinical infection and may develop acutely shortly before the onset of clinical signs.

The second cohort consisted of three 1-year-old animals (6B, 6D, and 6F) which were inoculated in the bone marrow with virus. All three developed typical phases 1 and 2 of infection. 6D developed intractable diarrhea and was euthanized at 4 weeks postinoculation. This animal had no virus in brain and had not produced antiviral antibodies. The other two animals are 6 months into their infection and are healthy. Both have produced anti-viral antibodies.

These data indicate the difficulty of reproducing encephalopathy using classical concepts and methods of neuroadaptation of the virus. However, the data are consistent with the prediction that continuous replication of macrophage-tropic virus in animals that failed to develop any antiviral immune responses would probably result in encephalopathy provided other types of life-threatening AIDS did not develop. The correlation between development of encephalitis and poor antiviral antibody responses has been reported previously (ZHANG et al. 1988). In most of the animals that developed antiviral antibodies, the virus remained latent in brain for prolonged periods and could not be activated by TNF-α and IL-6 infused into the brain. Thus, there was no evidence of a gradually increasing growth curve of virus replication in the brain of these animals, a phenomenon that occurs normally with the other neuroadapted RNA viruses. Similar results had been obtained with visna virus infection in sheep. It is possible that brain homogenate contained large amount of viral antigen which could have induced antiviral immune response and that animals developing immunoprecipitating antibodies also developed antiviral $CD8^+$ cells which could have repressed virus replication in brain.

Since the infectious viruses in the brain homogenate undoubtedly contained a virus swarm consisting of many genotypes or quasispecies, we attempted to determine whether an infectious, molecularly cloned virus that could replicate in brain macrophages, could cause encephalitis. We used PCR to amplify the env gene of the virus from DNA of 17E brain, cloned the gene using molecular genetic techniques, and substituted this env gene for the corresponding env gene of infectious $SIV_{mac}239$ DNA (ANDERSON et al. 1993). The rationale for this approach was that sequences in the *env* gene are the major determinant for cell and tissue tropism of the virus (CORDONNIER et al. 1989). Examination of the replication potential of this chimeric virus 239/17E showed that it replicated productively in macaque macrophages (ANDERSON et al. 1993; STEPHENS et al. 1995). We then tested the ability of this chimeric virus to cause neurological disease. To ensure that the viral genome did not undergo variation by passage in CEMx174 cells, we transfected bone marrow cells of four macaques with viral DNA, after which the cells were reinfused back into the respective animals (JOAG 1995). All four macaques became infected as determined by PCR evaluation of PBMC DNA and later by rescue of infectious virus by cocultivation of mitogen-treated PBMC with CEMx174 (proving infection in lymphocytes) and in cultures of blood-derived macrophages (proving infection in macrophage lineage cells). However, cytopathic virus was only rescuable from these animals during the first few weeks postinoculation. Beyond 12 weeks, virus could be detected only by PCR and p27 assay on the PBMCs and cells from lymph nodes and spleen. Unexpectedly, all four animals developed potent neutralizing antibodies to the virus by 1 month postinoculation and titers at 6 months exceeded 1 in 10 000. The major distinguishing feature of this infection compared to infection caused by $SIV_{mac}239$ and the R71 brain homogenate was that there was no true phase 1 of infection in the infected animals. There were no infected activated T cells in peripheral blood, no plasma viremia, no lymphoid hyperplasia, and no infected cells in the CSF. Brain biopsies were obtained from all four animals and viral DNA found in brain of three of them. However, no viral antigens or lesions were found in the brain and virus was not rescuable from macrophages cultivated from the biopsies. One animal developed immunosuppression and $CD4^+$ T cell depletion by 8 months postinoculation and was euthanized at 12 months after T cell anergy had developed. Necropsy showed AIDS, characterized by cryptosporidia-associated gastritis and enteritis. Despite this, the animal did not develop encephalitis. One possible reason for this was that by the time the animal had become immunosuppressed, the virus that was present was no longer capable of productive replication in macrophages. The other three animals have remained healthy at one year postinoculation. They remain persistently infected with a virus that is no longer cytopathic in the lymphocytes and monocytes and they have high neutralizing antibodies to the virus. Thus, despite the tropism of the virus for macrophages and the fact that the *env* gene of the virus was derived from encephalitic brain, this agent, which seemed to have the ideal genotypic credentials for causing encephalitis, failed to cause neurological disease. The major failing of the virus was that it was highly immunogenic in eliciting large amounts of neutralizing antibodies and parenthetically, failed to cause immunosuppression early during infection.

Our present data suggest that neurovirulent lentiviruses cannot be derived by neuroadaptation procedures that have been applied classically to other RNA viruses. Neurovirulent lentiviruses are macrophage-tropic and do not have tropism for neuroectodermal cells. Further, neurovirulence of these lentiviral agents seems to require a combination of different factors, two cardinal candidates being, first, presence of replication-competent, macrophage-tropic virus and, second, thorough suppression of the immune system. $SIV_{mac}239$ caused immunosuppression but it is not macrophage-tropic. In contrast, brain homogenates and bone marrow contained both potential factors and caused encephalitis in some, but not all, animals. The chimeric virus had the appropriate cell tropism but failed to cause immunosuppression. Our data suggest that mechanistically, the syndrome requires at least two hits. First, immunosuppression, possibly caused by a lymphocyte-tropic virus, and second, replication of a macrophage-tropic virus in the brain.

Acknowledgments. Supported by grants NS-32203, NS-12127, AI-29382, and RR-06753 from the National Institutes of Health. We also thank Dr. David Pinson (University of Kansas Medical Center) for the photomicrographs in Fig. 1.

References

Anderson MG, Hauer D, Sharma DP, Joag SV, Narayan O, Zink MC, Clements JE (1993) Analysis of envelope changes acquired by $SIV_{mac}239$ during neuroadaptation in rhesus macaques. Virology 195: 616–626

Brew BJ, Bhalla RB, Paul M, Gallardo H, McArthur JC, Schwartz MK, Price RW (1990) Cerebrospinal fluid neopterin in human immunodeficiency virus type 1 infection. Ann Neurol 28: 556–560

Brew BJ, Bhalla RB, Paul M, Sidtis JJ, Keilp JJ, Sadler AE, Gallardo H, McArthur JC, Schwartz MK, Price RW (1992) Cerebrospinal fluid beta 2-microglobulin in patients with AIDS dementia complex: an expanded series including response to zidovudine treatment. AIDS 6: 461–465

Budka H (1986) Multinucleated giant cells in brain: a hallmark of the acquired immune deficiency syndrome (AIDS). Acta Neuropathol (Berl) 69: 253–258

Chakrabarti L, Hurtrel M, Maire MA, Vazeux R, Dormont D, Montagnier L, Hurtrel B (1991) Early viral replication in the brain of SIV-infected rhesus monkeys. Am J Pathol 139: 1273–1280
Cordonnier A, Montagnier L, Emerman M (1989) Single amino-acid changes in HIV envelope affect viral tropism and receptor binding. Nature 340: 571–574
Cork LC, Hadlow WJ, Crawford TB, Gorham JR, Pyper RC (1974a) Infectious leukoencephalomyelitis of goats (CAEV). J Infect Dis 129: 134–141
Cork LC, Hadlow WJ, Gorham JR, Piper RC, Crawford TB (1974b) Pathology of viral leukoencephalomyelitis of goats. Acta Neuropathol (Berl) 29: 281–292
DeMartini JC, Brodie SJ, de la Concha Bermejillo A, Ellis JA, Lairmore MD (1993) Pathogenesis of lymphoid interstitial pneumonia in natural and experimental ovine lentivirus infection. Clin Infect Dis 17 [Suppl 1]: S236–S242
Desrosiers RC, Hansen Moosa A, Mori K, Bouvier DP, King NW, Daniel MD, Ringler DJ (1991) Macrophage-tropic variants of SIV are associated with specific AIDS-related lesions but are not essential for the development of AIDS. Am J Pathol 139: 29–35
Gabuzda DH, Hess JL, Small JA, Clements JE (1989) Regulation of the visna virus long terminal repeat in macrophages involves cellular factors that bind sequences containing AP-1 sites. Mol Cell Biol 9: 2728–2733
Gendelman HE, Orenstein JM, Baca LM, Weiser B, Burger H, Kalter DC, Meltzer MS (1989) The macrophage in the persistence and pathogenesis of HIV infection. AIDS 3: 475–495
Georgsson G, Houwers DJ, Palsson P, Petursson G (1989) Expression of viral antigens in the central nervous system of visna-infected sheep: an immunohistochemical study on experimental visna induced by virus strains of increased neurovirulence. Acta Neuropathol (Berl) 77: 299–306
Ginsberg AH, Johnson KP (1976) Vaccinia virus meningitis in mice after intracerebral inoculation. Infect Immun 13: 1221–1227
Gorrell MD, Brandon MR,Sheffer D, Adams RJ, Narayan O (1992) Ovine lentivirus is macrophagetropic and does not replicate productively in T lymphocytes. J Virol 66: 2679–2688
Heyes MP, Brew BJ, Martin A, Price RW, Salazar AM, Sidtis JJ, Yergey JA, Mouradian MM, Sadler AE, Keilp J et al (1991) Quinolinic acid in cerebrospinal fluid and serum in HIV-1 infection: relationship to clinical and neurological status. Ann Neurol 29: 202–209
Hickey WF, Hsu BL, Kimura H (1991) T-Lymphocyte entry into the central nervous system. J Neurosci Res 28: 254–260
Houwers DJ, Pekelder JJ, Akkermans JW, van der Molen EJ, Schreuder BE (1988) Incidence of indurative lymphocytic mastitis in a flock of sheep infected with maedi-visna virus. Vet Rec 122: 435–437
Joag SV, Adams RJ, Foresman L, Galbreath D, Zink MC, Pinson DM, McClure H, Narayan O (1994a) Early activation of PBMC and appearance of antiviral CD8⁺ cells influence the prognosis of SIV-induced disease in rhesus macaques. J Med Primatol 23: 108–116
Joag SV, Adams RJ, Pinson DM, Adany I, Narayan O (1994b) Intracerebral infusion of TNF-α and IL-6 failed to activate latent SIV infection in the brain of macaques inoculated with macrophage-tropic neuroadapted SIVmac. J Leukoc Biol 56: 353–357
Joag SV, Stephens EB, Foresman L, Adams RJ, Narayan O (1994c) Pathogenesis of SIV_{mac} infection in Chinese and Indian rhesus macaques: effects of splenectomy on virus burden. Virology 200: 436–446
Joag SV, Stephens EB, Narayan O (1994d) Lentiviruses. In: Fields BN, Knipe DM, Howley PM et al (eds) Virology, 3rd edn. Raven, New York
Joag SV, Stephens EB, Galbreath D, Zhu GW, Li Z, Foresman L, Zhao L.-J, Pinson DM, Narayan O (1995) Simian Immunodeficiency virus SIV_{mac} Chimeric Virus whose env gene was derived from SIV-encephalitic brain is macrophage-tropic but not neurovirulent. J Virol 69: 1367–1369
Johnson RT (1980) Selective vulnerability of neural cells to viral infection. Brain 103: 447–472
Johnson RT (1982) Mumps virus. In: Johnson RT (ed) Viral infections of the nervous system. Raven, New York, pp 98–100
Kennedy PGE, Narayan O, Ghotbi Z, Hopkins J, Gendelman HE, Clements JE (1985) Persistent expression of Ia antigen and viral genome in visna-maedi virus-induced inflammatory cells: possible role of lentivirus-induced interferon. J Exp Med 162: 1970–1982
Kennedy PGE, Narayan O, Zink MC, Hess J, Clements JE, Adams RJ (1988) The pathogenesis of visna, or lentivirus induced immunopathologic disease of the central nervous system. In: Gilden DH, Lipton H (eds) Clinical and molecular aspects of viral illness of the nervous system. Nijhoff, Boston
Kennedy-Stoskopf S, Narayan O, Strandberg JD (1985) The mammary gland as a target organ for infection with caprine arthritis-encephalitis virus. J Comp Pathol 95: 609–617

Kennedy-Stoskopf S, Zink C, Narayan O (1989) Pathogenesis of ovine lentivirus-induced arthritis: phenotypic evaluation of T lymphocytes in synovial fluid, synovium, and peripheral circulation. Clin Immunol Immunopathol 52: 323–330

Kestler H, Kodama T, Ringler D, Marthas M, Pedersen N, Lackner A, Regier D, Sehgal P, Daniel M, King N et al (1990) Induction of AIDS in rhesus monkeys by molecularly cloned simian immunodeficiency virus. Science 248: 1109–1112

Kestler HW, Li Y, Naidu YM, Butler CV, Ochs MF, Jaenel G, King NW, Daniel MD, Desrosiers RC (1988) Comparison of simian immunodeficiency virus isolates. Nature 331: 619–622

Kodama T, Mori K, Kawahara T, Ringler DJ, Desrosiers RC (1993) Analysis of simian immunodeficiency virus sequence variation in tissues of rhesus macaques with simian AIDS. J Virol 67: 6522–6534

Kure K, Lyman WD, Weidenheim KM, Dickson DW (1990) Cellular localization of an HIV-1 antigen in subacute AIDS encephalitis using an improved double-labeling immunohistochemical method. Am J Pathol 136: 1085–1092

Lackner AA, Smith MO, Munn RJ, Martfeld DJ, Gardner MB, Marx PA, Dandekar S (1991) Localization of simian immunodeficiency virus in the central nervous system of rhesus monkeys. Am J Pathol 139: 609–621

Lackner AA, Vogel P, Ramos RA, Kluge JD, Marthas M (1994) Early events in tissues during infection with pathogenic (SIV_{mac}239) and nonpathogenic (SIV_{mac}IAII) molecular clones of SIV. Am J Pathol

Lairmore MD, Butera ST, Callahan GN, DeMartini JC (1988) Spontaneous interferon production by pulmonary leukocytes is associated with lentivirus-induced lymphoid interstitial pneumonia. J Immunol 140: 779–785

McEntee MF, Gorrell MD, Adams RJ, Narayan O, Pitha P (1992) Tumour necrosis factor and interleukin 6 production during interaction between activated $CD4^+$ lymphocytes and simian immunodeficiency virus-infected macrophages. J Gen Virol 73: 1107–1113

Monath TP (1990) Flaviviruses associated primarily with the encephalitic syndrome. In: Fields BN, Knipe DM (eds) Virology, 2nd edn. Raven, New York, pp 775–787

Mori K, Ringler DJ, Desrosiers RC (1993) Restricted replication of simian immunodeficiency virus strain 239 in macrophages is determined by env but is not due to restricted entry. J Virol 67: 2807–2814

Narayan O, Cork LC (1985) Lentiviral diseases of sheep and goats: chronic pneumonia leukoencephalomyelitis and arthritis. Rev Infect Dis 7: 89–98

Narayan O, Wolinsky JS, Clements JE, Strandberg JD, Griffin DE, Cork LC (1982) Slow virus replication: the role of macrophages in the persistence and expression of visna viruses of sheep and goats. J Gen Virol 59: 345–356

Narayan O, Kennedy-Stoskopf SW, Sheffer D, Griffin DE, Clements JE (1983) Activation of caprine arthritis-encephalitis virus expression during maturation of monocytes to macrophages. Infect Immun 41: 67–73

Narayan O, Strandberg JD, Griffin DE, Clements JE, Adams RJ (1984) Aspects of the pathogenesis of visna in sheep. In: Mimms CA, Cuzner ML, Kelly RE (eds) Symposium on viruses and demyelinating diseases. Academic, New York, pp 125–140

Narayan O, Sheffer D, Clements JE, Tennekoon G (1985) Restricted replication of lentiviruses: visna viruses induce a unique interferon during interaction between lymphocytes and infected macrophages. J Exp Med 162: 1954–1969

Navia BA, Cho ES, Petito CK, Price RW (1986) The AIDS dementia complex: II. Neuropathology. Ann Neurol 19: 525–535

Palsson PA (1976) Maedi and visna in sheep. In: Kimberlin RH (ed) Slow viral disease of animals and man. North-Holland, Amsterdam, pp 17–43

Panitch H, Petursson G, Georgsson G, Palsson PA, Nathanson N (1976) Pathogenesis of Visna. III. Immune responses to central nervous system antigens in experimental allergic encephalomyelitis and visna. Lab Invest 35: 452–460

Price RW (1994) Understanding the AIDS dementia complex. In: Price RW, Perry SW (eds) HIV, AIDS, and the brain. Raven, New York, pp 1–45

Price RW, Brew B, Sidtis J, Rosenblum M, Scheck AC, Cleary P (1988) The brain in AIDS: central nervous system HIV-1 infection and AIDS dementia complex. Science 239: 586–592

Rimstad E, East NE, Torten M, Higgins J, DeRock E, Pedersen NC (1993) Delayed seroconversion following naturally acquired caprine arthritis-encephalitis virus infection in goats. Am J Vet Res 54: 1858–1862

Sharer LR, Epstein LG, Cho ES, Joshi VV, Meyenhofer MF, Rankin LF, Petito CK (1986) Pathologic features of AIDS encephalopathy in children: evidence for LAV/HTLV-III infection of brain. Hum Pathol 17: 271–284

Sharer LR, Baskin GB, Cho E-S, Murphey-Corb M, Blumberg BM, Epstein LG (1988) Comparison of simian immunodeficiency and human immunodeficiency virus encephalitides in the immature host. Ann Neurol 23 [Suppl]: S108–S112

Sharma DP, Zink MC, Anderson M, Adams R, Clements JE, Joag SV, Narayan O (1992) Derivation of neurotropic simian immunodeficiency virus from exclusively lymphocytetropic parental virus: pathogenesis of infection in macaques. J Virol 66: 3550–3556

Sheffield WD, Narayan O, Strandberg JD, Adams RJ (1980) Visna-maedi-like disease associated with an ovine retrovirus infection in a Corriedale sheep. Vet Pathol 17: 544–552

Sheppard HW, Lang W, Ascher MS, Vittinghoff E, Winkelstein W (1993) The characterization of non-progressors: long-term HIV-1 infection with stable $CD4^+$ T-cell levels. AIDS 7: 1159–1166

Shih DS, Carruth LM, Anderson M, Clements JE (1992) Involvement of FOS and JUN in the activation of visna virus gene expression in macrophages through an AP-1 site in the viral LTR. Virology 190: 84–91

Sigurdsson B (1954) Observations on three slow infections of sheep: Maedi, paratuberculosis, rida, a slow encephalitis of sheep with general remarks on infections which develop slowly, and some of their special characteristics. Br Vet J 110: 255–270

Stephens EB, McClure HM, Narayan O (1995) The proteins of Lymphocyte- and macrophage-tropic strains of simian immunodeficiency virus are processed differently in macrophages. Virology 206: 535–544

Vitu C, Russo P, Vignoni M (1993) Caprine arthritis-encephalitis: trial of an adjuvant vaccine preparation. II. Study of the antibody response. Comp Immunol Microbiol Infect Dis 16: 137–144

Zhang JY, Martin LN, Watson EA, Montelaro RC, West M, Epstein L, Murphey-Corb M (1988) Simian immunodeficiency virus/delta-induced immunodeficiency disease in rhesus monkeys: relation of antibody response and antigenemia. J Infect Dis 158: 1277–1286

Zink MC, Gorrell MD, Narayan O (1991) The neuropathogenesis of visna virus infection in sheep. Neurosciences 3: 125–130

Feline Immunodeficiency Virus as a Model for Study of Lentivirus Infection of the Central Nervous System

S.J. HENRIKSEN[1], O. PROSPERO-GARCIA[1], T.R. PHILLIPS[1], H.S. FOX[1], F.E. BLOOM[1], and J.H. ELDER[2]

1 Introduction

A hallmark clinical feature of human acquired immunodeficiency syndrome (AIDS) is the neuropsychiatric and cognitive deficits attending progression of the illness (JANSSEN et al. 1991; NAVIA et al. 1986a). This severe neurologic disorder has been referred to as the acquired immunodeficiency syndrome dementia complex (ADC) (NAVIA et al. 1986a, PRICE and PERRY 1994). A recent task force report constituted by the executive board of the American Academy of Neurology proposed a more definitive nomenclature and research definitions for the neurologic manifestations of HIV-1 infection (JANSSEN et al. 1991). The participants suggested separating the complex syndrome into two categories: (1) a more severe form, HIV-1-associated dementia complex or HIV-1-associated myelopathy; and (2) a less severe form, HIV-1-associated neurocognitive/motor disorder (HMCD). Both classes, previously termed the AIDS dementia complex (ADC), are forms of a subcortical dementia comprised of a complex collection of neuropsychological and cognitive signs symptomatic of HIV-1 infection of the CNS.

[1] Department of Neuropharmacology, The Scripps Research Institute, 10666 North Torrey Pines Road, La Jolla, C 92037, USA
[2] Department of Molecular Biology. The Scripps Research Institute, 10666 North Torrey Pines Road, La Jolla, CA 92037, USA

From the middle stages of AIDS, some 40%–67% of patients present with deteriorating mental function and at autopsy, 75%–90% of HIV-1 infected brains reveal significant neuropathological changes (NAVIA et al. 1986bl ELDER and SEVER 1988; HO et al. 1989; JOHNSON et al. 1988; WIGDAHL and KUNSCH 1989; PRICE and PERRY 1994). Decline in neuropsychological measures and the appearance of other neurological signs and behavioral abnormalities are believed to parallel the early and persistent infection of the human CNS by HIV-1 (MALESSA et al. 1989; MCARTHUR 1994).

The precise temporal relationship between CNS viral infection and the onset of specific neuropsychological and cognitive changes remains unclear (GOODIN et al. 1990; PERRY 1990; HANDELSMAN et al. 1992; HELMSTAEDTER et al. 1992; KORALNIK et al. 1990; MCARTHUR 1994). Individual neuronal cell bodies located in virally penetrated areas of the CNS rarely, if ever, are invaded *directly* by competent HIV-1 in "detectable amounts" (TENG 1990; DAWSON et al. 1991). However, it has recently been reported that in severely debilitated HIV-1-infected patients, neurons may show evidence of viral sequestration (NUOVO et al. 1994).

The most likely primary cellular targets are believed to be CNS macrophages, vascular endothelial cells, and astrocytes, and/or microglia. These observations suggest a differentially enhanced CNS neuroglial tropism for the infecting virus (JOHNSON et al. 1988). In addition, postmortem evidence of AIDS-associated astrocytosis of subcortical white matter in mildly demented patients suggests that primary diencephalic gliosis may be the first proliferative CNS sign of the disease and appears to antedate other neuropathological abnormalities. How these proposed indirect actions of the virus might cause the specific "subcortical" form of the observed dementia remains unknown (NAVIA et al. 1986a). Virally induced release of cytokines from either CNS cellular compartments or via invading macrophages from the periphery; a possible "functional" neuronal toxicity induced by locally released viral glycoprotein products (i.e., gp120); or elevated levels of quinolinic acid have been suggested as inducible vectors that may perturb CNS function (JOHNSON et al. 1988; BRENNEMAN et al. 1988; HEYES et al. 1989). The interesting observation that 3-azido-2,3-dideoxythymidine (AZT) may both decrease levels of quinolinic acid in the CSF and reverse, at least temporarily, ADC in infected individuals (YARCHOAN et al. 1988) is consistent with the idea that neurons are not irreversibly damaged by HIV-1. These findings also suggest that alteration in neuronal function may be due to a continuous but reversible release of some unidentified neuromodulatory agent.

Despite impressive advances, very little is known about the consequences of HIV-1 infection or, for that matter, infection by any other lentiviruses, on CNS neuronal function. Lesions observed in postmortem material taken from AIDS infected brains have generally shown a predisposition for reactive and non-neuronal cellular pathology distributed in early disease in the subcortical white matter, dorsal and ventral striatal regions, thalamus, and pons. Although subcortical white matter pallor and astrogliosis are the most common histological findings, microglial and macrophage infiltration of thalamic and basal ganglia structures are frequently observed (JOHNSON et al. 1988; MCARTHUR 1994). Specific brain areas, including the dorsal and ventral striatal complex, the hippocampus, and their major efferent and afferent projection sites are important for behavioral arousal and goal-directed motor and mnemonic functions. Further, subtle sensory and motor deficits of peripheral and central origin suggest nerve conduction deficits in several modalities (MALESSA et al. 1989; HO et al. 1985; LIPKIN et al. 1985). Therefore, the observed histopathological changes, whatever the initiating mechanism(s), are consistent with sensory, attentional, motivational, memory and motor dysfunctions indicative of a subcortical dementia (NAVIA et al. 1986a).

It is clear from the above observations that there are an abundance of hints and hypotheses to pursue regarding the mechanisms by which HIV-1 induces dementia. It is equally clear that investigation is severely limited by the system itself. There is limited availability of tissues from individuals showing signs of dementia during early stages of infection. Also, inconsistencies and unknowns in documentation of the infection makes it impossible to accurately assess the temporal relationship between viremia, immunological responsiveness, and dementia (GRANT and MARTIN 1994). An animal model that exhibits pathology similar to human AIDS, but is amenable to experimental manipulation, is sorely needed. As will be detailed below, the feline/feline immunodeficiency virus (FIV) system serves as a valuable and economic model for study of lentivirus infection of the CNS.

2 The Feline/FIV Model

FIV is a lentivirus found associated with a debilitating immunodeficiency disease of cats (PEDERSEN et al. 1987, 1989). Similar to HIV-1 infection of humans, FIV infection is associated with a decrease in CD4$^+$ T lymphocytes, which in turn contributes to the increase in opportunistic infections, resulting in upper respiratory illness, gingivitis, and wasting, observed with the disease (ACKLEY et al. 1990; TORTEN et al. 1991; HOFFMANN-FEZER et al. 1992; MATSUMURA et al. 1993). In spite of differences in genomic structure between FIV and HIV-1 (*vidante*; also see ELDER and PHILLIPS 1994), they exhibit similarities in the general properties of their life cycles, cell types infected, and pathology. Both lentiviruses primarily infect T cells and cells of the monocyte/macrophage lineage, even though evidence indicates that the receptor used by HIV-1, the CD4 molecule (MADDON et al. 1986) is not the receptor for FIV (BROWN et al. 1991;WILLETT et al. 1994). Both viruses cause an initial acute phase syndrome, characterized by a sharp early decline in CD4$^+$ T lymphocytes, with symptoms of lymphadenopathy, febrile episodes, and general malaise. This period is followed by a long "latent" period of several years, during which humoral antibody remains high the antigenemia is low, with infected individuals appearing essentially normal during this phase. The length of this latent period is highly variavle in both cats and humans and is influenced by other factors, such as infection by other viral pathogens (PEDERSEN et al. 1990; LIN et al.

1992; Witt et al. 1989; Waters et al. 1993). With cats, use of specific pathogen free (SPF) animals in experimental infections results in consistent acute phase disease, but very slow progression to end stage disease. Random source cats appear to progress more readily, particularly when challenged by secondary infections. Both humans and cats proceed into an AIDS-related complex (ARC) phase, followed by full-blown immunodeficiency. During this phase, antibody production declines, antigenemia increases, and individuals generally succumb to opportunistic infections.

Important to the subject at hand, FIV also invades the CNS, resulting in neurological abnormalities and observed behavior pathologies (Dow et al. 1990, 1992; Egberink et al. 1989; Hurtrel et al. 1992; Lafrado et al. 1993; Podell et al. 1993; Phillips et al. 1994). As such, the cat offers an excellent natural model system for study of lentivirus infection of the CNS, particularly in the acute phase of the infection, for which very little data are available for HIV-1 infection of humans. In the following, we detail the findings to date and, where possible, compare and contrast results of CNS infection by FIV with those reported for HIV-1 infected people.

3 Comparison of the Genomic Organization of FIV and HIV-1

A comparison of the genomic structure of FIV and HIV-1 is shown in Fig. 1 (for review, see Elder and Phillips 1994). Both lentiviruses contain the major gene regions gag, pol and env, typical of all the retroviruses. Also, they both are bordered by long terminal repeats, which encode regulators of transcription and facilitate entry of the viral genome into the host cell DNA. Typical of lentiviruses, both FIV and HIV-1 contain additional coding capacity for regulatory elements that influence the relative level and success of the infection. These elements include viral infectivity factor (Vif, Tomonaga et al. 1993) and the regulator of virion expression (Rev, Phillips et al. 1992). However, striking differences are also evident between the human and cat lentiviruses. FIV encodes a unique enzyme, deoxyuridine triphosphatase (DU, Elder et al. 1992) between reverse transcriptase (RT) and integrace (IN) of *pol*, that is lacking in HIV-1. FIV also possesses a short open reading frame (Orf 2) between *vif* and *env*, which is unrelated to *vpu* encoded at this location in HIV-1. Conversely, HIV-1 encodes a transactivator, TAt, as well as two additional proteins Vpr and Nef, that have not been demonstrated for FIV.

At the nucleotide sequence level, the cat and human viruses are also divergent, with approximately 50% identity within the most conserved domains of *pol* (Talbott et al. 1989; Olmsted et al. 1989; Phillips et al. 1990). This structural

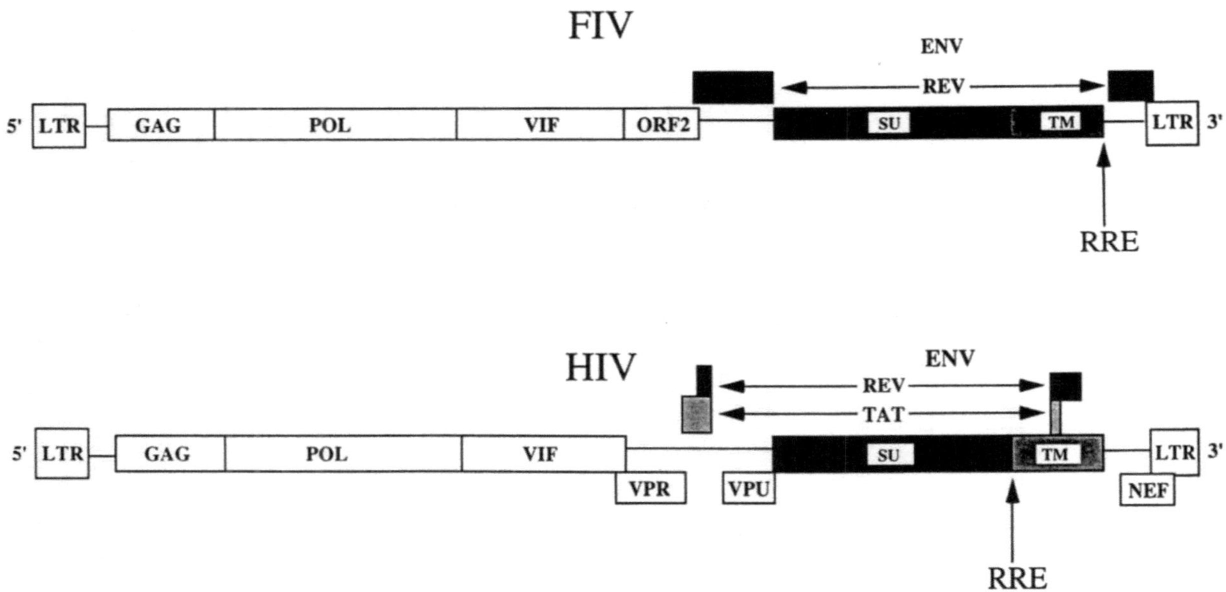

Fig 1. Genomic organization of FIV and HIV. The genomes of the two lentiviruses are shown as integrated proviruses bordered by 5' and 3' long terminal repeats (*LTRs*). Relative location of the Rev responsive element (*RRE*) present as a *cis* element in the viral RNA is also indicated. Both HIV and FIV are approximately 10 kb in length and encode common, but diverse, *gag, pol, vif,* and *env* genes. The *gag* gene encodes, from NH_2- to COOH-terminal, matrix, capsid and nucleocapsid proteins in both lentiviruses. HIV has, in addition, a p6 protein expressed at the COOH-terminal of Gag, whereas FIV encodes a p2 peptide at this location. The *pol* gene of FIV encodes a polyprotein consisting of, from NH_2- to COOH-terminal, protease, reverse transcriptase, deoxyuridine triphosphatase, and integrase proteins. HIV encodes protease, reverse transcriptase, and integrase proteins, but lacks deoxyuridine triphosphatase. Both FIV and HIV encode regulators of virion expression (*REV*) elements, but only HIV encodes the transactivator, *TAT*. FIV encodes a short open reading frame, *ORF* 2, at the approximate location of the *VPR* element of HIV, but lacks ORFs corresponding in location to *VPU* and *NEF*. There is no indication that ORF 2 and VPR share any common function. In fact, the functions of all the short Orfs, including ORF 2 of FIV and *VPR, VPU,* and *NEF* of HIV, as well as *VIF* in both viruses, have yet to be defined. However, it is likely that all are required for unrestricted growth and replication in vivo

divergence between FIV and HIV-1 makes the similarities in etiology all the more striking. The differences also serve to highlight areas of the genome and viral functions that are most likely to be involved directly in producing the observed phenotypes.

4 Neurological Findings in FIV-Infected Cats

The feline model provides an opportunity to study a broad spectrum of electrophysiological events shown to be perturbed during human HIV-1 infection. Most importantly, these events can be examined in the acute phase of infection, in animals in whom the history of infection is precisely known. FIV has an unidentified CNS cellular tropism, but has exhibited lymphocytic actions identical to what has been observed in HIV-1 infected individuals (PEDERSON et al. 1987; YAMAMOTO et al. 1988, 1989). Preliminary analysis of brain tissues from infected cats has revealed the presence of gliosis, glial nodules, and infiltration of perivascular endothelia, similar to what has been observed in humans infected with HIV-1, vida ante. Although the viruses from the cat and human differ somewhat in genomic organization, as discussed above, the fact that they cause similar clinical disease patterns suggests that the study of the feline model will rapidly yield answers directly applicable to understanding the human form of the disease.

Both animal and human data indicate functional pathology of both brainstem and cortically derived sensory evoked potentials that are early signs of CNS involvement (MCALLISTER et al. 1992; PODELL et al. 1991; WHEELER et al. 1991). WHEELER and colleagues (1991) observed electrophysiologic abnormalities in evoked spinal cord potentials from cats infected for 11 months with the Petaluma strain of FIV. Some 50% of their FIV positive cats had nervous system dysfunction, comprised of abnormally slow sensory and motor nerve conduction velocities consistent with polyneuropathy. PODELL and colleagues (1993) and LAFRADO and colleagues (1993) have also presented neurodiagnostic evaluations of felines infected with the Mt. Airy FIV strain (FIV-MA) 12–14 weeks after infection. These authors reported that brainstem auditory evoked responses (BAERS) and visual evoked potentials (VEPS) were abnormal in the FIV infected group compared to age matched control cats. Delays in interpeak BAER latencies and a prolongation in peaktime of specific positive waveforms were observed 3 months after infection and prolonged sequential positive peak waveforms were also observed.

These findings suggest pathophysiological processes of both sensory modalities. Finally, EEG analysis of two of five FIV-infected cats demonstrated a higher percentage of suppression-burst and paroxysmal events in the EEGs of infected cats. All of these results mirror findings reported by others regarding HIV-1-

infected individuals (GOODIN et al. 1990; MALESSA et al. 1989; PARISI et al. 1989a, b; ITIL et al. 1990).

Our series of FIV-infected cats also have demonstrated prominent neurological signs (see PHILLIPS et al. 1994). Disease signs from all of our FIV studies are summarized in Table 1. These data include cats that were infected with FIV-MA, FIV-PPR, and FIV-PPR lacking functional DU. FIV-infected animals had substantially more abnormalities in both the physical and neurologic examinations. It is interesting to note that the neurologic abnormalities (delayed pupillary reflex, delayed righting reflex, anisocoria, and posterior paresis) were confined to the FIV-infected animals. All four of these neurologic abnormalities spontaneously reverted to normal in some animals, indicating that the abnormalities were reversible.

BAERs and VEPs were obtained from infected and control felines from 6 months to 13 months postinfection (three separate assessments made). We observed modest delays in both auditory and visually evoked cortically recorded potential in all infected cats compared to controls.

A second cohort of FIV-MA-infected felines was similarly studied with a further focus on arousal/sleep-related abnormalities. This analysis group consisted of 13 SPF female cats, four controls and nine experimental animals. Cats were infected at 10 weeks of age with 1000 $TCID_{50}$ of the FIV-MA strain inoculate i.v. Controls were sham-infected using tissue culture supernatant from uninfected cells. The cohort was assessed for electrophysiological (E.P./EEG) changes 6 months postinfection. Significant delays in conduction velocity were observed for both auditory and visual evoked potentials for this groups of infected cats. Figure 2 illustrates the averaged data taken from the auditory and visual model. These data demonstrate FIV-related delays in the sequential positive peaks of these evoked potentials. Tables 2 and 3 illustrate the quantitive data for these results. Strong individual differences in the latencies of some wave form peaks were observed. Significant differences were noted in the auditory P1, P3, and P4 waves and the amplitude and latency of the visual P3 wave.

Table 1. Distribution of clinico-neurological signs for FIV-infected vs control subjects compiled across three studies

Clinical signs	FIV cats (n=21)	Control cats (n=12)
Lymphadenopathy	21	1
Rough hair coat	11	
Anisocoria	10	
Delayted righting reflex	7	
Oral ulcers	6	1
Delayed pupillary reflex	4	
Hind limb paresis	2	
Conjunctivitis	2	
Abdominal distension	1	
Feline acne	1	
Cutanteious ulcers	1	1
Eczema	1	

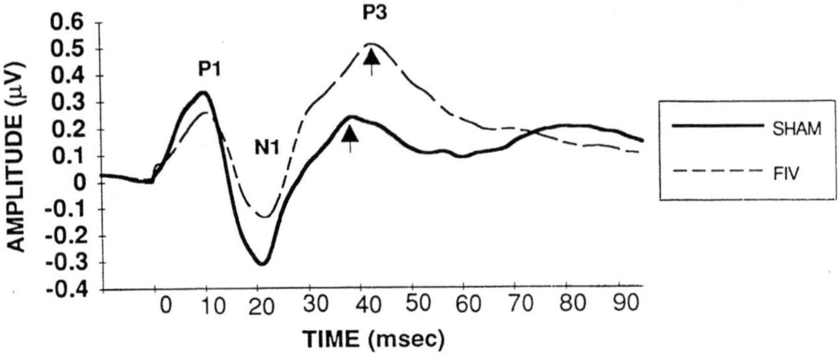

a

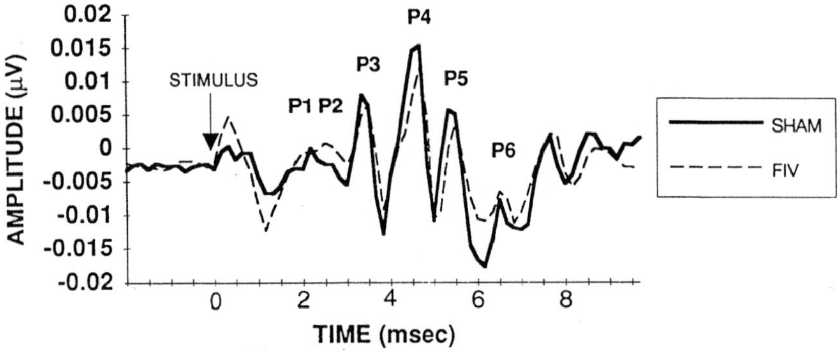

b

Fig. 2a,b. Cortically recorded auditory **(a)** and visual **(b)** evoked potentials for control (sham) and FIV-treated cats. Mean brainstem visual evoked events are illustrated for a group of four control and nine FIV-MA infected cats using the paradigms described below. Felines were fitted with scalp needle electrodes (subcutaneous) to differentially record both brainstem auditory evoked **(a)** and visual **(b)** evoked potentials. A computer program was used to generate the stimuli and to collect the data. On-line averaging of the signals allowed us to repeat trials within a recording session to determine stability of the evoked responses. For brainstem auditory (BAER) and visual evoked events (VEPs) simulation was accomplished in ketamine anesthetized cats. Responses to 100 stimuli (visual stimuli) and 1024 (auditory stimuli) were averaged for each subject in each recording session. One replication was done for each experiment. Significant delays were observed in specific components of the FIV-infected cats in both auditory and visual evoked responses compared to controls. Particularly striking is the delay in the auditory P1, P3 and P4 waves and the visually evoked P3 wave (*arrows*) in the FIV-infected group. Similar findings have been reported by PODELL et al. (1993)

In addition, when comparing individual traces, it was clear that the felines with the most prominent clinical neurological signs (e.g., anisocoria, loss of righting reflex and behavioral signs) had the most delayed evoked response peaks.

We have also infected a third cohort of felines with a molecular clone of FIV-PPR, isolated from a cat in the San Diego area (PHILLIPS et al. 1990). This viral isolate has been cloned and sequenced, and molecularly defined defective variants have been produced (WAGAMAN et al. 1993). Four felines have been infected with the wild-type cloned virus isolate and four with a dUTPase (DU)

Table 2. Quantitative data for Brainstem auditory evoked potentials: FIV-infected subjects vs controls

	P1		P2		P3		P4		P5	
	Lat (ms)	Amp (μV)	Lat (ms)	Amp (μV)	Lat (ms)	Amp (μV)	Lat (ms)	Amp (μV)	Lat (ms)	Amp (μV)
Normal	2.6 ±0.03	5.6 ±1.9	3.3 ±0.1	1.1 ±0.9	4.1 ±0.1	14.5 ±5.2	5.6 ±0.1	28.2 ±11.3	6.6 ±0.1	18.2 ±7.5
FIV	2.99* ±0.2	4.51 ±2.6	3.54 ±0.2	1.94 ±2.2	4.32* ±0.1	9.87 ±5.3	5.72* ±0.03	19.66 ±12	6.68 0.1	17.58 ±9

*$p < 0.05$.

Table 3. Quantitative data for visual evoked potentials: FIV-infected subjects vs controls

	P1		N1		P3	
	Lat (ms)	Amp (μV)	Lat (ms)	Amp (μV)	Lat (ms)	Amp (μV)
Normal	11.9 ±0.7	0.33 ±0.2	26.0 ±0.9	0.67 ±0.09	44.95 ±1	0.59 ±2
FIV	13.4* ±0.5	0.22 ±0.09	26.69 ±0.7	0.43* ±0.2	50.87* ±0.7	0.67 ±0.3

*$p < 0.05$.

mutant variant. Mutation in DU results in a strain that fails to establish a productive infection in primary macrophages (WAGAMAN et al. 1993). DU⁻ FIVs do infect macrophages, but only low levels of infectious virus are released from the cells. Since macrophages are thought to be a primary route of infection of the CNS, it was postulated that DU^- FIVs might exhibit attenuated neurovirulence. Compared to age matched controls, both FIV-infected groups (wild type and the DU mutant) demonstrated abnormalities in sensory evoked potentials in a subgroup of the infected cats at 5 months postinfection, but only in the visual modality. However, sensory-motor conduction velocities assessed during collaborative studies with David Wheeler, Colorado State University, have indicated substantial abnormalities in three out of the four animals infected with the wild-type virus and no changes in the DU^- FIV infected animals. Figure 3 demonstrates potentials for both wild-type and DU^- FIV treatment groups compared to the uninfected cats. These early alterations in neurobiological signs suggest that some sensory modality selectivity may occur in the progression of pathophysiological signs following FIV infection. In addition, it also suggests that individual differences exist in the penetration of neurological signs between infected subjects, thereby mimicking HIV-1 infection of humans. The results also show that DU^- FIVs are attenuated to a degree but are still able to influence CNS function, in spite of their reduced macrophage replication.

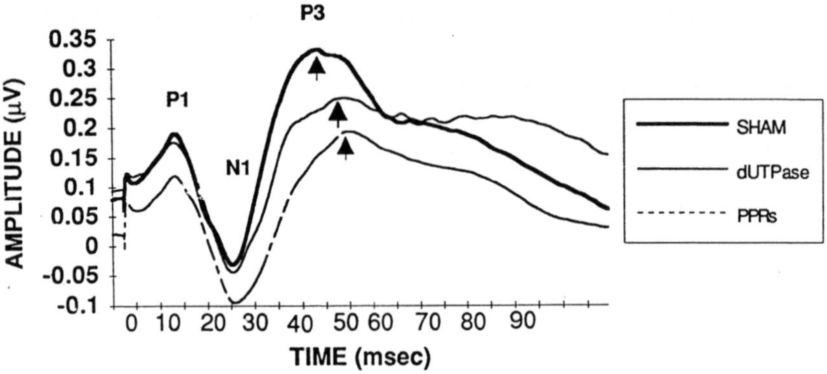

Fig. 3. Visual evoked responses in cats as a function of infection by FIV-PPR or FIV-PPR deficient in dUTPase (DU⁻ FIV-PPR). Data were collected as described in Fig. 2. Note delayed P3 wave latency for cats infected with either FIV-PPr or DU⁻ FIV-PPR, as compared to controls. Unlike brainstem auditory evoked responses which represent short latency peripheral and early brainstem synaptic events, these cortically recorded visual potentials are measures of long latency pathophysiology generated from diencephalic and cortical sites

5 Sleep and Arousal Patterns and FIV Infection

Several recent reports have suggested that alteration in the distribution and quantity of specific sleep states is an early marker for HIV-1 pathology in humans (Aldrich et al. 1988; St. Kubicki et al. 1988; Rothenberg et al. 1990; Norman et al. 1988, 1990, 1992). Indeed, our own human sleep studies have generally confirmed these findings and added a new interpretation to these data, suggesting a selective effect of HIV-1 infection on slow frequencies in the EEG (Darko and colleagues, personal communication). Other findings include a distortion in non-REM (NREM)/REM sleep cycles, and increase in the total percent of slow wave sleep (SWS) and the displacement of SWS toward the last half of the recording period. Given the proposed relationships between SWS and immune function (Krueger et al. 1989; Moldofsky et al. 1986, 1989; Toth and Krueger 1990), these findings support a potential interaction between cytokines and sleep processes (see also Krueger and Majde 1994).

In order to evaluate potential similarities between the reported human sleep disturbance with the FIV model, we studied the sleep/wakefulness patterns of a selected group of our FIV-infected cats. Complete sleep analysis was accomplished in nine infected and four age-matched control cats. Our results have confirmed our early observations of significant sleep abnormalities in these FIV-infected cats (Phillips et al. 1994). As seen in Table 4, results indicate that the FIV-infected cats spent more time awake than the sham-inoculated controls. In addition, these infected cats showed a substantial reduction of rapid eye movement (REM) sleep. The latency to sleep and REM sleep onset was significantly delayed as well in FIV-infected cats. Figure 4 shows representative

Table 4. Table of quantitative sleep measures obtained in control (n=4) and FIV infected (n=9) felines Mean ± SEM of sleep parameters

	Sham-inoculated	FIV-inoculated
Waking(%)	27.36 ± 3.3	53.22 ± 4.33*
SWS1 (%)	13.02 ± 4.24	11.58 ± 2.25
SWS2 (%)	40.76 ± 3.91	32.58 ± 8.13
REM sleep (%)	12.82 ± 4.79	6.82 ± 2.31*
SLEEP latency (min)	47.87 ± 26.3	86.62 ± 21.312*
REM latenay (min)	114.9 ± 35.87	176.28 ± 48.97*
REM latenay from sleep onset (min)	68.45 ± 15.0	90.16 ± 34.5
REM duration (min)	6.16 ± 1.52	3.71 ± 0.9*
REM frequency (total number/7.5 h)	11.0 ± 0.0	7.0 ± 2.15

Note large increase in waking and sleep (and REM) latency in the FIV-infected group. In addition, FIV-infected felines demonstrated reduced REM time and REM as a percent of total sleep.
*p<0,01; sws, slow wave sleep.

hypnograms of a control and FIV-infected cat. These hypnograms demonstrate the marked sleep architecture changes accompanying FIV infection. In addition, the FIV-infected cats demonstrated a remarkable increase in cortically recorded spindle activity during nonrapid eye movement (NREM) sleep, which was analyzed by spectral techniques (Fig. 5).

It has been suggested that most, if not all, of the neurophysiological alterations observed in humans and in animal models of HIV-1 may be due to the toxic manifestations of the virus envelope proteins, i.e., gp120 (SU) (Lipton 1992). Indeed, despite the observation that viral infection can be very host-specific, the action of SU seems to have no restriction of host sensitivity and has been shown to have the ability to damage the brain of different species (Lipton 1991, 1992). For example, picomolar concentrations of HIV-1 SU induces cellular death of ciliary neurons from chicken embryos in tissue culture and retinal ganglion cells from rats (Lipton 1991, Lipton et al. 1991). In addition, i.c.v. administration of HIV-1 SU impairs subsequent performance of adult rats in the Morris water maze (Glowa et al. 1992) and reduces glucose utilization in the suprachiasmatic nucleus (Raymond et al. 1989).

In order to determine if FIV envelope proteins can perturb central nervous system function of other species, we have purified and administered FIV SU protein to rats in order to assess its effect on the sleep-wake cycle. An Epstein-Barr virus envelope glycoprotein (gp105; Wickham and Nemerow 1993) was employed as a negative control for recombinant glycoprotein expressed and purified in an identical manner from insect cells.

Compared to saline and a fragment of the Epstein-Barr virus envelope glycoprotein (gp105) serving as controls, administration of 200 ng of FIV SU resulted in a significant increase of rodent wakefulness and a decrease in REM sleep throughout the 4 h of postinjection recordings (Prospero-Garcia et al. 1994; Table 5). These findings support and extend the sleep studies we have reported

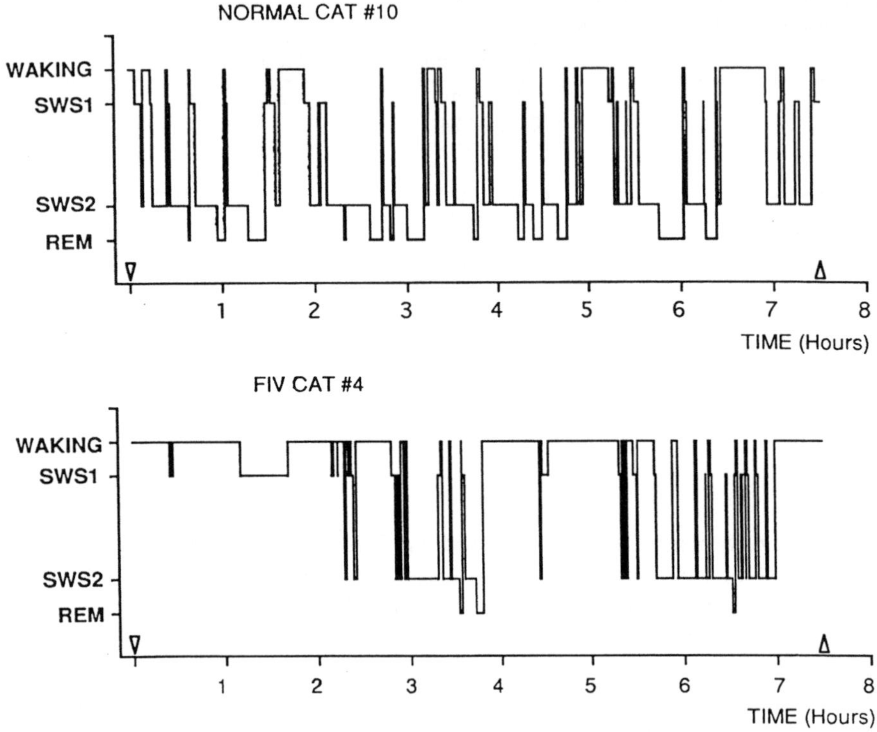

Fig. 4. Differences in sleep architecture for a representative control (# 10) and FIV-infected (# 4) feline. Some 10–12 months after the FIV inoculation, cats were surgically implanted under halothane anesthesia with a standard electrode array to record the sleep-wake cycle. Following 2 weeks from surgical recovery cats were acclimatized to the recording conditions for 5 days. Sleep recordings (8:00 am to 3:30 pm) were collected for 3 consecutive days in each subject. The sleep hypnograms were obtained from the third day of recording. Note the displacement of sleep to the last half of the recording period for the FIV-infected subject and the occurrence of more state changes compared to the control subject. See Table 4 for quantitative sleep/wakefulness measures

in the FIV-infected cats (discussed above), in which we have observed similar abnormalities associated with viral infection.

6 Neuropathological Signs

Although the target cells and the mechanism(s) of viral injury within the CNS are not known for FIV, it is clear that infectious virus can be recovered from the brain and CSF (Dow et al. 1990; Phillips et al. 1994). In the study by Dow and colleagues (1990) virus was isolated from the CNS from five out of ten naturally infected cats, with the presence of FIV antibodies in the CSF in nine of the ten animals. Dow and coworkers were the first to report that cats experimentally infected with FIV

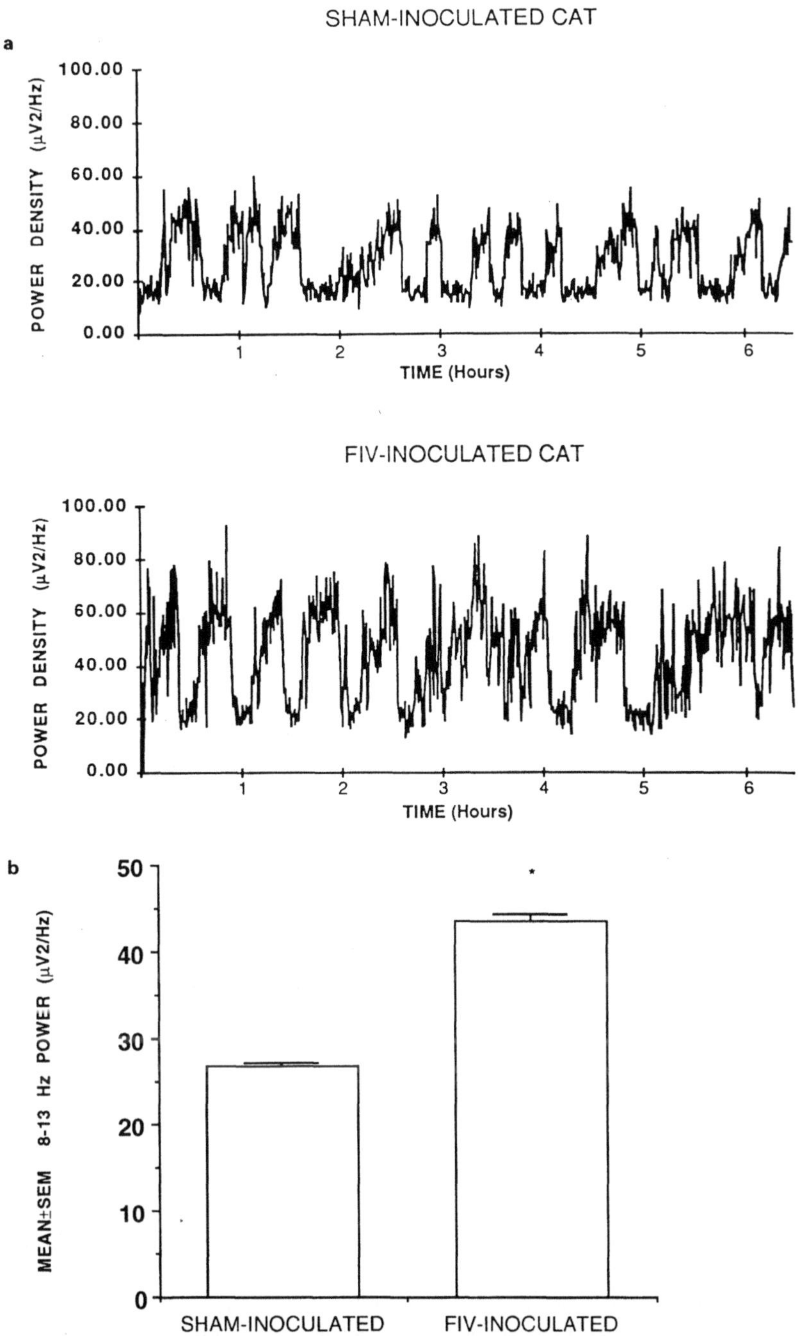

Fig. 5a,b. EEG power spectral band time series **(a)** of cortical EEG in sham inoculated and a FIV-infected cat. Note the increase in relative EEG power in the 8–13 Hz frequency band during sleep in the infected subject. Although the periods of waking-related EEG (*troughs*) were about the same in both groups, the FIV-infected cat had significantly larger and more prolonged periods of 8–13 Hz activity during drowsy and sleep-related EEG (*peaks*). **b** Quantitative changes occurring in the EEG power in this frequency band for these two cats. These EEG changes were characterized by a large increase in sleep spindle frequency indentified in the raw EEG during slow wave sleep in most infected subjects recorded to date

Table 5. Quantitative comoparisons between the sleep measures of rodents injected intraventricularly with either saline (injection control), EB gp105 (protein control), or the FIV SU-Env proteinmean±sem of sleep parameters

	Saline	FIV SU-Env	EB gb105
Waking (%)	24 ± 1.52	40 ± 4.4*	23.04 ± 3.0
SWS1 (%)	13.46 ± 2.74	11.46 ± 3.0	12.64 ± 3.22
SWS2(%)	51.28 ± 2.17	49.12 ± 6.08	55.68 ± 4.14
REM sleep(%)	10.42 ± 3.6	2.82 ± 1.21*	8.28 ± 1.16
Sleep latency (min)	8.82 ± 2.6	33.5 ± 7.58*	20.14 ± 3.8*
REM latency (min)	22.1 ± 4.1	80.7 ± 23.9*	44.4 ± 2.44*
REM latency from sleep onset (min)	13.76 ± 1.8	71.2 ± 15.2*	25.9 ± 4.0
REM duration (min)	1.84 ± 0.16	1.0 ± 0.3*	1.45 ± 0.12
REM frequency (total number/4h)	14.2 ± 1.43	4.6 ± 1.96*	15.4 ± 1.6

As compared to both controls, the FIV SU-Env injection elicited a significant increase in total waking and REM sleep latency and a decrease in REM sleep percentage and duration. These changes in quantitive sleep measures are similare to our observed changes in feline sleep following FIV infection and comparable to reported changes in the sleep of human HIV-infected subjects.
*$p < 0.01$; sws, slow wave sleep

developed a productive infection of the brain when inoculated by either the intracerebral or peripheral route (Dow et al. 1990). Virus was isolated from tissue explants from the cerebral cortex, caudate nucleus, midbrain, cerebellum, rostral and caudal brainstem, but not from the choroid plexus (Dow et al. 1990). Since the virus was not isolated from the choroid plexus, it is unlikely that contaminating blood was the source of the virus. Primary cultures of microglia and astrocytes were productively infected with FIV in vitro. However, similar cultures of neurons, oligodendrocytes and choroid plexus cells showed no evidence of infection. FIV was cytopathic to the astrocytes in culture, with syncytia forming after 3 days. However, no cytotoxicity was observed in the productively infected microglia cultures (Dow et al. 1990). HIV-1 also infects, in vitro, both astrocyte (Dewhurst et al. 1987; Erfle et al. 1991) and microglia (Sharpless et al. 1992) cultures. Similar to HIV-1 (Sharpless et al. 1992), the ability of FIV to infect CNS cell types is clone-specific, as we have shown that the 34TF10 clone of FIV productively infects an astrocyte cell line whereas the PPR clone does not (Phillips et al. 1990).

Pathologic examination of two cats, at 7 months postinoculation, revealed perivascular infiltrates of mononuclear cells, diffuse gliosis, glial nodules and white matter pallor. These lesions were primarily located in the caudate, midbrain, and rostral brain stem (Dow et al. 1990). Similarly, in our examination of infected, healthy animals, histopathological changes have been found in the CNS. Perivascular mononuclear cell infiltrates, and focal glial nodules were present in diffuse regions of the brain (Phillips et al. 1994). A recent more comprehensive neuropathological study examined the time course of the FIV-induced CNS lesions and confirmed many of the above findings (HURTREL et al. 1992). This study demonstrated that CNS lesions were seen as early as 1 month postinfection, when the virus was given intracerebrally, and as early as 2 months postinfection when the virus was inoculated intravenously. These results indicate that CNS entry for FIV is an early event, even when inoculated peripherally. Similarly, early penetration of HIV-1 into the human CNS (DAVIS et al. 1992) and SIV into monkey CNS (CHAKRABARTI et al. 1991; LACKNER et al. 1991) also occurs.

In cats that were coinfected with feline leukemia virus, the neurologic lesions were more frequent and severe (HURTREL et al. 1992). This supports a previous finding that feline leukemia virus enhances FIV disease (PEDERSON et al. 1990), similar to the interactions of HIV-1 with human T cell leukemia virus or cytomegalovirus. Multinucleated giant cells were found in only one cat. Generally, the neurologic lesions in the cats were less severe than lesions reported from the human AIDS autopsies. The authors suggested these differences may reflect the early stage of FIV disease, late asymptomatic to early ARC, compared to terminal AIDS stage for most of the human data. Primarily, FIV-infected cats die or are euthanized in the early ARC stage of the disease, as aggressive treatment of opportunistic infections is usually not attempted. People, however, often go through multiple treatments for opportunistic infections prior to death, thus allowing for continual advancement of their neurologic disease. Neuropathologic data are available from only a few human cases in the late asymptomatic or early ARC stages of the disease. From these cases it is clear that the neurologic lesions are less severe and resemble the reported pathologic findings with FIV: mild chronic lymphocyte meningitis (DAVIS et al. 1992; LENHARDT et al. 1988), perivascular inflammation (DAVIS et al. 1992; GRAY et al. 1992), myelin pallor, and gliosis (GRAY et al. 1992; MCARTHUR et al. 1989). Thus, the difference in the severity between neurologic lesions of FIV and HIV-1 are likely to result from the stage of the disease at the time of neuropathologic examination.

7 Behavioral Signs

In addition to the neurological signs reported for FIV infection discussed above, there have been several accounts of behavioral abnormalities associated with infection. These reports have included signs of motor disturbances such as "compulsive roaming" and circling behaviors (PEDERSON et al. 1987; WHEELER et al. 1991; YAMAMOTO et al. 1988). In our own studies, behavioral assessments of FIV-infected cats have been made in an informal fashion. Felines obtained at 5 months of age lived in separate group housing. These rooms, approximately 120 sq. ft. each, are fitted with appropriately placed Plexiglass shelves, scratching posts and various play articles. In addition there are numerous large animal "transport" cages that the cats may use for sleeping if they wish. The rooms are on a 12 h light/dark cycle. The doors to the rooms are fitted with a small window in order to allow observation of the cats from outside the rooms. During weekly physical and neurological exams and during periods of electrophysiological assessment, we have noted that the neurological measures have been, to a degree, paralleled by behavioral changes best be described as a growing neophobia and reticence of handling in the FIV-infected cats. This has been seen to a greater or lesser degree in all the infected cats. Lack of affection, fearfulness, and nonaggressive escape behavior remains to be noted for the infected cats. More sophisticated behavioral tests have yet to be employed to assess cognitive function, either by us or other groups investigating FIV-infected cats.

8 Summary

Feline immunodeficiency virus infects the CNS and results in predictable pathophysiology strikingly similar to that seen with HIV-1 infection of humans. The observed pathophysiology is mimicked in several physiologically assessed modalities, further supporting the validity of the feline model. Peripheral and control evoked potential findings and the occurrence of the sleep architecture changes in both cat and human disease provide an intriguing focus for further investigation. Although structurally diverse in an absolute sense, FIV and HIV-1 share basic structural features and commonalities of their life cycle. It is likely that by understanding the common mechanisms by which these lentiviruses influence CNS function, a more complete understanding of the neurological deficits seen in HIV-1-infected patients will be obtained.

The cat model is particularly valuable for study of CNS disease, since it allows detailed analyses of events during the acute phase of infection, under circumstances in which the nature and timing of the infection are carefully controlled. The availability of molecular clones for mutational analysis will facilitate mapping of genomic regions critical to the perturbation of CNS function. It is suggested that development of intervention strategies in the cat model will yield treatment modalities directly applicable to HIV-1 infection of humans.

Acknowledgments. The authors wish to thank P. Wagaman, D. Pauoi, D. Lerner, and D. Wheeler for efforts related to work contributing to this chapter. This work was supported by grant 1P50 MH47680 from the National Institute of Mental Health, the National Institutes of Health.

References

Ackley CD, Yamamoto JK, Levy N, Pedersen NC, Cooper MD (1990) Immunologic abnormalities in pathogen-free cats experimentally infected with feline immunodeficiency virus. J Virol 64: 5652–5655

Aldrich MS, Rogers AE, Angell K (1988) Excessive daytime sleepiness as a presenting manifestation of HIV infection. Sleep Res 17: 137

Brenneman DE, Westbrook GL, Fitzgerald SP, Ennist DL, Elkins KL, Ruff MR, Pert CB (1988) Neuronal cell killing by the envelope protein of HIV and its prevention by vasoactive intestinal peptide. Nature 335: 639–642

Brown WC, Bissey L, Logan KS, Pedersen NC, Elder JH, Collisson EW (1991) Feline immunodeficiency virus infects both $CD4^+$ and $CD8^+$ T lymphocytes. J Virol 65: 3359–3364

Chakrabarti L, Hurtrel M, Maire M-A, Vaseux R, Dormont D, Montagnier L, Hurtrel B (1991) Early viral replication in the brain of SIV-infected rhesus monkeys. Am J Pathol 139: 1273–1279

Davis LE, Hjelle BL, Mille VE, Palmer DL, Llewellyn AL, Merlin TL, Young SA, Mills RG, Wachsman W, Wiley CA (1992) Early viral brain invasion in iatrogenic human immunodeficiency virus infection. Neurology 42: 1736–1739

Dawson S, Smyth NR, Bennett M, Gaskell RM, McCracken CM, Brown A, Gaskell CJ (1991) Effect of primary-stage feline immunodeficiency virus infection on subsequent feline calicivirus vaccination and challenge in cats. AIDS 5: 747–750

Dewhurst S, Dakai K, Bresser J, Stevension M, Evinger-Hodges MJ, Volsky DJ (1987) Persistent productive infection of human glia cells by human immunodeficiency virus (HIV) and by infectious molecular clones of HIV. J Virol 61: 3774–3782

Dow SW, Poss ML, Hoover EA (1990) Feline immunodeficiency virus: a neurotropic lentivirus. J AIDS 3: 658–668

Dow SW, Dreitz MJ, Hoover EA (1992) Feline immunodeficiency virus neurotropism: evidence that astrocytes and microglia are the primary target cells. Vet Immunol Immunopathol 35: 23–35

Egberink HF, Berrocal A, Bax HA, van den Ingh TS, Walter JH, Horzinek MC (1989) Feline immunodeficiency virus infection. Vet Immunol Immunopathol 21: 111–129

Elder JH, Phillips TR (1994) Molecular properties of feline immunodeficiency virus (FIV). Infect Agents Dis 2: 361–374

Elder GA, Sever JL (1988) AIDS and neurological disorders: an overview. Ann Neurol 23 [Suppl]: S4–6

Elder JH, Lerner DL, Hasselkus-Light CS, Fontenot DJ, Hunter E, Luciw PA, Montelaro RC, Phillips TR (1992) Distinct subsets of retroviruses encode dUTPase. J Virol 66: 1791–1794

Erfle V, Stoeckbauer P, Kleinschmidt A, Kohleisen B, Mellert W, Stavrou D, Brack-Werner R (1991) Target cells for HIV in the central nervous system: macrophages or glial cells? Res Virol 142: 139–144

Gendelman HE, Baca LM, Husayni H, Turpin JA, Skillman D, Kalter DC, Orenstein JM, Hoover DL, Meltzer MS (1990) Macrophage-HIV interaction: viral isolation and target cell tropism. AIDS 4: 221–228

Glowa JR, Panlilio LV, Brenneman DE, Gozes I, Fridkin M, Hill JM (1992) Learning impairment following intracerebral administration of the HIV envelope protein gp120 or a VIP antagonist. Brain Res 570: 341–346

Goodin DS, Aminoff MJ, Chernoff DN, Hollander H (1990) Long latency event-related potentials in patients infected with human immunodeficiency virus. Ann Neurol 27: 414–419

Grant I, Martin A (eds) (1994) Neuropsychology of HIV infection. Oxford University Press, New York

Gray F, Lescs MC, Keihane C, Parair F, Marc B, Durigon M, Gherardi R (1992) Early brain changes in HIV infection: neuropathological study of 11 HIV seropositive, non-AIDS cases. J Neuropathol Exp Neurol 51: 177–185

Handelsman L, Aronson M, Maurer G, Wiener J, Song IS, Jacobson J, Bernstein D, Ness R, Herman S, Losonczy M, Holloway K, Horvath T, Donnelly N, Hirschowitz J, Rowan AJ (1992) Neuropsychological and neurological manifestations of HIV-1 dementia in drug users. J Neuropsych Clin Neurosci 4: 21–28

Helmstaedter C, Hartmann A, Niese C, Brackmann H, Sass R (1992) Stage independent and individual courses of neurocognitive deficits in HIV: a follow-up study of 62 HIV positive hemophiliacs. Nervenarzt 62: 88–94

Heyes MP, Rubinow D, Lane C, Markey SP (1989) Cerebrospinal fluid quinolinic acid concentrations are increased in Acquired Immune Deficiency Syndrome. Ann Neurol 26: 275–277

Ho D, Rota J, Schooley R (1985) Isolation of HTLV-III from cerebralspinal fluid and neural tissues of patients with neurologic syndromes related to the acquired immunodeficiency syndrome. N Engl J Med 313: 278–286

Ho DD, Bredesen DE, Vinters HV (1989) The acquired immunodeficiency syndrom (AIDS) dementia complex. Ann Intern Med 111: 400–410

Hoffmann-Fezer G, Thum J, Ackley C, Herbold M, Mysliwietz J, Thefeld S, Hartmann K, Kraft W (1992) Decline in $CD4^+$ cell numbers in cats with naturally acquired feline immunodeficiency virus infection. J Virol 66: 1484–1488

Hurtrel M, Ganiere JP, Guelfi JF, Chakrabarti L, Maire M-A, Gray F, Montagnier L, Hurtrel B (1992) Comparison of early and late feline immunodeficiency virus encephalopathies. AIDS 6: 399–406

Itil T, Ferracuti S, Freedman AM, Sherer C, Mehta P, Itil KZ (1990) Computer-analyzed EEG (CEEG) and dynamic brain mapping in AIDS and HIV related syndrome: a pilot study. Clin Electroencephalogr 21: 140–144

Janssen RS, Cornblath D, Epstein L, Foa R, McArthur J, Price R, Asbury AK, Beckett A, Benson DF, Bridge TO, Leventhal CM, Satz P, Soykin AJ, Sidtis JJ, Tross S (1991) Nomenclature and research case definitions for neurologic manifestations of human immunodeficiency virus-type 1 (HIV-1) infection. Neurology 41: 778–785

Johnson RT, McArthur JC, Narayan O (1988) The neurobiology of human immunodeficiency virus infections. FASEB J 2: 2970–2981

Koralnik IJ, Beaumanoir A, Hausler R, Kohler A, Safran AB, Delacoux R, Vibert D, Mayer E, Burkhard P, Nahory A, Magistris M, Sanches J, Myers P, Paccolat F, Quoëx F, Gabriel V, Perrin L, Mermillod B, Gauthier G, Waldvogel F, Hirshel B (1990) A controlled study of early neurologic abnormalities in men with asymptomatic human immunodeficiency virus infection. N Engl J Med 13: 864–870

Krueger J, Majde J (1994) Microbial products and cytokines in sleep and fever regulation. Crit Rev Immunol 14: 355–379

Krueger J, Obal F, Johanssen L, Cady A, Toth L (1989) Endogenous slow wave sleep substances: a review. In: Wacquier A, Dugovic C, Radulovacki M (eds) Slow wave sleep: physical, pathophysiological and functional aspects. Raven, New York, pp 75–89

Lackner AA, Smith MO, Munn RJ, Martfeld DJ, Gardner MB, Marx PA, Dandekar S (1991) Localization of simian immunodeficiency virus in the central nervous system of rhesus monkeys. Am J Pathol 139: 609–621

Lafrado L, Podell M, Krakowka S, Hayes K, Hanlon M, Mathes L (1993) FIV: a model for retrovirus-induced pathogenesis. AIDS Res Rev 3: 115–150

Lenhardt TM, Super MA, Wiley CA (1988) Neuropathological changes in an asymptomatic HIV seropositive man (letter). Ann Neurol 23: 209–210

Lin DS, Bowman DD, Jacobson RH (1992) Immunological changes in cats with concurrent Toxoplasma gondii and feline immunodeficiency virus infections. J Clin Microbiol 30: 17–24

Lipkin W, Parry G, Kiprov D (1985) Inflammatory neuropathy in homosexual men with lymphadenopathy. Neurology 35: 1479–1493

Lipton SA (1991) Calcium channel antagonists and human immunodeficiency virus coat protein-mediated neuronal injury. Ann Neurol 30: 110–114

Lipton SA (1992) Models of neuronal injury in AIDS: another role for NMDA receptor? TINS 15: 75–79

Lipton SA, Sucher NJ, Kaiser PK, Dreyer EB (1991) Synergistic effects of HIV coat protein and NMDA receptor-mediated neurotoxicity. Neuron 7: 111–118

Maddon PJ, Dagleish AG, McDougal JS, Clapham PR, Weiss RA, Axeel R (1986) The T4 gene encodes the AIDS virus receptor and is expressed in the immune sysetm and the brain. Cell 47: 333–348

Malessa R, Heuser L, Brockenmeyer N (1989) Evoked potentials in neurologically asymptomatic persons during the early stages of HIV infection. EEG-EMG 20: 257–266

Matsumura S, Ishida T, Washizu T, Tomoda I, Nagata S, Chiba J, Kurata T (1993) Pathologic features of acquired immunodeficiency-like syndrome in cats experimentally infected with feline immunodeficiency virus. J Vet Med Sci 55: 387–394

McAllister RH, Herns MV, Harrison MJG, Newman SP, Connolly S, Fowler CJ, Fell M, Durrance P, Manji H, Kendall BE, Valentine AR, Weller IVD, Adler M (1992) Neurological and neuropsychological performance in HIV seropositive men without symptoms. J Neurol Neurosurg Psychiatry 55: 143–148

McArthur JC (1994) Neurological and neuropathological manifestations of HIV infection. In: Grant I, Martin A (eds) Neuropsychology of HIV infection. Oxford University Press, New York, pp 56–107

McArthur JC, Becker PS, Parisi JE, Trapp B, Selnes OA, Cornblath DR, Balakrishnan J, Griffin JW, Price D (1989) Neuropathological changes in early HIV-1 dementia. Ann Neurol 26: 681–684

Moldofsky H, Lue F, Eisen J, Keystone E, Gorczynski R (1986) The relationship of interleukin-1 and immune functions to sleep in humans. Psychosom Med 48: 309–318

Moldofsky H, Lue F, Davidson J, Gorczinski R (1989) Effects of sleep deprivation on human immune functions. FASEB J 3: 1972–1977

Navia BA, Jordan BD, Price RW (1986a) The AIDS dementia complex: I. clinical features. Ann Neurol 19: 517–524

Navia BA, Cho ES, Petito CK (1986b) The AIDS dementia complex: II. Neuropathology. Ann Neurol 19: 525–535

Norman SE, Resnick L, Cohn MA, Duara R, Herbst J, Berger JR (1988) Sleep disturbances in HIV-seropositive patients (letter). JAMA 260(7): 922

Norman S, Demirozu M, Chediak A (1990) Disturbed sleep architecture (NREM/REM sleep cycles) in HIV-infected healthy men. Sleep Res 19: 339

Norman SE, Chediak AD, Freeman C, Kiel M, Mendez A, Duncan R, Simoneau J, Nolan B (1992) Sleep disturbances in men with asymptomatic human immunodeficiency (HIV) infection. Sleep 15(2): 50–55

Nuovo GJ, Gallery F, MacConnell P, Braun A (1994) In situ detection of polymerase chain reaction-amplified HIV-1 nucleic acids and tumor necrosis factor-α RNA in the central nervous system. Am J Pathol 144: 659–666

Olmsted RA, Hirsch VM, Purcell RH, Johnson PR (1989) Nucleotide sequence analysis of feline immunodeficiency virus: genome organization and relationship to other lentiviruses. Proc Natl Acad Sci USA: 86 8088–8092

Parisi A, Di PG, Strosselli M (1989a) Usefulness of computerized electroencephalography in diagnosing, staging and monitoring of AIDS-dementia complex. AIDS 3(4): 209–213

Parisi A, Strosselli M, and DiPerri G (1989b) Electroencephalography in the early diagnosis of HIV-related subacute encephalities: analysis of 185 patients. Clin Electroencephologr 20: 1–5

Pedersen NC, Ho EW, Brown ML, Yamamoto JK (1987) Isolation of a T-lymphotropic virus from domestic cats with an immunodeficiency-like syndrome. Science 235: 790–793

Pedersen NC, Yamamoto JK, Ishida T, Hansen H (1989) Feline immunodeficiency virus infection. Vet Immunol Immunopathol 21: 111–129

Pedersen NC, Torten M, Rideout B, Sparger E, Tonachini T, Luciw PA, Ackley C, Levy N, Yamamoto J (1990) Feline leukemia virus infection as a potentiating cofactor for the primary and secondary stages of experimentally induced feline immunodeficiency virus infection. J Virol 64: 598–606

Perry SW (1990) Organic mental disorders caused by HIV-1: update on early diagnosis and treatment. Am J Psychiatry 147: 696–710

Phillips TR, Talbott RL, Lamont C, Muir S, Lovelace K, Elder JH (1990) Comparison of two host cell range variants of feline immunodeficiency virus. J Virol 64: 4605–4613

Phillips TR, Lamont C, Konings DA, Shacklett BL, Hamson CA, Luciw PA, Elder JH (1992) Identification of the Rev transactivation and Rev responsive elements of feline immunodeficiency virus. J Virol 66: 5464–5471

Phillips TR, Prospero-Garcia O, Puaoi DL, Lerner DL, Fox HS, Olmsted RA, Bloom FE, Henriksen SJ, Elder JH (1994) Neurological abnormalitites associated with feline immunodeficiency virus infection. J Gen Virol 75: 979–987

Podell M, Olmsted R, Krakowka S, Mathes L, Oglesbee M, Lafrado L (1991) Neurodiagnostic evaluation in cats experimentally infected with feline immunodeficiency virus. In: 1st International Conference of Feline Immunodeficiency Virus Researchers. Davis, California, USA, p 36

Podell M, Oglesbee M, Mathes L, krakowka S, Olmsted R, Lafrado L (1993) AIDS-associated encephalopathy with experimental feline immunodeficiency virus infection. J AIDS 6: 758–771

Price RW, Perry S (eds) (1994) HIV-1, AIDS, and the brain. Raven, New York

Prospero-Garcia O, Herold N, Waters AK, Phillips TR, Elder JH, Henriksen SJ (1994) Intraventricular administration of a FIV-envelope induces sleep architecture changes in rats. Brain Res

Raymond LP, Kimes AS, Tabakoff B, London ED (1989) AIDS and sleep disorders: effect of gp120 on cerebral clucose metabolism. CRS Soc Biol 183: 407–418

Rothenberg S, Zozula R, Funesti J, McAuliffe V (1990) Sleep habits in asymptomatic HIV-seropositive men. Sleep Res 19: 342

Sharpless NE, O'Brien WA, Verdin E, Kufta CV, Chen IS, Dubois-Dalcq M (1992) Human immunodeficiency virus type 1 tropism for brain microglial cells is determined by a region of the env glycoprotein that also controls macrophage tropism. J Virol 66: 2588–2593

St Kubicki H, Henkes H, Terstegge K, Ruf B (1988) AIDS related sleep disturbances–a preliminary report. In: St Kubicki H, Henkes J, Beinzle U, Pohle HD (eds) HIV and nervous system. Fischer, Stuttgart, pp 97–105

Talbott RL, Sparger EE, Lovelace KM, Fitch WM, Pedersen NC, Luciw PA, Elder JH (1989) Nucleotide sequence and genomic organization of feline immunodeficiency virus. Proc Natl Acad Sci USA 86: 5743–5747

Teng CS (1990) Induction of feline immunodeficiency syndrome by feline leukemia virus: pituitary and adrenocortical dysfunctions. AIDS 4(12): 1219–1224

Tomonaga K, Miyazawa T, Sakuragi J, Mori T, Adachi A, Mikami T (1993) The feline immunodeficiency virus ORF-A gene facilitates efficient viral replication in established T-cell lines and peripheral blood lymphocytes. J Virol 67: 5889–5895

Torten M, Franchini M, Barlough JE, George JW, Mozes E, Lutz H, Pedersen NC (1991) Progressive immune dysfunction in cats experimentally infected with feline immunodeficiency virus. J Virol 65: 2225–2230
Toth LA, Krueger JM (1990) Infectious disease, cytokines and sleep. In: Mancia GM M (ed) The diencephalon and sleep. Raven, New York, pp 331–341
Wagaman PC, Hasselkus-Light CS, Henson M, Lerner DL, Phillips TR, Elder JH (1993) Molecular cloning and characterization of deoxyuridine triphosphatase from feline immunodeficiency virus (FIV). Virology 196: 451–457
Waters L, Hopper CD, Gruffydd-Jones TJ, Harbour DA (1993) Chronic gingivitis in a colony of cats infected with feline immunodeficiency virus and feline calicivirus. Vet Rec 132: 340–342
Wheeler DW, Mitchell TW, Gasper PW, Barr MC, Whalen LR (1991) FIV infection associated with neurologic abnormalities. In: 1st International Conference of Feline Immunodefiency Virus Researchers. Davis, California, USA
Wickham TJ, Nemerow GR (1993) Optimization of growth methods and recombinant protein production in BT1-Tn-5B1-4 insect cells using the baculovirus expression system. Biotechnol Prog 9: 25–30
Wigdahl B, Kunsch C (1989) Minireview: role of HIV in human nervous system dysfunction. AIDS Res Hum Retroviruses 5: 369–374
Willett BJ, Hosie MJ, Jarrett O, Neil JC (1994) Identification of a putative cellular receptor for feline immunodeficiency virus as the feline homologue of CD9. Immunology 81: 228–233
Witt CJ, Moench TR, Gittelsohn AM, Bishop BD, Childs JE (1989) Epidemiologic observations on feline immunodeficiency virus and Toxoplasma gondii coinfection in cats in Baltimore, MD. J Am Vet Med Assoc 194: 229–233
Yamamoto JK, Sparger E, Ho EW, Andersen PR, O'Connor TP, Mandell CP, Lowenstine L, Munn R, Pedersen NC (1988) Pathogenesis of experimentally induced feline immunodeficiency virus infection in cats. Am J Vet Res 49: 1246–1258
Yamamoto JK, Hansen H, Ho EW, Morishita TY, Okuda T, Sawa TR, Nakamura RM, Pedersen NC (1989) Epidemiologic and clinical aspects of feline immunodeficiency virus infection in cats from the continental United States and Canada and possible mode of transmission. J Am Vet Med Assoc 194: 213–220
Yarchoan R, Thomas R, Grafman J, Whichman A, Dalkas M, McAtee N, Berg G, Fischl M, Perno CF, Klecker RW, Buchbinder A, Tay S, Larson SM, Myers CE, Broder S (1988) Long-term administration of 3'-azido-2', 3" dideoxythymidine to patients with AIDS-related neurological disease. Ann Neurol 23: S82–S87

Transgenic Models to Assess the Neuropathogenic Potential of HIV-1 Proteins and Cytokines

L. Mucke[1], E. Masliah[2], and I.L. Campbell[1]

1 HIV-1 Associated Nervous System Damage

Human immunodeficiency virus type 1 (HIV-1) infection continues to spread at a rapid rate and the clinical syndromes associated with it present increasing medical and socioeconomic problems. A significant number of people infected with HIV-1 develop neurological complications, including impairments of cognition and motor performance that can culminate in dementia and paralysis (Working Group of the American Academy of Neurology AIDS Task Force 1991). These functional alterations are accompanied at the structural level by loss of neuronal subpopulations in combination with more widespread degeneration of neuronal dendrites and synapses, reactive astrocytosis, microglial changes and alterations of the blood-brain barrier (BBB) (for review see Sharer 1992; Budka 1992; Spencer and Price 1992; Price and Perry 1994). The effective prevention and treatment of these abnormalities, impossible at the present time, will likely require a better

[1] Department of Neuropharmacology, The Scripps Research Institute, 10666 North Torrey Pines Road, La Jolla, CA 92037, USA
[2] Departments of Neurosciences and Pathology, University of California at San Diego, LA Jolla, CA 92093-0624, USA

understanding of the molecular and cellular processes that underlie HIV-1 associated CNS damage.

2 AIDS Dementia Complex—A Multifactorial Disease Manifestation

Brain damage and neurological impairment is seen in only a proportion of patients with HIV-1 infection of the CNS. This interesting restriction of disease manifestation to a subpopulation of infected patients is not unique to HIV-1 but is seen in a variety of infectious CNS diseases, including those associated with herpes simplex, mumps, measles and polio virus. The intensity of CNS damage in such infections can in general be expected to depend on both neurovirulence-determining properties of the infecting viral strain and specific susceptibilities of the infected host.

Neuroinvasive strains of HIV-1 appear to enter the CNS via infected monocytoid cells and replicate within the brain primarily within macrophages/microglia. While a causal involvement of HIV-1 CNS damage is strongly suggested by the topographic correlation between viral burden and synapto-dendritic damage in the brains of HIV-1 infected patients (MASLIAH et al. 1992a), direct infection of neurons by HIV-1 appears to play an insignificant, if any, role in the pathogenesis of AIDS dementia complex (see PRICE and PERRY 1994, for review). Mechanisms by which HIV-1 could alter neuronal functions indirectly include the infection and impairment of glial cells that provide crucial support for neurons as well as the generation of soluble microbial and host-derived neurotoxins which could diffuse and affect neurons at a distance (for review see MUCKE and EDDLESTON 1993; EPSTEIN and GENDELMAN 1993; PRICE and PERRY 1994).

3 Microbial Products

In vitro studies indicate that the HIV-1 envelope glycoprotein gp120 has detrimental effects on neurons (BRENNEMAN et al. 1988; LIPTON 1992a; GENIS et al. 1992; LO et al. 1992; MÜLLER et al. 1992; SAVIO and LEVI 1993; LEVI et al. 1993; DAWSON et al. 1993; USHIJIMA et al. 1993; SWEETNAM et al. 1993). Notably, macrophage/microglia-derived factors appear to be essential mediators of gp120-induced neurotoxicity (LIPTON 1992b; GIULIAN et al. 1993). In vivo, systemic injection of gp120 has been reported to induce neuronal loss/dendritic damage in the CNS and to retard behavioral development in neonatal rats (HILL et al. 1993). Injection of gp120 into the cerebral ventricles decreased cerebral glucose utilization (KIMES et al. 1991) and impaired learning ability (GLOWA et al. 1992) in adult rats. Peptides representing portions of HIV-1 Tat or Rev have been reported to be lethal when injected intracerebroventricularly into rodents (SABATIER et al. 1991; MABROUK et al. 1991). However, the latter studies provide no information on neuropathological CNS alterations or on the exact cause of death in the injected animals.

Due to the immunodeficiency that characterizes AIDS, patients with advanced HIV-1 infection are prone to opportunistic infections and frequently harbor other microbes, in variable combinations, in their brains including viruses, bacteria, parasites and fungi. While this article focuses on the neuropathogenic potential of HIV-1 products, it is important to bear in mind that a number of these other microbes may also generate factors that could directly or indirectly contribute to the neurological impairment seen in patients with AIDS.

4 Host Factors

Infection of the CNS by HIV-1 is accompained by a compound host response involving elements of both the immune system and the local milieu. At the cellular level this response is characterized respectively, by the presence of T cells and macrophages in the meninges, the CSF and, variably, the parenchyma and by prominent gliosis with activation of both astrocytes and microglia (reviewed in DAL CANTO 1989; SHARER 1992; SPENCER and PRICE 1992; PRICE abd PERRY 1994). When activated, each of these different cell types has the potential to produce an array of soluble factors (such as cytokines, proteases and short lived but highly reactive free radical species) that may help coordinate the antiviral host response and contribute to the repair of tissue injury. However, as these factors are also extremely potent modifiers of both cellular structure and function, besides their beneficial actions they may also possess considerable pathogenic potential.

Dysregulated production of cytokines is a hallmark of HIV-1 infection both in the periphery (ROSENBERG and FAUCI 1990; LEVY 1993) and in the CNS (MERRILL and CHEN 1991; GELEZIUNAS et al. 1992; EPSTEIN and GENDELMAN 1993). Direct assays of CSF demonstrated elevated levels of interleukin (IL)-1, IL-6 and tumor necrosis factor (TNF)-α, while immunocytochemical staining of autopsy tissue revealed increased levels of IL-1, IL-6, TNF-α, and transforming growth factor (TGF)-β (GALLO et al. 1989; WAHL et al. 1991; TYOR et al. 1992; PERRELLA et al. 1992; ACHIM et al. 1993).

Notably, HIV-1 infection and HIV-1 products, particularly gp120, may induce the production of cytokines such as IL-1, IL-6, TGF-β and TNF-α from T cells, B cells, macrophages, microglia and/or astrocytes (ROSENBERG and FAUCI 1990; EMILIE et al. 1990; VITKOVIC et al. 1990; WAHL et al. 1991; SUNDAR et al. 1991; MERRILL et al. 1992; GENIS et al. 1992). Conversely, cytokines may be potent regulators of HIV-1 gene transcription and translation with IL-6 and TNF-α having primarily enhancing (ROSENBERG and FAUCI 1990) and TGF-β suppressing (KEKOW et al. 1990) effects. Thus, important regulatory loops exist between HIV-1 and the host cytokine network that may strongly influence the development of damage in the HIV-1 infected CNS.

In addition to their possible influence on the replication of HIV-1 in the brain, cytokines are likely mediators of many of the clinical and pathological manifestations seen in HIV-1 associated neurological disease (MERRILL and CHEN 1991; GELEZIUNAS et al. 1992; EPSTEIN and GENDELMAN 1993; PRICE and PERRY 1994). The potential role of cytokines in neurodegeneration has been the focus of detailed reviews (e.g., MERRILL 1992; BENVENISTE 1992). While the current article focuses on the neuropathogenic effects of cytokines produced within the brain parenchyma, there is ample evidence that elevations in systemic cytokine levels may also contribute to a number of CNS-related signs/symptoms commonly seen during HIV-1 infections, including headache, fever, fatigue and somnolence (BOCCI 1988; BLATTEIS 1990; NISTICO and DE SARRO 1991; ROTHWELL 1991; PLATA-SALAMAN 1991; SPENCER and PRICE 1992).

5 The Transgenic Approach

Since the neurodegenerative changes that accompany HIV-1 infection evolve slowly, postmortem material often reflects the intricate combination of causative and compensatory processes. This makes it difficult to differentiate crucial cause-effect relationships from less important associations by conventional neuropathological approaches. Many features of neurological diseases are also difficult to analyze in vitro because of the extensive interactions between different types of predominantly non-dividing cells that characterize the adult CNS. Nonetheless, both clinicopathological investigations and in vitro studies have identified a variety of viral and host-derived factors that could contribute to the development and progression of AIDS dementia complex (see above and SPENCER and PRICE 1992, and PRICE and PERRY 1994, for review). However, for the design of effective therapeutic strategies it is desirable to determine more directly to what extent specific factors can contribute to nervous system damage in vivo. Fusion gene constructs expressed in transgenic animals can be helpful in this respect because they allow for the prolonged, reproducible delivery of selected proteins (devoid of any contaminations) to specific areas/cells within the intact CNS. They also circumvent a number of problems associated with the direct injection of suspected etiological factors into the CNS, e.g., secondary responses to mechanical trauma and gross variations in the amount of protein delivered. A number of constructs have been developed which direct the in vivo expression of fusion genes predominantly to specific types of CNS cells. Well characterized examples of such constructs include those derived from the neuron specific enolase (NSE; targets neurons) (FORSS-PETTER et al. 1990), myelin basic protein (MBP; targets oligodendrocytes) (TURNLEY et al. 1990), and glial fibrillary acidic protein (GFAP; targets astrocytes) (MUCKE et al. 1991) genes. Astrocytes constitute particularly appropriate targets for the expression of cytokine encoding transgenes in the CNS because these cells are highly plastic and respond vigorously to diverse CNS insults (EDDLESTON and MUCKE 1993; MUCKE and EDDLESTON 1993). In addition, in response to appropriate stimuli, astrocytes are known to produce a variety of cytokines including IL-1, IL-6, TNF-α and TGF-β (HATTEN et al. 1991; EDDLESTON and MUCKE 1993) indicating that these cells possess the necessary machinery for the correct synthesis, processing and secretion of such proteins.

The following criteria help assess the specificity of findings obtained in transgenic models. Since transgenes usually integrate into different sites of the genome when microinjected into separate oocytes, the presence of similar CNS changes in distinct lines of transgenic mice expressing the same fusion gene indicates that these alterations are not caused by an insertional mutation with neurodevelopmental consequences. The absence of CNS alterations in mice expressing unrelated proteins from the same promoter demonstrates that these changes do not result nonspecifically from the aberrant expression of any protein within the tartgeted cell type. Such transgenic controls also help exclude other potential transgene-related artifacts, e.g., diversion of critical transcription factors by the transgene promoter. In combination with these important negative controls, a positive correlation between the intensity of CNS alterations and CNS levels of specific transgene products across different lines of transgenic mice and across different brain regions provides strong in vivo support for a cause-effect relationship between the suspected pathogenic factor and the induction of CNS damage. As described in detail elsewhere (TOGGAS et al. 1994; CAMPBELL et al. 1993a), the GFAP-gp120 and GFAP-IL6 transgenic models summarized below meet all of the specificity criteria outlined above.

6 HIV-1 Transgenic Models

A number of transgenic models have been generated which carry complete HIV-1 proviral sequences, subgenomic fragments or reporter genes linked to the HIV-1 LTR (reviewed in KOCH and RUPRECHT 1992). While these models are interesting in their own right, the reports that have been published on them provide no conclusive information on the effect of specific HIV-1 proteins on the CNS.

Since the HIV-1 envelope glycoprotein gp120 has detrimental effects on nerve cells in vitro (see above), we wanted to assess the neuropathogenic potential of this molecule in vivo by expressing it in the CNS of transgenic mice. While monocytoid cells are the primary source of HIV-1 proteins in the human brain (for review see SPENCER and PRICE 1992, and PRICE and PERRY 1994), there was no promoter available at the time that would allow for the effective constitutive expression of transgenes in brain microglia/macrophages. We therefore decided to use regulatory sequences from the glial fibrillary acidic protein (GFAP) gene to target the expression of gp 120 to astrocytes. Astroglial expression of GFAP-driven fusion genes has been shown to be an effective means of delivering diverse proteins to the CNS parenchyma (MUCKE et al. 1991, 1992; MUCKE and ROCKENSTEIN 1993; CAMPBELL et al. 1992, 1993a, b) and increasing evidence suggests that astrocytes are involved in the pathogenesis of HIV-1 associated neurologic disease (LIPTON 1992a; GENIS et al. 1992; LEVI et al. 1993; PULLIAM et al. 1993; MUCKE and EDDLESTON 1993; TORNATORE et al. 1994; SAITO et al. 1994).

As described in detail elsewhere (TOGGAS et al. 1994), the env gene of HIV-1_{LAV} was truncated to encode soluble gp120 (lacking the gp41 transmembrane anchor) and inserted into exon 1 of a modified murine GFAP gene. In vitro studies confirmed that astroglial cell lines expressing this construct effectively secrete gp120 into the culture media (TOGGAS et al. 1994). By in situ hybridization, GFAP-gp120 transgenic mice showed widespread astroglial expression of gp120 mRNA with the highest concentrations of radioactively labeled astrocytes being found in neocortex, olfactory bulb, hippocampus, tectum, selected white matter tracts, and along the glia limitans. The quantitative neuropathological analysis of adult GFAP-gp120 transgenic mice (TOGGAS et al. 1994) revealed extensive vacuolar degeneration of neuronal dendrities (Fig. 1d), a decrease in the number of presynaptic terminals, a drop-out of large cortical neurons and a widespread astrocytosis (Figs. 2 and 3d) and microgliosis. Notably, these changes were absent in nontransgenic mice as well as in transgenic controls expressing α1-antichymotrypsin (Figs. 1a, b; 2; 3a,b) or β-galactosidase (TOGGAS et al. 1994) from the same GFAP regulatory sequences. More recently, we have evaluated gp120 transgenic mice during their early postnatal development. Whereas brains of gp120 transgenic neonates were indistinguishable at the structural level from brains of nontransgenic littermate controls, vacuolar degeneration of MAP-2 immunoreactive neuronal dendrites was evident in gp120 transgenic mice but not in nontransgenic littermate controls by 7 days postnatally (Mucke et al. 1995). This early development of brain damage in GFAP-gp120 transgenic mice is consistent with the postnatal increase in the expression directed by the GFAP promoter (TARDY et al. 1989). Notably, it also correlates well with the developmental expression of NMDA receptors MACDONALD et al. 1993), which appear to play an important role in gp120-induced neurotoxicity (LIPTON et al. 1991; LIPTON 1992a).

In summary, our studies demonstrate that the GFAP-driven expression of gp120 within the otherwise unmanipulated CNS of transgenic mice is sufficient to induce pathological effects on neurons, astrocytes and microglia that resemble changes seen in brains of patients with AIDS dementia complex (see Table 1 and further below). Experiments are now underway to: (a) determine whether gp120 acts directly or indirectly in this model, (b) identify and manipulate the cellular and molecular mediators of gp120-induced toxicity in vivo, and (c) explore the CNS effects of other HIV-1 proteins in related models.

7 Cytokine Transgenic Models

While diverse cytokines have been expressed in a great variety of transgenic mice (for review see LUBBERT et al. 1990, and TAVERNE 1993), transgene expression in these models was not targeted primarily to the CNS. In addition, many of the published reports on these models provide no information on the CNS; notable

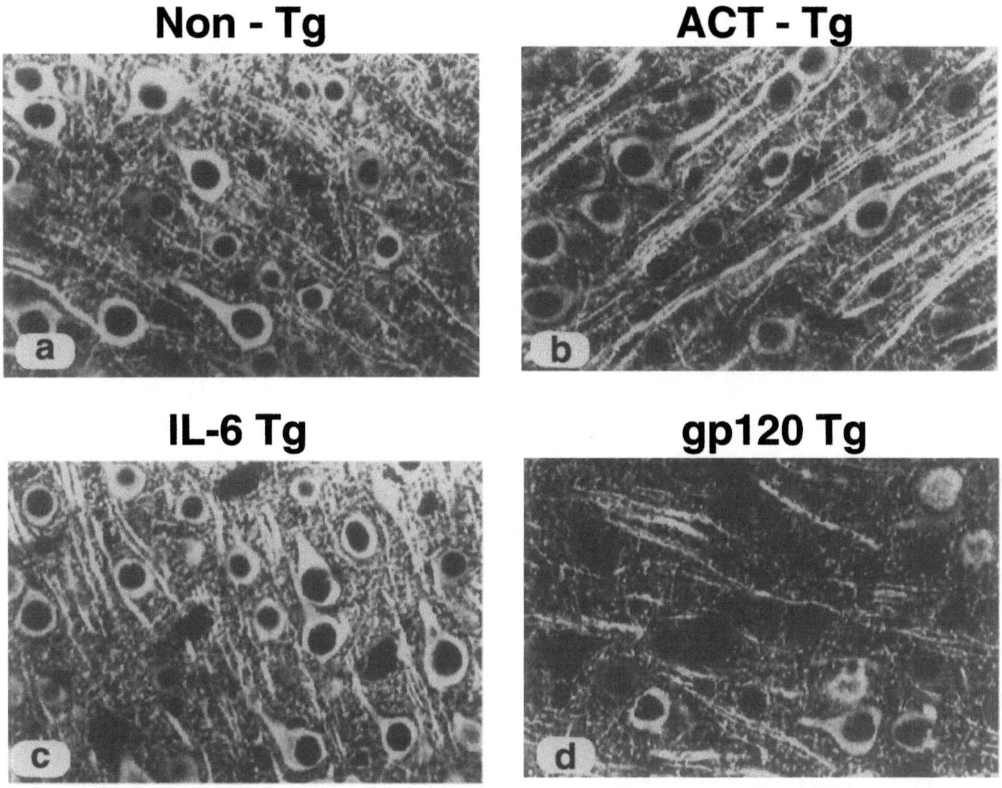

Fig. 1a–d. Prominent damage of MAP-2 immunoreactive (MAP2-IR) neurons in the neocortex of gp120 transgenic mice. MAP2-IR pyramidal neurons in the neocortex of **a** nontransgenic mice (*Non-Tg*); **b** transgenic controls expressing α1-antichymotrypsin (ACT) from the glial fibrillary acidic protein (GFAP) promoter (*ACT-Tg*); and **c** GFAP-interleukin-6 transgenic mice (*IL-6 Tg*) displayed abundant dendritic arborizations with no or minimal vacuolizations. In contrast, **d** GFAP-gp120 transgenic mice showed extensive loss and vacuolization of MAP2-IR neuronal cell bodies and dendrites. Vibratome sections were immunolabeled with a monoclonal antibody against MAP-2 and imaged with a laser scanning confocal microscope as previously described (TOGGAS et al. 1994)

exceptions are briefly summarized below. Overexpression of granulocyte macrophage–colony stimulating factor (GM-CSF) driven by the Moloney murine leukemia virus long terminal repeat (LTR) in transgenic mice resulted in significant infiltration of the eyes by macrophages and blindness associated with diffuse degeneration of the outer nuclear and photoreceptor layers of the retina, retinal astrocytosis as well as corneal and vitreal neovascularization (CUTHBERTSON et al. 1990). Transgenic mice expressing both human IL-2 and the L chain of the IL-2 receptor under the control of the H-2K^d promoter were runted, developed motor abnormalities and revealed a selective loss of Purkinje cells in the cerebellum (ISHIDA et al. 1989). A preliminary report indicates that mice transgenic for the human IL-1β gene placed downstream of the metallothionein-1 promoter are also runted and develop prominent CNS inflammation (CHEN et al. 1994). In contrast to these detrimental cytokine effects, CNS expression of bovine basic fibroblast

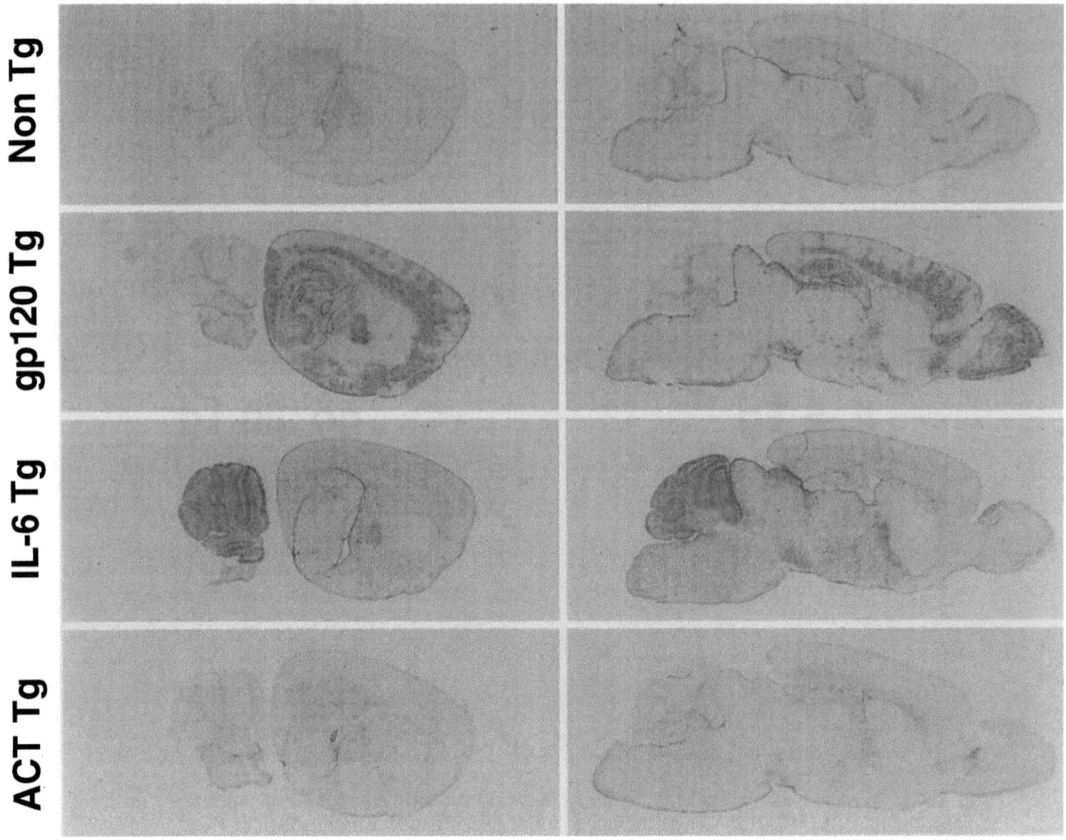

Fig. 2. Regions of maximal reactive astrocytosis in glial fibrillary acidic protein (GFAP)-gp120 and GFAP-interleukin-6 (IL-6) transgenic mice. GFAP-immunolabeled sagittal hemibrain sections representing two distinct anatomical levels reveal low levels of baseline staining in nontransgenic mice and GFAP-α1-antichymotrypsin (ACT) transgenic controls. Areas of dense immunostaining in transgenic mice expressing gp120 or IL-6 identify brain regions with maximal levels of astrotytosis. Note that in gp120 transgenic mice the astrocytosis is most prominent in the neocortex and hippocampus, whereas in IL-6 transgenic mice it is maximal in the cerebellum and brainstem. Hemibrains were fixed with Bouin's solution, paraffin-embedded, sectioned at 9 μm and stained with GFAP antibodies as described (MUCKE et al. 1991) Note that the GFAP immunoreactivity in transgenic mice reflects the expression of endogenous GFAP, because no GFAP mRNA/protein can be derived from GFAP-driven transgenes (TOGGAS et al. 1994; CAMPBELL et al. 1993a)

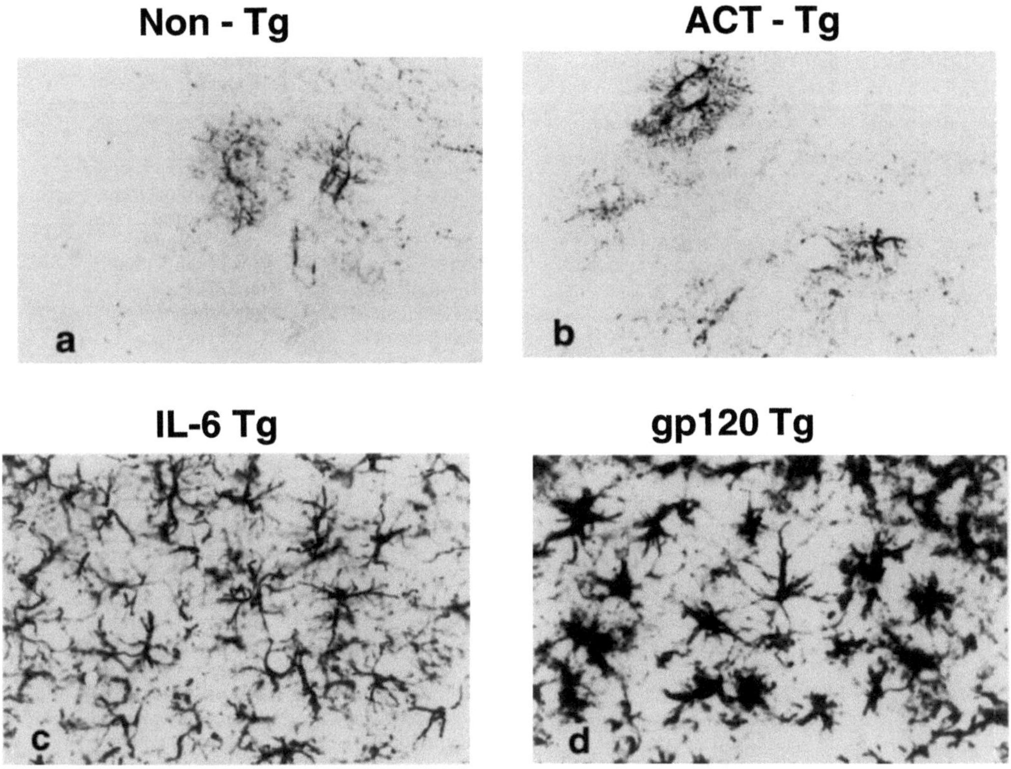

Fig. 3a–d. Difference in the morphology of reactive astrocytes in glial fibrillary acidic protein-interleukin-6. In the neocortex of **a** nontransgenic (Non-Tg) and **b** GFAP-ACT transgenic mice occasional finely ramified GFAP immunoreactive fibrillary astrocytes were seen, primarily around blood vessels. In contrast, the neocortex of transgenic mice expressing **c** IL-6 or **d** gp120 showed a prominent astrocytosis. Note that the astrocytes of gp120 transgenic mice are more hypertrophic than the astrocytes of IL-6 transgenic mice. It is unclear if this difference in morphology reflects differences in the type or degree of activation, in the specific subpopulation of astrocytes activated, or in some other glial property. Vibratome sections were immunolabeled with a monoclonal antibody against GFAP as described (TOGGAS et al. 1994)

growth factor (bFGF) driven by the Rous sarcoma virus LTR in transgenic mice decreased their susceptibility to hypoxic-ischemic cerebral damage (MACMILLAN et al. 1993).

In the first reported study to target the expression of cytokines primarily to the CNS, the modified GFAP gene (see above) was used to overexpress murine IL-6 in astrocytes of transgenic mice (CAMPBELL et al. 1992, 1993a). By in situ hybridization, expression of transgene-derived IL-6 was localized predominantly to subcortical regions such as the thalamus, cerebellum and brain stem, with highest expression in cerebellar astrocytes. Production of bioactive IL-6 in these mice was

Table 1. CNS alteration in human AIDS patients and in transgenic models expressing HIV-1 gp120 or IL-6

CNS alterations	HIV-1 infected humans[a]	GFAP-gp120 transgenic mice	GFAP-IL6 transgenic mice
Functional	Cognitive impairment dementia	(Behavioral analysis in progress)	(Behavioral analysis in progress)
	Motor incoordination	No overt signs	Motor incoordination
	Paralysis	No paralysis	Hind limb paralysis in older mice
	Seizures and/or EEG/AER abnormalities	No overt seizures (electrophysiological analysis in progress)	Motor seizures; lack of theta rhythm on EEG
	Developmental delay	Growth retardation of delayed onset in Balb /cByJ strain	Early runting
Structural	Distribution of changes: basal ganglia > neocortex > other brain regions	Distribution of changes: neocortex > hippocampus > other brain regions	Distribution of changes: cerebellum > brain stem > other brain regions
	Neuronal loss (pyramidal cells and interneurons)	Neuronal loss (primarily large neocortical pyramidal cells)	Neuronal loss (primarily neurons in cerebellar cortex and hippocampal interneurons)
	Vacuolar degeneration of neuronal dendrites and loss of dendritic spines	Vacuolar degeneration of neuronal dendrites and loss of dendritic spines	Vacuolar degeneration of neuronal dendrites and loss of dendritic spines
	Astrocytosis	Astrocytosis	Astrocytosis
	Microglial activation including nodules	Microglial activation including nodules	Microglial activation including nodules
	Multinucleated giant cells	None	None
	White matter changes	White matter changes	White matter changes
	BBB disruption	BBB intact	BBB disruption
	Inflammatory infiltrates	No infiltrates	Inflammatory infiltrates
	No angiogenesis	No angiogenesis	Angiogenesis
Molecular	Increased expression of cytokines (see text)	Normal cytokine pattern at mRNA level[b]	Increased expression of cytokine mRNAs[b]

GFAP, glial fibrillary acidic protein; IL-6, interleukin-6; AER, auditory evoked responses; BBB, blood-brain barrier.
[a] See PRICE and PERRY (1994) for review and note that these alterations are found to variable degrees and in variable combinations in only a proportion of HIV-1 infected patients.
[b] See Fig. 5

demonstrated from isolated astrocyte cultures. IL-6 expression in the CNS of transgenic mice was associated with the development of a neurological syndrome in which both the clinical severity and neuropathology correlated positively with the level and distribution of transgene expression (CAMPBELL et al. 1993a; CHIANG et al. 1994).

High levels of cerebral IL-6 expression were fatal with mice dying by 2 months of age exhibiting developmental retardation with pronounced runting, ataxia, tremors, motor impairment and seizures. Transgenic mice with low or moderate levels of IL-6 expression could be bred and have been used to establish perma-

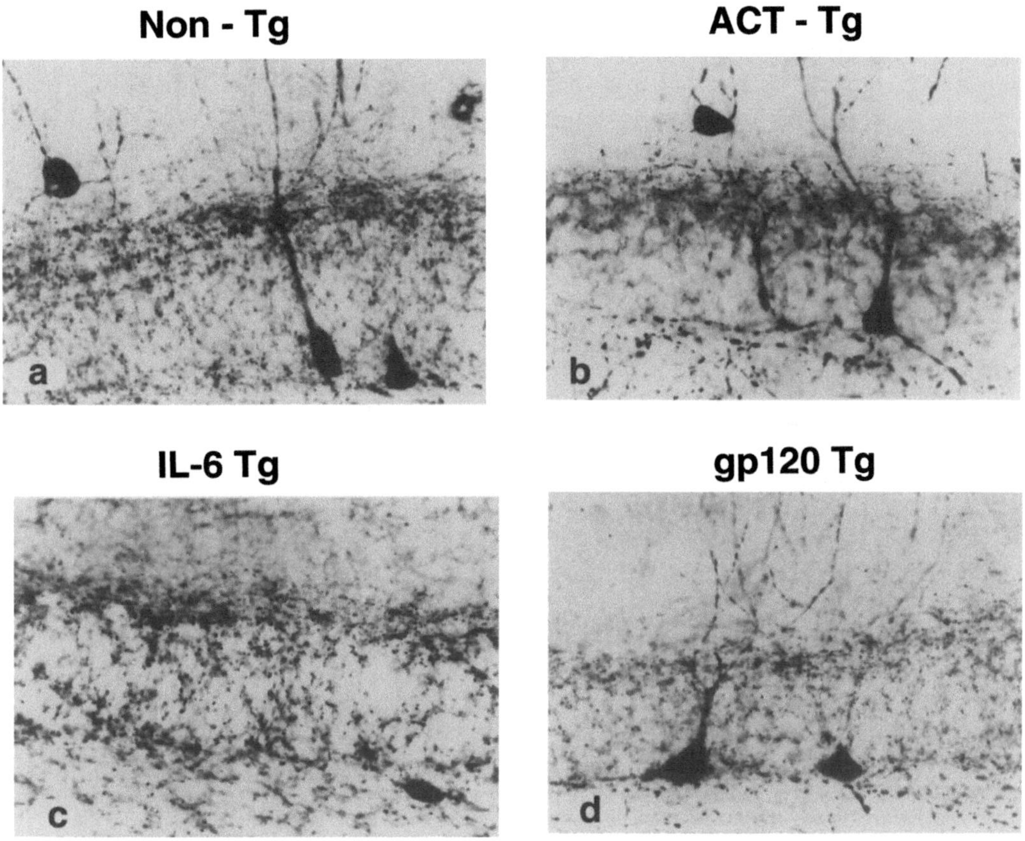

Fig. 4 a-d. Prominent damage of parvalbumin immunoreactive (PV-IR) Interneurons in Interleukin-6 (IL-6) transgenic mice. In the dentate gyrus, **a** nontransgenic mice and **b** glial fibrillary acidic protein-α1-antichymotrypsin (GFAP-ACT) transgenic mice showed prominent multibranched PV-IR intrinsic neurons. **c** In contrast, transgenic mice expressing IL-6 displayed a significant loss of PV-IR interneurons in the dentate region. **d** Transgenic mice expressing gp120 did not show significant alterations of this neuronal population. Vibratome sections were immunolabeled with monoclonal antibodies against PV as previously described (MASLIAH et al. 1992b; CAMPBELL et al. 1993a)

nent lines. Transgenic offspring from such GFAP-IL6 mice develop a more slowly progressive neurological disease which in its end stages is characterized by severe motor impairment, tremors and seizures. Electrophysiological studies in these mice revealed hippocampal pathophysiology with an abnormal EEG pattern characterized by rhythmic power bursts and an absence of theta wave activity (STEFFENSEN et al. 1994). This EEG profile is consistent with the clinical seizures seen in these animals and may reflect an increased excitatory activity in the hippocampus. GFAP-IL6 mice also developed progressive neuropathological alterations of the CNS (CAMPBELL et al. 1993a; CHIIANG et al. 1994). Neurodegeneration was conspicuous, consisting of dendritic vacuolization and reduced

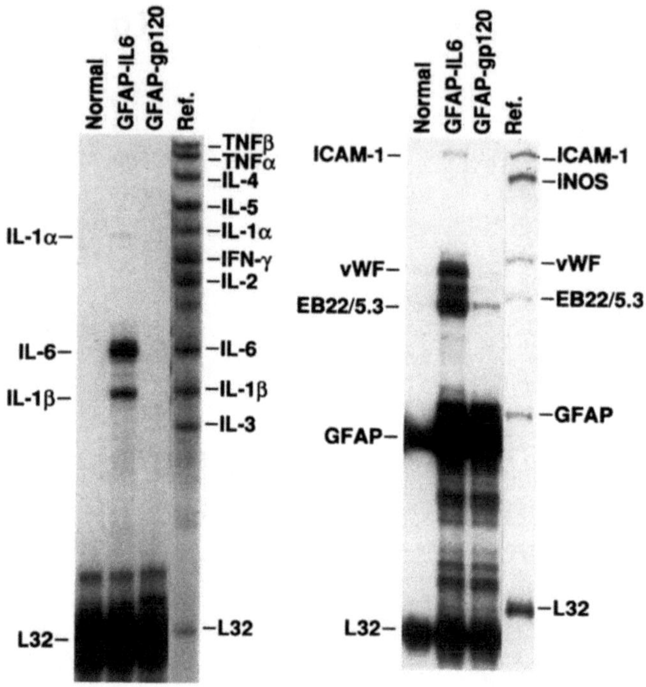

Fig. 5. Levels of cytokine and cellular response protein mRNAs in brains of interleukin-6 (IL-6) and gp120 transgenic mice and nontransgenic controls. PolyA$^+$-enriched RNA was isolated from brains and analyzed by RNase protection assay, as described previously for cytokines (5 µg RNA per lane; *left panel*) (CAMPBELL et al. 1994b; HOBBS et al. 1992) and for cellular response genes (0.5 µg RNA per lane; *right panel*) (CHIANG et al. 1994). The lane marked *Ref* corresponds to the labeled probe set used in the assay. In IL-6 transgenic mice, IL-6, IL-α and IL-1β RNAs were increased as well as transcripts encoding intercellular adhesion molecule-1 (*ICAM-1*), von Willebrand factor (*vWf*) and the acute phase response gene *EB22/5.3*. In contrast, gp120 mice showed no significant increase in the expression of these mRNAs. Both IL-6 and gp120 mice showed increased expression of GFAP mRNA which, as outlined previously (TOGGAS et al. 1994; CAMPBELL et al. 1993a), represents an increase in the expression of the endogenous GFAP gene. The lowest band seen in all the lanes represents mRNA encoding the ribosomal protein L32 and was used as a marker for RNA loading. *GFAP*, glial fibrillary acidic protein; *TNF*, tumor necrosis factor; *IFN*, interferon; *iNOS*, inducible form of nitric oxide synthase

branching of dendritic spines in CA1 neurons, loss of parvalbumin-immunoreactive hippocampal interneurons (Fig. 4c) as well as atrophy and loss of neurons in the molecular and granular layers of the cerebellum (CAMPBELL et al. 1993a). At the ultrastructural level, there was spongiosis and axonal dystrophy (throughout the cerebellum) as well as demyelination and oligodendrocyte injury (L. POWELL and I.L. CAMPBELL, unpublished observation). GFAP-IL6 mice also showed a pronounced gliosis with activation of astrocytes (Fig. 2c, 3) and microglia. Despite its hematopoietic and proimmunoinflammatory actions in the periphery, expression of IL-6 in the CNS of GFAP-IL6 mice was not associated with prominent infiltration of the brain with proinflammatory cells. However, discrete perivascular accumulations of mononuclear cells were seen around vessels of the cerebellar sulci and the brain stem. While IL-6 has not been reported to have significant angiogenic properties, GFAP-IL6 mice exhibited a progressive cerebrovascular genesis that resulted in a massive increase in cerebral blood vessel density. Additional changes of the cerebrovascular endotheliun in these mice included an increased expression of von Willebrand's factor (Fig. 5) and a disruption of the blood-brain barrier (CAMPBELL 1994). Consistent with its actions in the periphery, IL-6 expression in the CNS of GFAP-IL6 mice resulted in a chronic activation of acute-phase response gene expression, e.g., the murine homologue of $\alpha 1$ antichymotrypsin, EB22/5.3 (Fig. 5), and the C3 complement gene (S. Barnum and I.L. Campbell, unpublished observation). Finally brains of GFAP-IL6 mice showed increased expression of endogenous cytokine mRNAs encoding IL-1α, IL-1β and TNF-α (Fig. 5). These studies in mice indicate that the cerebral expression of IL-6, a cytokine implicated in the pathogenesis of HIV-1-associated neurologic impairment (GALLO et al. 1989; TYOR et al. 1992), can recapitulate many of the molecular, cellular and functional alterations seen in the human disease (see Table 1 and further below), providing strong in vivo support for a significant neuropathogenic potential of this cytokine.

We have recently used the GFAP promoter to express a number of other cytokines in the CNS of transgenic mice including IFN-α (CAMPBELL et al. 1993b), IL-3 (C.-S. CHIANG and I.L. CAMPBELL, unpublished) and TGF-β (WYSS-CORAY et al. 1995). Initial characterization of these animals indicates that, like IL-6, expression of each of these cytokines at high levels results in a severe neurologic syndrome with lethal consequences. The preliminary characterization of the various GFAP-cytokine transgenic mice indicates that these models show overlapping (e.g., gliosis) as well as unique (e.g., angiogenesis, up-modulation of acute phase or extracellular matrix proteins, prominent inflammatory infiltrates) features. There are few reports concerning the expression of cytokines in the CNS of transgenic mice driven by other neural promoters. In a preliminary report (RENNO et al. 1992), transgenic mice carrying the interferon (IFN)-γ gene under the control of the MBP (oligodendrocyte targeted) promoter, showed, surprisingly, no dicernible clinical or pathological phenotype. While CNS expression of the transgene at the RNA level was confirmed in these mice, evidence for the production of bioactive IFN-γ protein was not given. In contrast to this study, expression of the IFN-γ gene under the control of the rhodopsin promoter resulted in blindness in transgenic mice (GEIGER et al. 1994). Histopathologic examination of the eyes in these animals revealed intraocular inflammation, loss of photoreceptors, cataract formation and gliosis. Finally, expression of the insulin-like growth factor 1 gene under the control of the MBP promoter was found to increase brain growth and CNS myelination in transgenic mice (CARSON et al. 1993).

8 Comparison of Structural and Molecular Alterations in Brains of Transgenic Models and HIV-1 Infected Humans

As outlined above, a multitude of microbes and dysregulated host responses may contribute to nervous system damage in HIV-1 infected patients. Transgenic mice expressing specific microbial proteins and/or cytokines in their CNS allow the assessment of the relative contributions these factors make to the development of CNS damage in vivo. Differences between distinct transgenic models (Table 1, Figs. 1–5) likely reflect a combination of the specific properties of the transgene products expressed and the selective vulnerability of the various cell populations in the brain. For example, while transgenic mice expressing gp120 display extensive damage to neocortical pyramidal neurons and minimal involvement of interneurons that contain calcium-binding proteins, transgenic mice expressing IL-6 present significant disruption of such interneurons and little, if any, alterations of neocortical, pyramidal neurons (Figs. 1, 4). Our preliminary comparative analysis further indicates that gp120 and IL-6 also induce a different host response at the molecular level (Fig. 5). Hence, differences in gp120 and IL-6 induced brain damage could be due to the induction of different neurotoxic activities. It should be noted that the in vivo expression directed by a given promoter can be influenced by the coding sequence with which it has been fused (PALMITER et al. 1991). The brain regions showing maximal damage in GFAP-gp120 and GFAP-IL6 transgenic mice, neocortex versus cerebellum, respectively, are also the regions showing maximal levels of transgene expression in the two distinct models (CAMPBELL et al. 1993a; TOGGAS et al. 1994). However, differences in the distribution of transgene expression are not sufficient to account for the neuropathological differences between the GFAP-gp120 and GFAP-IL6 models because (a) the distribution of transgene expression in these models overlapped widely, (b) similar model-specific changes were found across several transgenic lines showing different overall levels of transgene expression and (c) brain regions that are pathologically altered in both models, for example the hippocampal formation, show distinct structural alterations. The susceptibility of different neuronal subpopulations to specific neurotoxins may be critically influenced by differences in the expression of endogenous molecules (e.g., cell surface receptors, calcium binding proteins, antioxidants, proteases, protease inhibitors and cytokines) across brain regions. It should also be noted in this context that the neurodegenerative changes induced by gp120 and IL-6 in the CNS of transgenic mice are progressive and that the mice included in the preliminary comparative analysis presented here were all under 6 months of age. It is possible that in older mice some of the "model-specific" features will become more distinct, whereas others may become obscured by progressive nonspecific injury responses.

It is interesting that the neuropathological changes found in gp120 and IL-6 transgenic mice resemble a number of alterations seen in patients with AIDS dementia complex (Table 1). The loss of specific neuronal subpopulations in combination with more widespread vacuolar degeneration of neuronal dendrites in transgenic mice is clearly consistent with neuropathological effects seen in HIV-1-infected human brains (ARTIGAS et al. 1989; KETZLER et al. 1990; EVERALL et al. 1991; WILEY et al. 1991; MASLIAH et al. 1992a). Microglial cells in the transgenic models had an activated morphological appearance and were occasionally clustered in groups (TOGGAS et al. 1994), reminiscent of microglial nodules seen in HIV-1 encephalitis and other neurodegenerative disorders (BUDKA 1991; DICKSON et al. 1993). In addition, both transgenic models developed a prominent reactive astrocytosis (Figs. 2, 3) and reactive astrocytosis is one of the earliest and most consistent alterations in the HIV-1 infected human CNS (DE LA MONTE et al. 1987; MCARTHUR et al. 1989; BUDKA 1991; SHARER 1992; DICKSON et al. 1993). It should be noted that the CNS responds to diverse forms of damage with similar structural alterations. This is clearly illustrated by the neuropathological changes associated with HIV-1 infection in humans which, in large part, can also be found in a variety of other neurological diseases (MASLIAH et al. 1992a; SHARER 1992; DICKSON et al. 1993; EDDLESTON and MUCKE 1993). The majority of CNS changes found in the transgenic models described above are also not unique to gp120 or IL-6 induced neurotoxicity but overlap widely with alterations found in other neurological disease models, for example, infections with scrapie (HOGAN et al. 1987; CAMPBELL et al. 1994a) or murine retroviruses (BASZLER and ZACHARY 1990; NAGRA et al. 1993) and injections of excitotoxins into the CNS (IACOPINO et al. 1992).

In conclusion, our studies show that HIV-1 products such as gp120 and host factors such as IL-6 are sufficient to induce significant CNS damage in vivo and, hence, should be considered as potentially important therapeutic targets in HIV-1 associated neurological disease. Possible synergistic effects between these factors are currently being explored more directly in gp120/IL-6 bigenic mice.

9 Outlook

From a therapeutic perspective, one can only hope that the diverse processes which may contribute to the development of AIDS dementia complex converge into one or a few final common pathogenetic pathway(s), for example, the dysregulation of the neuronal calcium homeostasis (LIPTON and ROSENBERG 1994). Notably, preliminary data obtained in the GFAP-gp120 model indicate that neuronal overexpression of the amyloid precursor protein can effectively prevent gp120-induced neuronal injury in vivo (MUCKE et al. 1994, 1995). IL-6 transgenic mice are currently being tested in similar experimental paradigms.

In combination with the examination of human patients and the analysis of other experimental models, transgenic mice can be expected to advance our understanding of HIV-1 associated neurological disease, help identify critical targets for therapeutic interventions and facilitate the preclinical evaluation of therapeutic strategies.

Acknowledgments. We thank M. Ruppe and A. Samimi for excellent technical assistance. This work was supported by NIH grants MH-47680 (LM) and MH-50546 (ILC) and by a Harry Weaver Neuroscience Scholarship Award from the National Multiple Sclerosis Society (LM).

References

Achim CL, Heyes MP, Wiley CA (1993) Quantitation of human immunodeficiency virus, immune activation factors, and quinolinic acid in AIDS brains. J Clin Invest 91: 2769–2775

Artigas J, Niedobitek F, Grosse G, Heise W, Gosztonyi G (1989) Spongiform encephalopathy in AIDS dementia complex: report of five cases. J Acquir Immune Defic Syndr 2: 374–381

Baszler TV, Zachary JF (1990) Murine retroviral-induced spongiform neuronal degeneration parallels resident microglial cell infection: ultrastructural findings. Lab Invest 63: 612–623

Benveniste EN (1992) Inflammatory cytokines within the central nervous system: sources, function, and mechanism of action. Am J Physiol 263: 1–16

Blatteis CM (1990) Neuromodulatory actions of cytokines. Yale J Biol Med 63: 133–146

Bocci V (1988) Central nervous system toxicity of interferons and other cytokines. J Biol Regul Homeost Agents 2: 107–118

Brenneman DE, Westbrook GL, Fitzgerald SP, Ennist DL, Elkins KL, Ruff MR, Pert CB (1988) Neuronal cell killing by the envelope protein of HIV and its prevention by vasoactive intestinal peptide. Nature 335: 639–642

Budka H (1991) Neuropathology of human immunodeficiency virus infection. Brain Pathol 1: 163–175

Budka H (1992) Cerebral pathology in AIDS: a new nomenclature and pathogenetic concepts. Curr Opin Neurol Neurosurg 5: 917–923

Campbell IL(1994) Proliferative angiopathy in the brain induced by the cerebral overexpression of interleukin-6 in transgenic mice. J Cell Biochem 18A: 283

Campbell IL, Oldstone MBA, Mucke L (1992) Neurologic disease induced in transgenic mice by the astrocyte-specific expression of interleukin-6. Soc Neurosci Abstr 18: 482

Campbell IL, Abraham CR, Masliah E, Kemper P, Inglis JD, Oldstone MBA, Mucke L (1993a) Neurologic disease induced in transgenic mice by cerebral overexpression of interleukin 6. Proc Natl Acad Sci USA 90: 10061–10065

Campbell IL, Mucke L, Sandberg K (1993b) Cerebral overexpression of interleukin-6 or interferon-α1 induces distinct neuropathology in transgenic mice. Soc Neurosci Abstr 19: 982

Campbell IL, Eddleston M, Kemper P, Oldstone MBA, Hobbs MV (1994a) Activation of cerebral cytokine gene expression and its correlation with onset of reactive astrocyte and acute-phase response gene expression in scrapie. J Virol 68: 2383–2387

Campbell IL, Hobbs MV, Kemper P, Oldstone MBA (1994b). Cerebral expression of multiple cytokine genes in mice with lymphocytic choriomeningitis. J Immunol 152: 716–723

Carson MJ, Behringer RR, Brinster RL, McMorris FA (1993) Insulin-like growth factor I increases brain growth and central nervous system myelination in transgenic mice. Neuron 10: 729–740

Chen HY, Trumbauer ME, Grabiec CC, Hofmann KJ, Zheng H, Stevens KA, Sirinathsingji D, Van der Ploeg LHT, Shaw AR (1994) Transgenic mice expressing human interleukin-1β exhibited runted phenotype and pathology in various tissues. J Cell Biochem 18A: 283

Chiang CS, Stalder A, Samimi A, Campbell IL (1994) Reactive gliosis as a consequence of interleukin- 6 expression in the brain. Studies in transgenic mice. Dev Neurosci 16: 212–221

Cuthbertson RA, Lang RA, Coghlan JP (1990) Macrophage products IL-1α, TNF-α and bFGF may mediate multiple cytopathic effects in the developing eyes of GM-CSF transgenic mice. Exp Eye Res 51: 335–344

Dal Canto MC (1989) AIDS and the nervous system: current status and future perspectives. Hum Pathol 20: 410–417

Dawson VL, Dawson TM, Uhl GR, Snyder SH (1993) Human immunodeficiency virus type 1 coat protein neurotoxicity mediated by nitric oxide in primary cortical cultures. Proc Natl Acad Sci USA 90: 3256–3259

de la Monte SM, Ho DD, Schooley RT, Hirsch MS, Richardson EPJ (1987) Subacute encephalomyelitis of AIDS and its relation to HTLV-III infection. Neurology 37: 562–569

Dickson DW, Lee SC, Mattiace LA, Yen S-HC, and Brosnan C (1993) Microglia and cytokines in neurological disease, with special reference to AIDS and Alzheimer's disease. Glia 7: 75–83

Eddleston MP, Mucke L (1993) Molecular profile of reactive astrocytes–implications for their role in neurologic disease. Neuroscience 54: 15–36

Emilie D, Peuchmaur M, Maillot MC, Crevon MC, Brousse N, Delfraissy JF, Dormont J, Galanaud P (1990) Production of interleukins in human immunodeficiency virus-1-replicating lymph nodes. J Clin Invest 86: 148–159

Epstein LG, Gendelman HE (1993) Human immunodeficiency virus type 1 infection of the nervous system: pathogenetic mechanisms. Ann Neurol 33: 429–436

Everall IP, Luthert PJ, Lantos PL (1991) Neuronal loss in the frontal cortex in HIV infection Lancet 337: 1119–1121

Forss-Petter S, Danielson PE, Catsicas S, Battenberg E, Price J, Nerenberg M, Sutcliffe JG (1990) Transgenic mice expressing beta-galactosidase in mature neurons under neuron-specific enolase promoter control. Neuron 5: 187–197

Gallo P, Frei K, Rordorf C, Lazdins J, Tavolato B, Fontana A (1989) Human immunodeficiency virus type 1 (HIV-1) infection of the central nervous system: an evaluation of cytokines in cerebrospinal fluid J Neuroimmunol 23: 109–116

Geiger K, Howes E, Gallina M, Huang XJ, Travis GH, Sarvetnick N (1994). Transgenic mice expressing IFN-g in the retina develop inflammation of the eye and photoreceptor loss. Invest Opthalmol Vis Sci 35: 2667–2681

Geleziunas R, Schipper HM, Wainberg MA (1992) Pathogenesis and therapy of HIV-1 infection of the central nervous system. AIDS 6: 1411–1426

Genis P, Jett M, Bernton EW, Boyle T, Gelbard HA, Dzenko K, Keane RW, Resnick L, Mizrachi Y, Volsky DJ, Epstein LG, Gendelman HE (1992) Cytokines and arachidonic acid metabolites produced during human immunodeficiency virus (HIV)-infected macrophage-astroglia interactions: implications for the neuropathogenesis of HIV disease. J Exp Med 176: 1703–1718

Giulian D, Wendt E, Vaca K, Noonan CA (1993) The envelope glycoprotein of human immunodeficiency virus type 1 stimulates release of neurotoxins from monocytes. Proc Natl Acad Sci USA 90: 2769–2773

Glowa JR, Panlilio LV, Brenneman DE, Gozes I, Fridkin M, Hill JM (1992) Learning impairment following intracerebral administration of the HIV envelope protein gp120 or a VIP antagonist. Brain Res 570: 49–53

Hatten ME, Liem RK, Shelanski ML, Mason CA (1991) Astroglia in CNS injury. Glia 4: 233–243

Hill JM, Mervis RF, Avidor R, Moody TW, Brenneman DE (1993) HIV envelope protein-induced neuronal damage and retardation of behavioral development in rat neonates. Brain Res 603: 222–233

Hobbs MV, Weigle WO, Noonan DJ, Torbett BE, McEvilly RJ, Koch RJ, Cardenas GJ, Ernst DN (1992) Patterns of cytokine gene expression by $CD4^{+}$ T cells from young and old mice. J Immunol 150: 3602–3614

Hogan RN, Baringer JR, Prusiner SB (1987) Scrapie infection diminishes spines and increases varicosities of dendrites in hamsters: a quantitative Golgi analysis J Neuropathol Exp Neurol 46: 461–473

Iacopino A, Christakos S, German D, Sonsalla PK, Altar CA (1992) Calbindin-D_{28K}-containing neurons in animal models of neurodegeneration: possible protection from excitotoxicity. Mol Brain Res 13: 251–261

Ishida Y, Nishi M, Taguchi O, Inaba K, Hattori M, Minato N, Kawaichi M, Honjo T (1989) Expansion of natural killer cells but not T cells in human interleukin 2/interleukin 2 receptor (Tac) transgenic mice. J Exp Med 170: 1103–1115

Kekow J, Wachsman W, McCutchan JA, Cronin M, Carson D, Lotz M (1990) Transforming growth factor β and noncytopathic mechanisms of immunodeficiency in human immunodeficiency virus infection. Proc Natl Acad Sci USA 87: 8321–8325

Ketzler S, Weis S, Haug H, Budka H (1990) Loss of neurons in the frontal cortex in AIDS brains. Acta Neuropathol (Berl) 80: 92–94

Kimes AS, London ED, Szabo G, Raymon L, Tabakoff B (1991) Reduction of cerebral glucose utilization by HIV envelope glycoprotein gp120. Exp Neurol 112: 224–228

Koch JA, Ruprecht RM (1992) Animal models for anti-AIDS therapy. Antiviral Res 19: 81–109

Levi G, Patrizio M, Bernardo A, Petrucci TC, Agresti C (1993) Human immunodeficiency virus coat protein gp120 inhibits the β-adrenergic regulation of astroglial and microglial functions. Proc Natl Acad Sci USA 90: 1541–1545

Levy JA (1993) Pathogenesis of human immunodeficiency virus infection. Microbiol Rev 57: 183–289

Lipton SA (1992a) Models of neuronal injury in AIDS: another role for the NMDA receptor? Trends Neurosci 15: 75–79

Lipton SA (1992b) Requirement for macrophages in neuronal injury induced by HIV envelope protein gp120. NeuroReport 3, 913–915
Lipton SA, Rosenberg PA (1994) Mechanisms of disease: excitatory amino acids as a final common pathway for neurologic disorders. N Engl J Med 330: 613–622
Lipton SA, Sucher NJ, Kaiser PK, Dreyer EB (1991) Synergistic effects of HIV coat protein and NMDA receptor-mediated neurotoxicity. Neuron 7: 111–118
Lo T-M, Fallert CJ, Piser TM, Thayer SA (1992) HIV-1 envelope protein evokes intracellular calcium oscillations in rat hippocampal neurons. Brain Res 594: 189–196
Lubbert M, Jonas D, Herrmann F (1990) Animal models for the biological effects of continuous high cytokine levels. Blut 61: 253–257
Mabrouk K, Van Rietschoten J, Vives E, Darbon H, Rochat H, Sabatier J-M (1991) Lethal neurotoxicity in mice of the basic domains of HIV and SIV rev proteins. FEBS Lett 289: 13–17
MacDonald MC, Robertson HA, Wilkinson M (1993) Age- and dose-related NMDA induction of Fos-like immunoreactivity and c-fos mRNA in the arcuate nucleus of immature female rats. Dev Brain Res 73: 193–198
Macmillan V, Judge D, Wiseman A, Settles D, Swain J, Davis J (1993) Mice expressing a bovine basic fibroblast growth factor transgene in the brain show increased resistance to hypoxemic-ischemic cerebral damage. Stroke 24: 1735–1739
Masliah E, Achim CL, Ge N, DeTeresa R, Terry RD, Wiley CA (1992a) Spectrum of human immunodeficiency virus-associated neocortical damage. Ann Neurol 32: 321–329
Masliah E, Ge N, Achim CL, Hansen LA, Wiley CA (1992b) Selective neuronal vulnerability in HIV encephalitis. J Neuropathol Exp Neurol 51: 585–593
McArthur JC, Becker PS, Parisi JE, Trapp B, Selnes OA, Cornblath DR, Balakrishnan J, Griffin JW, Price D (1989) Neuropathological changes in early HIV-1 dementia. Ann Neurol 26: 681–684
Merrill JE (1992) Tumor necrosis factor alpha, interleukin 1 and related cytokines in brain development: normal and pathological. Dev Neurosci 14: 1–10
Merrill JE, Chen IS (1991) HIV-1, macrophages, glial cells, and cytokines in AIDS nervous system disease. FASEB J 5: 2391–2397
Merrill JE, Koyanagi Y, Zack J, Thomas L, Martin F, Chen IS (1992) Induction of interleukin-1 and tumor necrosis factor alpha in brain cultures by human immunodeficiency virus type 1. J Virol 66: 2217–2225
Mucke L, Eddleston M (1993) Astrocytes in infectious and immune-mediated diseases of the central nervous system. FASEB J 7: 1226–1232
Mucke L, Rockenstein EM (1993) Prolonged delivery of transgene products to specific brain regions by migratory astrocyte grafts. Transgenics 1: 3–9
Mucke L, Oldstone MBA, Morris JC, Nerenberg MI (1991) Rapid activation of astrocyte-specific expression of GFAP-lacZ transgene by focal injury. New Biol 3: 465–474
Mucke L, Forss-Petter S, Goldgaber D, Johnson W, Picard E, Rockenstein E, Abraham C (1992) Transgenic models to study the pathogenic role of mutated and non-mutated forms of human amyloid proteins in the development of Alzheimer's disease (AD). Neurobiol Aging 13 [Suppl I]: S101
Mucke L, Abraham CR, Masliah E (1994) Transgenic model of HIV-induced brain damage: successful therapy and implications for AIDS dementia complex. Clin Res 42: 154a
Mucke L, Abraham CR, Ruppe MD, Rockenstein EM, Toggas SM, Mallory M, Alford M, Masliah E (1995) Protection against HIV-1 gp120-induced brain damaged by neuronal expression of human amyloid precursor proteins (HAPP). J Exp Med 181: 1551–1556
Müller WEG, Schröder HC, Ushijima H, Dapper J, Bormann J (1992) gp120 of HIV-1 induced apoptosis in rat cortical cell cultures: prevention by memantine. Eur J Pharmacol 226: 209–214
Nagra RM, Masliah E, Wiley CA (1993) Synaptic and dendritic pathology in murine retroviral encephalitis. Exp Neurol 124: 283–288
Nistico G, de Sarro G (1991) Is interleukin 2 a neuromodulator in the brain? Trends Neurosci 14: 146–150
Palmiter Rd, Sandgren EP, Avarbock MR, Allen MR, Alen DD, Brinster RL (1991) Heterologous introns can enhance expression of transgenes in mice. Proc Natl Acad Sci USA 88: 478–482
Perrella O, Carrieri PB, Guarnaccia D, Soscia M (1992) Cerebrospinal fluid cytokines in AIDS dementia complex. J Neurol 239: 387–388
Plata-Salaman CR (1991) Immunoregulators in the nervous system. Biobehav Rev 15: 185–215
Price RW, Perry SW (eds) (1994) HIV, AIDS, and the brain. Raven, New York
Pulliam L, West D, Haigwood N, Swanson RA (1993) HIV-1 envelope gp120 alters astroycytes in human brain cultures. AIDS Res Hum Retroviruses 9: 439–444

Renno T, Antel J, Sekaly R, Owens T (1992) Transgenic expression of IFN-g in murine CNS. Soc Neurosci Abstr 18: 767

Rosenberg ZF, Fauci AS (1990) Immunopathogenetic mechanisms of HIV infection: cytokine induction of HIV expression. Immunol Today 11: 176–180

Rothwell NJ (1991) Functions and mechanisms of interleukin 1 in the brain. Trends Pharmacol 12: 430–437

Sabatier J-M, Vives E, Mabrouk K, Benjouad A, Rochat H, Duval A, Hue B, Bahraoui E (1991) Evidence for neurotoxic activity of tat from human immunodeficiency virus type 1. J Virol 65: 961–967

Saito Y, Sharer LR, Epstein LG, Michaels J, Mintz M, Louder M, Golding K, Cvetkovich TA, Blumberg BM (1994) Overexpression of nef as a marker for restricted HIV-1 infection of astrocytes in postmortem pediatric central nervous tissues. Neurology 44: 474–481

Savio T, Levi G (1993) Neurotoxicity of HIV coat protein gp120, NMDA receptors, and protein kinase C: a study with rat cerebellar granule cell cultures. J Neurosci Res 34: 265–272

Sharer LR (1992) Pathology of HIV-1 infection of the central nervous system. A review. J Neuropathol Exp Neurol 51: 3–11

Spencer DC, Price RW (1992) Human immunodeficiency virus and the central nervous system. Annu Rev Microbiol 46: 655–693

Steffensen SC, Campbell IL, Henriksen SJ (1994) Site-specific hippocampal pathophysiology due to cerebral overexpression of interleukin-6 in transgenic mice. Brain Res 652: 149–153

Sundar SK, Cierpial MA, Kamaraju LS, Long S, Hsieh S, Lorenz C, Aaron M, Ritchie JC, Weiss JM (1991) Human immunodeficiency virus glycoprotein (gp120) infused into rat brain induces interleukin 1 to elevate pituitary-adrenal activity and decrease peripheral cellular immune responses. Proc Natl Acad Sci USA 88: 11246–11250

Sweetnam PM, Saab OH, Wroblewski JT, Price CH, Karbon EW, Ferkany JW (1993) The envelope glycoprotein of HIV-1 alters NMDA receptor function. Eur J Neurosci 5: 276–283

Tardy M, Fages C, Riol H, LePrince G, Rataboul Charriere-Bertrand C, Nunez J (1989) Developmental expression of the glial fibrillary acidic protein mRNA in the central nervous system and in cultured astrocytes. J Neurochem 52: 162–167

Taverne J (1993) Transgenic mice in the study of cytokine function. Int J Exp Pathol 74: 525–546

Toggas SM, Masliah E, Rockenstein EM, Rall GF, Abraham CR, Mucke L (1994) Central nervous system damage produced by expression of the HIV-1 coat protein gp120 in transgenic mice. Nature 367: 188–193

Tornatore C, Chandra R, Berger JR, Major EO (1994) HIV-1 infection of subcortical astrocytes in the pediatric central nervous system. Neurology 44: 481–487

Turnley AM, Morahan G, Okano H, Bernard O, Mikoshiba K, Allison J, Bartlett PF, Miller JFAP (1991) Dysmyelination in transgenic mice resulting from expression of class I histocompatibility molecules in oligodendrocytes. Nature 353: 566–569

Tyor WR, Glass JD, Griffin JW, Becker PS, McArthur JC, Bezman L, Griffin DE (1992) Cytokine expression in the brain during the acquired immunodeficiency syndrome. Ann Neurol 31: 349–360

Ushijima H, Ando S, Kunisada T, Schröder HC, Klöcking H-P, Kijjoa A, Müller WEG (1993) HIV-1 gp120 and NMDA induce protein kinase C translocation differentially in rat primary neuronal cultures. J Acquir Immune Defic Syndr 6: 339–343

Vitkovic L, Kalebic T, da Cunha A, Fauci AS (1990) Astrocyte-conditioned medium stimulates HIV-1 expression in a chronically infected promonocyte clone. J Neuroimmunol. 30: 153–160

Wahl SM, Allen JB, McCartney-Francis N, Morganti-Kossmann MC, Kossmann T, Ellingsworth L, Mai UE, Mergenhagen SE, Orenstein JM (1991) Macrophage- and astrocyte-derived transforming growth factor beta as a mediator of central nervous system dysfunction in acquired immune deficiency syndrome. J Exp Med 173: 981–991

Wiley CA, Masliah E, Morey M, Lemere C, DeTeresa R, Grafe M, Hansen L, Terry R (1991) Neocortical damage during HIV infection. Ann Neurol 29: 651–657

Working Group of the American Academy of Neurology AIDS Task Force (1991) Nomenclature and research case definitions for neurologic manifestations of human immunodeficiency virus-type 1 (HIV-1) infection. Neurology 41: 778–785

Wyss-Coray T, Feng L, Masliah E, Ruppe MD, Lee HS, Toggas SM, Rockenstein EM, Mucke L (1995) Increased central nervous system production of extracellular matrix components and development of hydrocephalus in transgenic mice overexpressing transforming growth-factor-β1. Am J Pathol (in Press)

C Tissue Culture and Animal Models to Generate Basic Concepts on Viral Immune Cell: Central Nervous System Interactions

Virus Entry and Release in Polarized Epithelial Cells

R.W. COMPANS

1 Introduction

Many viruses initiate their infection processes by interacting with cells in tight epithelial layers, which line the surfaces of the body that are exposed to the external environment. The cell biology of this virus-cell interaction may be one of the important factors which play a role in viral pathogenesis. The individual cells in epithelial layers are tightly connected by junctional complexes, which form a barrier to diffusion of molecules across the cell layer and also divide the cell surface into two distinct plasma membrane domains: the apical domain which faces the lumen, and the basolateral domain which faces the interior of the body. In addition to their role in providing a permeability barrier, the junctional complexes restrict the lateral diffusion of membrane proteins and lipids between the apical and basolateral plasma membrane domains. As a result of the restricted diffusion of membrane components as well as the differential targeting of distinct sets of lipids and proteins to apical vs basolateral membranes, the epithelial cells in such tissues are highly polarized, with each plasma membrane domain having a distinct lipid and protein composition (reviewed by RODRIGUEZ-BOULAN and NELSON 1989).

It has now become apparent that both the entry and the release of many viruses from epithelial cells is also highly polarized, occurring selectively at either the apical or the basolateral plasma membrane domains (TUCKER and COMPANS 1993). Virus-infected epithelial cells have also been extensively studied in many

Emory University School of Medicine, Department of Microbiology and Immunology, Atlanta, GA 30322, USA laboratories as model systems for studying the mechanisms which determine protein sorting and cell polarity (SABATINI et al. 1988; SIMONS and WANDINGER-NESS 1990; RODRIGUEZ-BOULAN and POWELL 1992). In the present chapter, we will summarize some of the recent findings concerning virus entry and virus release from polarized epithelial cells, with particular emphasis on studies of HIV-1.

The interaction of viruses with polarized cells may be an important determinant of the pathogenesis of viral infection. As depicted schematically in Fig. 1, viruses of some families enter epithelial cells at apical surfaces and their progeny virions may be released again exclusively at the apical surface. If the epithelial cell layer remains intact, the progeny virus can spread to adjacent cells but may never traverse the epithelial cell layer, with the consequence that the infection process remains localized to the epithelial surface. In contrast, viruses of other families may be released preferentially at the basolateral surface. As a result, virions will traverse the cell layer to reach underlying cells and tissues, with the possibility of subsequently spreading to other organs. Thus, the patterns of virus entry and release from polarized epithelial cells may be a determinant of the spread of virus within an infected host.

Similar patterns of polarized release of certain enveloped viruses have been observed from cells of a variety of epithelial tissues including intestine (RINDLER and TRABER 1988), thyroid (NITSCH et al. 1985), endometrium (BALL et al. 1995) and neuroectodermal cells including ependymal cells and choroid plexus (KRISTENSSON et al. 1984). Polarized release of viruses has also been observed from endothelial

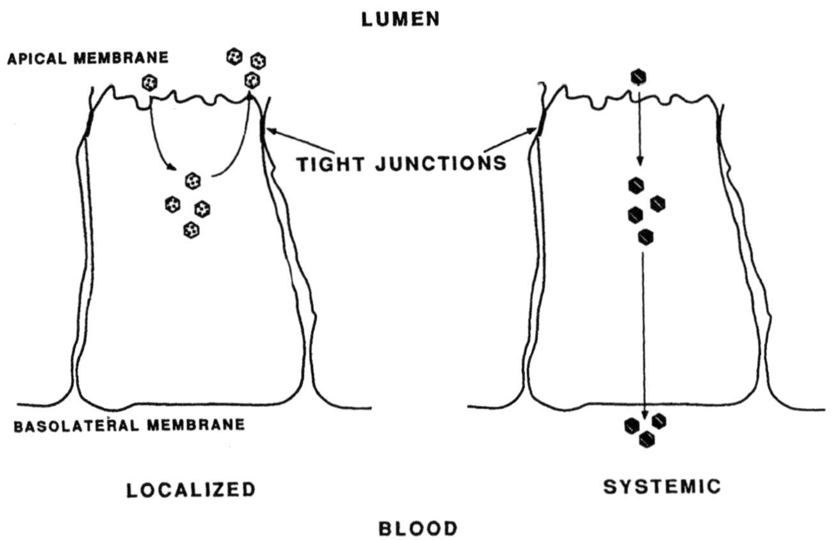

Fig. 1. Alternative interactions of viruses with polarized epithelial cells. On the *left*, virus entry and release at the apical surface would result in lateral spread of the virus from cell to cell, with the infection process remaining restricted to the epithelial surface. On the *right*, entry at the apical surface and release at the basolateral surface would enable a virus to traverse the epithelial barrier, with the possibility of leading to a systemic infection

Table 1. Alternate routes of viral transmission across epithelial cell layers

Paracellular passage
Free virus
Cell-associated virus
Transcytosis
Entry/release at opposite membrane domains
Epithelial cell destruction following infection
Physical damage to epithelial layers

cells (DROPULIC and MASTERS 1990). For two alphaviruses, the direction of virus release appears to vary depending on the tissue of origin of the cultured epithelial cell (ZURZOLO et al. 1992).

There are several alternative mechanisms (Table 1) by which viruses including HIV may traverse epithelial cell layers such as those forming the blood-brain barrier, and the cell biology of virus infection of the epithelial cells could play a role in this process. The infection process could involve passage of virus through the endothelium or, alternatively, through the neuroectodermal cells. Free virus or infected cells could traverse the epithelial or endothelial barrier by paracellular passage. Alternatively, virus could penetrate epithelial cell layers by transcytosis, a process that has been observed following interaction of several viruses (including HIV) with specialized transporting epithelial M cells which cover mucosal lymphoid tissues (WOLF et al. 1981; AMERONGEN et al. 1991). As another alternative, the virus could directly infect epithelial cells and be released by budding at the opposite surface. Finally, the epithelial cell layers could be disrupted as a result of the infection process, enabling the virus to traverse the barrier.

2 Virus Entry

The distribution of viral receptors on surfaces of epithelial cells determines the plasma membrane domains which can serve as sites for initiation of the infection process. For several viruses, it has been observed that entry is polarized, occurring exclusively at one or the other plasma membrane domain. Entry of vesicular stomatitis virus (VSV), Semliki Forest virus and vaccinia virus is restricted to the basolateral surfaces of epithelial cells (FULLER et al. 1984, 1985; RODRIGUEZ et al. 1991), whereas SV40 infection occurs only at the apical plasma membrane domain of polarized monkey kidney cells (CLAYSON and COMPANS 1988). The nature of the specific receptors for these viruses has not yet been established. Other viruses exhibit little or no preference for entry at a specific plasma membrane domain. Such nonpolarized entry may be mediated by virus binding to a single type of receptor molecule that is expressed on both the apical and basolateral surfaces. For example, influenza virus can initiate infection at either the apical or basolateral surfaces of polarized cells in culture, since membrane glycoproteins on both surfaces contain sialic acid residues which serve as the receptors for the viral hemagglutinin. Poliovirus, which binds to an immunoglobulin-like host cell receptor (MENDELSOHN et al. 1989), also can initiate infection at both the apical and basolateral surfaces of epithelial cell lines (TUCKER et al. 1993a). Virus entry at either surface was inhibited by prior incubation with a specific monoclonal antibody which is known to block viral attachment to the poliovirus cellular receptor (NOBIS et al. 1985). These results indicate that the poliovirus receptor is expressed on both the apical and basolateral surfaces at sufficient levels to mediate virus entry from either membrane domain. The presence of the same protein on both the apical and basolateral surfaces would be unusual, and it will be of interest to determine whether the poliovirus receptors present on each membrane domain represent different modifications of the receptor molecule.

Some viruses may bind to more than a single type of host cell receptor molecule, and the possibility exists that their entry may be mediated by binding to a different receptor species on each membrane domain. This appears to be the mechanism by which herpes simplex virus enters epithelial cells via either the apical or basolateral surface. The infectious entry of a glycoprotein C (gC)-deficient herpes simplex virus 1 (HSV-1) mutant occurs exclusively on the basolateral surface of canine kidney (MDCK) cells, whereas wild type HSV-1 was able to infect MDCK cells following adsorption to either the apical or basolateral surfaces (SEARS et al. 1991). It was thus concluded that HSV-1 infection is mediated by two different host cell receptor molecules; one expressed on the apical domain which must interact with gC to mediate infection, and a second located on the basolateral surface which binds to a different viral attachment protein (SEARS et al. 1991). As discussed below, alternative receptors have also been demonstrated in the case of HIV-1.

3 Virus Release

Early studies of the release of viruses from polarized epithelial cells were carried out by MURPHY and BANG (1952), who first observed the release of influenza virus by budding from the surface of the chorioallantoic membrane of embryonated eggs. They found that the release of virus was polarized, occurring exclusively at the free apical surface. More recently, a number of lines of cultured cells were found to retain many of the structural and functional properties of polarized cells from epithelial tissues, including the presence of junctional complexes and distinct apical and basolateral plasma membrane domains with distinct compositions. RODRIGUEZ-BOULAN and SABATINI (1978) observed the polarized budding of enveloped viruses from MDCK cells; VSV budded predominantly from the basolateral plasma membrane whereas influenza and Sendai virions were released exclusively from the apical domain. Polarized release appears to be a general feature of viruses that are released by budding at the plasma membrane (Table 2). Subsequent studies of these viruses have revealed that the glycoproteins of enveloped viruses which assemble at the plasma membrane are directionally transported to the same surface from which the virus buds even when they are expressed in the absence of other viral components (ROTH et al. 1983; JONES et al. 1985; STEPHENS et al. 1986). These observations have led to the hypothesis that the site of expression of the envelope glycoprotein(s) determines the site of viral assembly.

Several enveloped viruses which assemble at intracellular membranes (Table 2) are also asymmetrically released from polarized epithelial cells. HSV 1 and 2, which assemble and bud at the inner nuclear membrane (DARLINGTON and MOSS 1968), were found to be preferentially released at the basolateral plasma membrane (SRINIVAS et al. 1986). Since herpesviruses bud into the lumen of the nuclear envelope which is continuous with the endoplasmic reticulum, the virions are presumably transported by a mechanism similar to that responsible for the release of soluble, secreted polypeptides by these cells. Bunyavirus assembly usually occurs by budding at smooth surfaced membranes in the Golgi complex, after which virus is transported to the cell surface. Punta Toro virus, a member of the *Phlebovirus* genus of bunyaviruses, was found to be released almost exclusively from the basolateral surface of polarized epithelial cells following its assembly in the Golgi complex (CHEN et al. 1991). Immunoelectron microscopic analysis of hepatocytes infected with another bunyavirus, Rift Valley fever virus, also revealed an apparent selective release from the basolateral domain (ANDERSON

Table 2. Direction of virus release from polarized epithelial cells

Release process	Virus	Polarity	Reference
Enveloped viruses			
Budding at the cell surface	Influenza	Apical	RODRIGUEZ-BOULAN and SABATINI 1978
	Paramyxo	Apical	RODRIGUEZ-BOULAN and SABATINI 1978
	Vesicular stomatitis	Basolateral	RODRIGUEZ-BOULAN and SABATINI 1978
	Retro (C type)	Basolateral	ROTH et al., 1983
	HIV	Basolateral	OWENS et al., 1991
	Alpha, FRT cells	Apical	ZURZOLO et al., 1992
	Alpha, CaCo-2 cells	Basolateral	ZURZOLO et al., 1992
Budding at intracellular membranes and release by vesicular transport	Herpes	Basolateral	SRINIVAS et al., 1986
	Bunya	Basolateral	CHEN et al., 1991
Non-enveloped viruses			
Release without lysis	SV40	Apical	CLAYSON et al., 1989
	Polio, CaCo-2 cells	Apical	TUCKER et al., 1993b
	Polio, Vero cells	Bidirectional	TUCKER et al.,, 1993b
Release by lysis	?	?	

and SMITH 1987). The vectorial release of viruses that assemble at intracellular membranes suggests either that the virions express targeting signals recognized in a luminal context, or that an interaction occurs between virions and a targeted membrane bound factor(s). In this respect the release of such virions may resemble the transport of endogenous cellular secretory products which are released in a directional fashion.

The release of certain nonenveloped viruses has also been found to occur in a directional manner. Recent studies using SV40 and poliovirus have indicated that these nonenveloped viruses may also be targeted for release at a particular plasma membrane domain. SV40 is a nonenveloped DNA virus which is assembled in the nucleus of infected cells. SV40 virions were observed to accumulate on the apical surface of polarized Vero C1008 and primary African green monkey kidney epithelial cells prior to any detectable cell lysis (CLAYSON et al. 1989). Released virus was also detected almost exclusively in the apical culture medium. Treatment with monensin, an inhibitor of vesicular transport, resulted in almost complete inhibition of SV40 release but had no effect upon virus-specific protein synthesis or the assembly and intracellular accumulation of infectious virus, suggesting the release of SV40 may be mediated by a vesicular transport mechanism. Consistent with this hypothesis, cytoplasmic virions were found within membranous vesicular compartments during the period of virus release (CLAYSON et al. 1989). No information is available on the possible route by which virions could exit from the nucleus into cytoplasmic vesicular compartments.

Poliovirus assembles in the cytoplasm, and virions have been detected by electron microscopy in several forms; free within the cytoplasm, tightly packed in crystalline arrays and within membrane-enclosed bodies (DALES et al. 1965; DUNNEBACKE et al. 1969). Poliovirus was found to be released predominantly from the apical domain of Caco-2 intestinal cells (TUCKER et al. 1993b). Poliovirus release is generally considered to be mediated by cellular lysis, but several apparently nonlytic release processes have also been described (DUNNEBACKE et al. 1969; BIENZ et al. 1973). The mechanism of vectorial release of poliovirus from Caco-2 cells is not yet known. Possibilities include the targeting of poliovirus-containing vesicles to the apical plasma membrane or directional lysis of the apical plasma membrane. Interestingly, monensin had little or no effect upon the release of poliovirus from Caco-2 or Vero C1008 cells (TUCKER et al. 1993b) suggesting that the mechanism of release may differ from that used by SV40.

4 HIV Infection of Polarized Epithelial Cells

Several investigators have examined the interaction of HIV-1 with epithelial cells. Although HIV virions were not found to adsorb to enterocytes in explants of intestine (AMERONGEN et al. 1991), other studies have demonstrated the ability of HIV to infect intestinal and other types of polarized epithelial cells in culture (ADACHI et al. 1987; BOURINBAIAR and PHILLIPS 1991; YAHI et al. 1992a; TAN et al.

1993; MOYER and GENDELMAN 1991). Upon cocultivation of a line of intestinal epithelial cells with HIV-infected T cells or monocytes, the HIV-infected cells adhered to the surfaces of epithelial cells and a localized release of HIV virions occurred in regions of cell contact. The epithelial cells subsequently were productively infected and released progeny virions (BOURINBAIAR and PHILLIPS 1991; PHILLIPS and BOURINBAIAR 1992; TAN et al. 1993). Other investigators have demonstrated the infection of epithelial cells by cell-free virus. Galactosyl ceramide has been identified as a potential HIV-1 receptor, enabling virus infection of CD4-negative human intestinal epithelial cells lines (YAHI et al. 1992b; FANTINI et al. 1993). Following infection of several types of polarized epithelial cells, the release of virus was observed to occur preferentially at the basolateral surface (OWENS et al. 1991; FANTINI et al. 1991). Thus, the cell biology of the infection process provides a possible mechanism for HIV to transverse the epithelial barrier. In studies of a closely related primate model, it was found that SIV is able to establish infection and induce fatal immune deficiency by direct application of cell-free virus to the intact epithelial surface of the male or female genital tract (MILLER et al. 1989). Thus, a virus closely related to HIV can establish infection by interaction with an intact epithelium.

Evidence for the transepithelial transport of HIV-1 across an epithelial cell layer has also been obtained using mucosal explants from Peyer's patches of mice and rabbits (AMERONGEN et al. 1991). HIV virions bound to the luminal membranes of M cells of both species, were endocytosed, and delivered to the basolateral surfaces which adjoin mucosal lymphoid tissues. These results demonstrate the potential of HIV to transverse an intact epithelium by transcytosis, thus gaining exposure to underlying susceptible tissues.

Expression systems have been used to investigate the role of specific proteins of HIV-1 in determining the site of virus assembly and release in polarized epithelial cells. The HIV envelope protein, when expressed from a vaccinia virus recombinant, was preferentially expressed at basolateral membranes (OWENS and COMPANS 1989). HIV core proteins can be assembled into virus particles in the presence or absence of envelope proteins (HAFFAR et al. 1990; KARACOSTAS et al. 1989; SHIODA and SHIBUTA 1990), which provides an important advantage for these studies. To determine whether virus-like particles were released and to analyze polarity of release of core proteins in the presence or absence of envelope glycoproteins, the apical and basolateral media from recombinant vaccinia virus-infected epithelial cells were analyzed (OWENS et al. 1991). After metabolic labeling of filter-grown cells, HIV particles in media from the apical and basolateral chambers were individually pelleted for analysis of virion proteins. No proteins were observed in the samples from envelope-expessing cells, although such cells secrete gp120 into the culture media (OWENS and COMPANS 1989), indicating that the conditions used for analysis were sufficient to exclude soluble, nonvirion-associated envelope proteins. When medium from cells expressing core proteins was analyzed, a major protein band corresponding to p24 was observed in approximately equal amounts in both the basolateral and apical media samples, indicating that the transport of the capsid protein and its incorporation into released virus particles are nonpolarized under these conditions. In contrast, when the virion pellets released from cells coexpressing envelope and core proteins were analyzed, both the envelope glycoprotein and the core protein were found to be incorporated into particles and released predominantly at basolateral membranes. These results clearly indicate that in the presence of the envelope protein, HIV assembly and release preferentially occurs at basolateral membranes, whereas in the absence of *env* gene expression, nonpolarized budding and release of core particles occur. Thus, the site of expression of the envelope glycoprotein determines the site of release of core proteins, presumably via an interaction between these components during virus assembly. Evidence for a specific interaction between the matrix protein p17 and the intracytoplasmic domain of gp41 in determining polarized virus release was recently obtained by analysis of mutant forms of the envelope and matrix proteins (LODGE et al. 1994).

5 Polarized Entry and Release as a Determinant of Viral Pathogenesis

The distribution of viral receptors on epithelial surfaces is a primary determinant of the susceptibility of such cells to viral infection. If the receptors are present on apical surfaces, the epithelial cells can be infected by a virus present in the lumen. In contrast, if the receptors are expressed exclusively on the basolateral surfaces, a virus must first traverse the epithelial barrier by some mechanism before it can interact with appropriate receptors.

As discussed above, virus release by budding at specific plasma membrane domains may also play a role as a determinant of viral pathogenesis. A likely example is the parainfluenza viruses, which usually cause an infection that is restricted to the respiratory tract and are released by budding at apical surfaces of epithelial cells. Thus, the progeny virions are released into the lumen of the respiratory tract, and can infect adjacent epithelial cells. However, the virus is unable to reach underlying cells and tissues, and the infection process remains localized to the epithelial surface. In support of the concept that the size of virus release determines the ability of a virus to spread to other tissues, a mutant of murine parainfluenza (Sendai) virus, which was found to exhibit a bidirectional pattern of virus release from polarized epithelial cells, also caused a systemic infection (TASHIRO et al. 1990). Thus, the site of virus release appears to be a primary determinant of the pneumotropic infection by wild-type Sendai viruses vs the systemic infection caused by the mutant exhibiting bidirectional release.

In the case of HIV infection of the CNS, the virus must traverse an epithelial cell layer in the opposite direction from that which occurs at the initial stages of virus infection of a susceptible host. Virus particles in the blood have direct access to the basolateral surfaces of cells forming the blood brain barrier, and they must traverse the barrier from the basolateral to apical direction for penetration of the barrier to occur. Evidence has been reported for the presence of HIV receptors on basolateral membranes (YAHI et al. 1992a) as well as for the release of virus by budding at apical surfaces in some epithelial cell types (YAHI et al. 1992b). Thus, it is plausible that the cell biology of HIV entry and release could provide a mechanism for the virus to traverse the blood brain barrier.

6 Concluding Remarks

The interaction of viruses with polarized epithelial cells may be an important determinant of certain aspects of viral pathogenesis. The restriction of receptor molecules to particular plasma membrane domains is likely to determine whether a virus can initiate infection by direct adsorption to the luminal surface or whether traversal of the epithelial layer must first occur. The polarity of virus release may be important for viruses to gain access to the circulatory system for subsequent spread to target tissues. Virus maturation at restricted membrane domains may also limit the access of components of the immune system to viral antigens. In addition, the directional release of virus particles may determine the pattern of virus shedding from infected epithelial tissues, which influences subsequent viral transmission. Thus, the cell biology of viral entry and release from infected epithelial cells may play a role in multiple aspects of the pathogenesis of virus infections.

In order to provide further information about the mechanism of CNS infection by HIV, it will be of interest to explore the use of in vitro cell culture models to examine viral interactions with the cell types forming the blood brain barrier. Additional studies using primate models will also be of great value in investigating the mechanism of CNS infection by closely related lentiviruses.

Acknowledgments. Research by the author was supported by grants AI12680 from the National Institute of Allergy and Infectious Diseases, and CA18611 from the National Cancer Institute. I thank R.V. Srinivas and Lawrence R. Melsen for assistance in preparing Fig. 1.

References

Adachi A, Koenig S, Gendelman HE, Daugherty D, Gattoni-Celli S, Fauci AS, Martin MA (1987) Productive, persistent infection of human colorectal cell lines with human immunodeficiency virus. J Virol 61: 209–213

Amerongen H, Weltzin R, Farnet C, Michetti P, Haseltine W, Neutra M (1991) Transepithelial transport of HIV-1 by intestinal M cells: a mechanism for transmission of AIDS. J AIDS 4: 760–765

Anderson GWJ, Smith JF (1987) Immunoelectron microscopy of rift valley fever viral morphogenesis in primary rat hepatocytes. Virology 161: 91–100

Ball J, Moldoveanu Z, Melsen LR, Kozlowski PA, Jackson S, Mulligan MJ, Mestecky JF, Compans RW (1995) A polarized human endometrial cell line which binds and transports polymeric IgA In Vitro Cell. Dev Biol 31: 196–206
Bienz K, Egger D, Wolff DA (1973) Virus replication, cytopathology, and lysosomal enzyme response of mitotic and interphase hep-2 cells infected with poliovirus. J Virol 11: 565–574
Bourinbaiar AS, Phillips DM (1991) Transmission of human immunodeficiency virus from monocytes to epithelia. J AIDS 4: 56–63
Chen S-Y, Matsuoka Y, Compans RW (1991) Golgi complex localization of the Punta Toro virus G2 protein requires its association with the G1 protein. Virology 183: 351–365
Clayson ET, Compans RW (1988) Entry of SV40 is restricted to apical surfaces of polarized epithelial cells. Mol Cell Biol 8: 3391–339
Clayson ET, Brando LVJ, Compans RW (1989) Release of SV40 virions from epithelial cells is polarized and occurs without cell lysis. J Virol 63: 2278–2288
Dales S, Eggers HJ, Tamm I, Palade GE (1965) Electron microscopic study of the formation of poliovirus. Virology 26: 379–389
Darlington RW, Moss LJI (1968) Herpesvirus envelopment. J Virol 2: 48–55
Dropulic B, Masters CL (1990) Entry of neurotropic arboviruses into the central nervous system: an in vitro study using mouse brain endothelium. J Infect Dis 161: 685–691
Dunnebacke TH, Levinthal JD, Williams RC (1969) Entry and release of poliovirus as observed by electron microscopy of cultured cells. J Virol 4: 505–53
Fantini J, Baghdiguian S, Yahi N, Chermann J-C (1991) Selected human immunodeficiency virus replicates preferentially through the basolateral surface of differentiated human colon epithelial cells. Virology 185: 904–907
Fantini J, Cook DG, Nathanson N, Spitalnik SL, Gonzalez-Scarano F (1993) Infection of colonic epithelial cell lines by type 1 human immunodeficiency virus is associated with cell surface expression of galactosylceramide, a potential alternative gp120 receptor. Proc Natl Acad Sci USA 90: 2700–2704
Fuller SD, Bonsdorff C-H, Simons K (1984) Vesicular stomatitis virus infects and matures only through the basolateral surface of the polarized epithelial cell line, MDCK. Cell 38: 65–77
Fuller SD, von Bonsdorff C-H, Simons K (1985) Cell surface influenza haemagglutinin can mediate infection by other animal viruses. EMBO J 4: 2475–2485
Haffar O, Garrigues J, Travis B, Morgan P, Zarling J, Hu S-L (1990) Human immunodeficiency virus-like, nonreplicating, gag-env particles assemble in a recombinant vaccinia virus expression system. J Virol 64: 2653–2659
Jones LV, Compans RW, Davis AR, Bos TJ, Nayak DP (1985) Surface expression of the influenza neuraminidase, an amino-terminally anchored viral membrane glycoprotein, in polarized epithelial cells. Mol Cell Biol 5: 2181–2189
Karacostas V, Nagashima K, Gonda MA, Moss B (1989) Human immunodeficiency virus-like particles produced by a vaccinia virus expression vector. Proc Natl Acad Sci USA 86: 8964–8967
Kristensson K, Lundh B, Norrby E, Payne L, Orvell C (1984) Asymmetric budding of viruses in ependymal and choroid plexus epithelial cells. Neuropathol Appl Neurobiol 10: 209–219
Lodge R, Gottlinger H, Gabuzda D, Cohen EA, Lemay G (1994) The intracytoplasmic domain of gp41 mediates polarized budding of human immunodeficiency virus type 1 in MDCK cells. J Virol 68: 4857–4861
Mendelsohn C, Wimmer E, Racaniello VR (1989) Cellular receptor for poliovirus: molecular cloning, nucleotide sequence, and expression of a new member of the immunoglobulin superfamily. Cell 56: 855–865
Miller CJ, Alexander NJ, Sujipto S, Lackner AA, Gettie A, Hendrickx AG, Lowenstine LJ, Jennings M, Marx PA (1989) Genital mucosal transmission of simian immunodeficiency virus: animal model for heterosexual transmission of human immunodeficiency virus. J Virol 63: 4277
Moyer MP, Gendelman HE (1991) HIV replication and persistence in human gastrointestinal cells cultured in vitro. J Leucoc Biol 49: 499–504
Murphy JS, Bang FB (1952) Observations with the electron microscope on cells of the chick chorio-allantoic membrane infected with influenza virus. J Exp Med 95: 259
Nitsch L, Tramontano D, Ambesi-Impiombato FS, Quarto N, Bonatti S (1985) Morphological and functional polarity in an epithelial thyroid cell line. Eur J Cell Biol 35: 57–66
Nobis P, Zibirre R, Meyer G, Kuhne J, Warnecke G,Koch G (1985) Production of a monoclonal antibody against an epitope on the Hela cells that is the functional poliovirus binding site. J Gen Virol 66: 2563–2569
Owens RJ, Compans RW (1989) Expression of the HIV envelope glycoprotein is restricted to basolateral surfaces of polarized epithelial cells. J Virol 63: 978–982

Owens RJ, Dubay J, Hunter E, Compans RW (1991) The human immunodeficiency virus envelope protein determines the site of virus release in polarized epithelial cells. Proc Natl Acad Sci USA 88: 3987–3991

Phillips DM, Tan X (1992) Mechanism of trophoblast infection by HIV. AIDS Res Hum Retroviruses 9: 1697–1705

Phillips DM, Bourinbaiar AS (1992) Mechanism of HIV spread from lymphocytes to epithelia. Virology 186: 261–273

Rindler MJ, Traber MG (1988) A specific sorting signal is not required for the polarized secretion of newly synthesized proteins from cultured intestinal epithelial cells. J Cell Biol 107: 471–479

Rodriguez D, Rodriguez JR, Ojakian GK, Esteban M (1991) Vaccinia virus preferentially enters polarized epithelial cells through the basolateral surface. J Virol 65: 494–498

Rodriguez-Boulan E, Nelson J (1989) Morphogenesis of the polarized epithelial cell phenotype. Science 245: 718–725

Rodriguez-Boulan E, Pendergast M (1980) Polarized distribution of viral envelope proteins in the plasma membrane of infected epithelial cells. Cell 20: 45–54

Rodriguez-Boulan E, Powell SK (1992) Polarity of epithelial and neuronal cells. Annu Rev Cell Biol 8: 395–427

Rodriguez-Boulan E, Sabatini DD (1978) Asymmetric budding of viruses in epithelial monlayers: a model system for study of epithelial polarity. Proc Natl Acad Sci USA 75: 5071–5075

Roth MG, Compans RW, Guisti L, Davis AR, Nayak DP, Gething MJ (1983) Influenza virus hemagglutinin expression is polarized in cells infected with recombinant SV40 viruses carrying cloned hemagglutinin DNA. Cell 33: 435–442

Sabatini DD, Ivanov IE, Gottlieb TA, Compton T, Gonzales A, Beaudry G, Rindler MJ (1988) Use of cultured, virus-infected cells to study the biogenesis of polarity of epithelial cells. Ann Endocrinol 49: 270–286

Sears AE, McGwuire B, Roizman B (1991) Infection of polarized MDCK cells with herpes simplex virus 1: two asymmetrically distributed cell receptors interact with different viral proteins. Proc Natl Acad Sci USA 88: 5087–5091

Shioda T, Shibuta H (1990) Production of human immunodeficiency virus (HIV)-like particles from cells infected with recombinant vaccinia viruses carrying the gag gene of HIV. Virology 175: 139–148

Simons K, Wandinger-Ness A (1990) Polarized sorting in epithelia. Cell 62: 207–210

Srinivas RV, Balachandran N, Alonso-Caplen FV, Compans RW (1986) Expression of herpes simplex virus glycoproteins in polarized epithelial cells. J Virol 58: 689–693

Stephens EB, Compans RW, Earl P, Moss B (1986) Surface expression of viral glycoproteins in polarized epithelial cells using vaccinia virus vectors. EMBO J 5: 237–245

Tan X, Pearce-Pratt R, Phillips DM (1993) Productive infection of a cervical epithelial cell line with human immunodeficiency virus: implications for sexual transmission. J Virol 67: 6477–6452

Tashiro M, Yamakawa M, Tobita K, Seto JT, Klenk H-D, Rott R (1990) Altered budding site of a pantropic mutant of sendai virus, F1 -R, in polarized epithelial cells. J Virol 64: 4672–4677

Tucker SP, Compans RW (1993) Virus infection of polarized epithelial cells. In: Maramorosch K, Murphy FA, Shatkin AJ (eds) Advances in virus research, Vol 42. Academic, New York, pp 187–247

Tucker SP, Melsen LR, Compans RW (1993a) Bi-directional entry of poliovirus into polarized epithelial cells. J Virol 67: 29–38

Tucker SP, Thornton CL, Wimmer E, Compans RW (1993b) The vectorial release of the poliovirus from polarized human intestinal epithelial cells and the effect of Brefeldin A. J Virol 67: 4274–4282

Wolf JL, Rubin DH, Finberg R, Dambrauskas R, Trier JS (1981) Intestinal M cells: a pathway for entry of retrovirus into the host. Science 212: 471–472

Yahi N, Baghdiguian S,Bolmont C, Fantini J (1992a) Replication and apical budding of HIV-1 in mucous-secreting colonic epithelial cells. J AIDS 5: 993–1000

Yahi N, Baghdiguian S, Moreau H, Fantini J (1992b) Galactosyl ceramide (or a highly related molecule) is the receptor for human immunodeficiency virus type 1 on human colon epithelial HT-29 cells. J Virol 66: 4848–4854

Zurzolo C, Polistinia C, Saini M, Gentile R, Aloj L, Migliaccio G, Bonatti S, Nitsch L (1992) Opposite polarity of virus budding and of viral envelope glycoprotein distribution in epithelial cells derived from different tissues. J Cell Biol 117: 551–564

Traffic of Hematogenous Cells Through the Central Nervous System

K.C. Williams and W.F. Hickey

1 Introduction

As the interest in immunological events occurring in the central nervous system (CNS) has grown in the past quarter century, many investigators have repeatedly confirmed the relative absence of cellular elements of the immune system in the CNS of healthy mammals. This encompasses the lack of a significant number of T and B lymphocytes and low to nonexistent constitutive MHC class I and II antigen expression (Wekerle et al. 1986). Nevertheless, in numerous pathological conditions these cells are readily able to enter the CNS and contribute to salutary or destructive immunological processes. Scientific questions regarding the role and mechanisms of cellular entry are becoming clearer. These questions focus on the following concepts: What is the normal number of immunologically related cells types in the CNS? How do these cells gain access to the CNS under normal and pathological conditions? What important molecular signals and adhesion molecules do they require and how are these "signals" modulated in CNS pathology? In what processes do these hematogenous cells participate? And,

Department of Pathology, Dartmouth Medical School, Dartmouth Hitchcock Medical Center, Lebanon, NH 03756, USA how do their arrivals and departures vary in pathological conditions? From a variety of in vivo model systems and in vitro studies, the answers to these and related questions are now beginning to appear.

The discovery of HIV-1, and the well known tropism of this virus for $CD4^+$ T cells and macrophages, has given a somber but potent impetus for understanding cellular exchange between the vascular compartment and the nervous system. HIV produces pathological alterations in the CNS of approximately two thirds of those infected (JOHNSON et al. 1988; PRICE et al. 1988). Our understanding of which CNS immunological parameters are altered, thereby permitting the pathogenic changes to develop, is at this time rudimentary. Concern over which cells of the immune system participate in bringing the virus (and potentially other pathogens) to the CNS and whether such detrimental cellular traffic could be therapeutically altered provides an obvious link between the fields of neuroimmunology and AIDS research.

While HIV spread to the CNS is believed to occur very frequently early in the infection, it is in the later phases of the disease when the virus typically produces the neurologic syndromes of HIV encephalitis (HIVE) and the AIDS dementia complex (ADC) (PRICE et al. 1988; PRICE et al. 1991). Although patients with HIV infection do not die of HIVE or ADC, certainly the quality of life is severally affected, too commonly in a devastating manner. Additionally, there is considerable new evidence supporting the role that the CNS may play in augmenting or inhibiting functions of the immune system. Certainly, a compromised CNS may conceivably relate to the peripheral immune suppression noted in HIV infected individuals. There is no question that HIVE and ADC are a result of HIV infection in the CNS (HO et al. 1985; LEVY et al. 1985; KOENIG et al. 1986; MATHEZ et al. 1990). However, the contribution of the virus in directly mediating the observed CNS pathology, versus the contribution of endogenous glial cells and/or of infected, activated hematogenic cells trafficking into the CNS, is not defined.

Outside of the CNS the major cell types studied with regard to HIV infection are $CD4^+$ cells, macrophages and follicular dendritic cells (ROSENBERG and FAUCI 1989; MELTZER et al. 1990; PANTALEO et al. 1994; PANTALEO and FAUCI 1994). $CD4^+$T cells and macrophages are the major cell types in the body that are infected with HIV. T cells and macrophages both warrant consideration as a source of cell associated provirus routinely passing into the CNS. It is known that in the early stages of systemic HIV infection $CD4^+$T cells, monocytes, and macrophages exist in a highly activated state (PANTALEO and FAUCI 1994). It has also been demonstrated that monocyte/macrophages normally traffic into the CNS and replace a population of resident CNS macrophages (HICKEY and KIMURA 1988; HICKEY et al. 1992) and the rate of this described cell turnover is accelerated in CNS inflammation (LASSMAN et al. 1991b; LASSMANN et al. 1993). Additionally it has been demonstrated that both T cells (HICKEY 1991; Hickey et al. 1991) and B cells (TYOR et al. 1989; CSERR and KNOPF 1992; GRIFFIN et al. 1992; KNOPF et al. 1994) can enter the normal CNS parenchyma, a phenomenon that is thought to be dependent in part upon the level of T and B cell activation. Thus it would seem that as a direct result of early viral infection in the periphery, monocytes, T cells and B cells all might become activated to a sufficient level to cross the blood-brain barrier (BBB) and penetrate the CNS.

Macrophages and microglia are the major cell types demonstrated to be infected in the CNS. Autopsy studies of patients with HIV infection are usually representative of end stage AIDS and may not provide accurate information with regard to the early players in CNS infection. Additionally, the role of free virus, which exists in high concentrations in the cerebral spinal fluid (CSF) of HIV infected individuals (LEVY et al. 1985; HOLLANDER and LEVY 1987), cannot be entirely excluded as a source of CNS infection. In this review of cellular traffic between the blood and nervous system we will discuss the current state of knowledge regarding multiple cell types that traffic in to the CNS. Obviously, for an organ like the CNS which seems to have many mechanisms unique to itself, attention will be directed primarily toward experimental data which are CNS specific when such information is available.

2 General Considerations

The recirculation of cellular components of the immune system throughout the body on initial inspection appears to be a random, haphazard phenomenon. Yet, on closer investigation each cell type follows specific migratory rules relative to the cell surface molecules with which it is endowed, its state of activation, its phase of life and the nature of the vascular bed through which it passes (MACKAY et al. 1990; HARLAN and LIU 1992). Moreover, there are some general concepts and rules which are helpful in considering tissue entry by a hematogenous cell.

In general, cells circulating in the blood have few significant interactions with the vessel wall. When interactions do occur, they are orchestrated by molecules belonging predominantly to three general classes: integrins (HYNES 1992), selectins (LAWRENCE and SPRINGER 1991; BEVILACQUA and NELSON 1993), and members of the immunoglobulin supergene family (SPRINGER 1990). It appears that elements of each group participate in different phases of the cells impending emigration from the vascular compartment (SRINGER 1990; BUTCHER 1991; MCEVER 1992; SCHIMIZV et al. 1992). The initial interaction of a passing cell and the vessel wall produce a slowing of cell's normal velocity and motion. The cell interacts with P- and E-selectin on the endothelium producing a "rolling" motion of the cell along the vessel's inner surface (LAWRENCE and SPRINGER 1991; SLOAN et al. 1992). Next, the cell continues to slow and finally becomes adhesive to the wall, a process believed to be mediated via ICAM-1 and VCAM-1 on the endothelium (BOYD et al. 1988; SLOAN et al. 1992). Ultimately, the adherent cell identifies a suitable site of egress, and via a series of other cell-cell signals and the elaboration of basement membrane degrading enzymes (NAPARSTEK et al. 1984; IHRCKE et al. 1993), the hematogenous cell enters the organ (SLOAN et al. 1992). No complete list detailing the necessary adhesion molecules, intercellular signals, cytoskeletal reorganizations, and enzyme secretions exist for any circulating immunological cell. Nevertheless, this is one of the most active and fertile areas of current immunological research. At the outset it should be stated that there is no currently defined CNS specific adhesion molecules expressed on CNS endothelial cells. Although such specific CNS "addressins" may be discovered, their existence is merely a theoretical possibility at this time.

One series of factors with which all circulating cells must deal relates to the endothelium. The endothelium in the normal CNS capillary is most probably designed to be a surface at which only minimal numbers of circulating cell/endothelial interactions occur (MALE et al. 1990; HICKEY 1991; LASSMANN et al. 1991a). It has been postulated that CNS endothelial cells are less permissive to the traffic of T cells than their counterparts in other tissues (MALE et al. 1990). Reviews of the expression of many of the known cellular adhesion molecules by endothelia in various organs frequently note that endothelium of small CNS vessels express such molecules at low or undetectable levels (BARKER and BILLINGHAM 1977; HICKEY, 1991; LASSMANN et al. 1991a; ROSSLER et al. 1992; SLOAN et al. 1992). It would contribute to the oft described paucity of cells entering the CNS under normal conditions. This is in dramatic contrast with the adhesion molecule expression on the high endothelial venules of lymphoid tissue which appear to be designed to facilitate a maximum amount of transendothelial traffic for multiple types of cells. Yet, in pathological circumstances or following specific manipulations, the activated CNS endothelial bed can come to closely resemble that of high endothelial venules (Fig. 1) (CANNELLA et al. 1990; SOBEL et al. 1990; LASSMANN et al. 1991; ROSSLER et al. 1992; SASSEVILLE et al. 1992). Thus, the state of the endothelium, resting vs activated, must be taken into account when considering hematogenous cell migration. Similarly, the activation state of the migrating cell must be considered as will be noted below.

3 T Cells

In the normal mammalian CNS, T lymphocytes are very rare (WEKERLE et al. 1986; HICKEY 1991). Indeed, this is one of the pillars supporting the concept of "immunological privilege" of the CNS (BARKER and BILLINGHAM 1977; WEKERLE et al. 1986). Nevertheless, T cells do appear in the CNS during numerous conditions when an immunological stimulus is sufficient to produce inflammation. Such conditions include viral encephalitis, multiple sclerosis (MS), and experimental systems like experimental encephalomyelitis (EAE), Theiler's virus encephalomyelitis, lymphocytic choriomeningitis virus infection (LCMV), herpes virus encephalitis (HVE) and simian immunodeficiency virus (SIV) infection. What are the parameters guiding such lymphocyte entry?

Many of the critical observations come from the experimental systems just noted. When it was discovered that lymph node cells could be used to adoptively

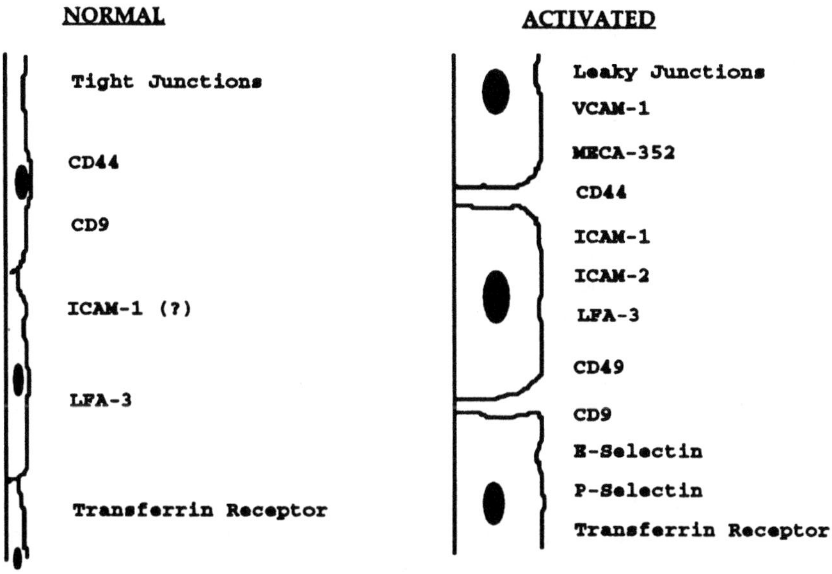

Fig. 1. CNS endothelial cell changes with activation. A comparison of the constitutive expression of immune accessory/adhesion molecules on normal, nonactivated, CNS endothelial cells compared to the de novo expression of accessory/adhesion molecules on activated CNS endothelium

transfer EAE to naive recipients whose CNS was totally normal (HOLDA et al. 1980), it became possible to investigate the mechanisms governing T cell entry into the CNS. One of the seminal observations was that EAE could not be produced unless the lymph node cells being transferred were stimulated! This stimulation could be effected by either the encephalitogenic protein (HOLDA et al. 1980; HINRICHS et al. 1981; TAKENAKA et al. 1986), or by a mitogenic lectin which had no similarity whatsover to the EAE inducing antigen (PANITCH et al. 1980; TAKENAKA et al. 1986). When EAE inducing T cell lines and clones became available which were specific for a single CNS antigen, again the phenomenon was observed: stimulated cells could produce disease and unstimulated ones could not (WEKERLE et al., 1986; HICKEY et al. 1991). These findings led Wekerle and coworkers to set forth the hypothesis that T lymphoblasts could gain access to the CNS while resting or naive T lymphocytes were excluded (WEKERLE et al. 1986). In their studies they also noted that activated T cells specific for ovalbumin could be detected in the rat CNS if given to the test animal in the activated state (WEKERLE et al. 1986). Thus, it remained to work out the parameters which permitted CNS entry and to define which molecules were active in the process.

Studies from this laboratory demonstrated that activated T cells could readily gain entry to the CNS while inactive ones were excluded (HICKEY 1991; HICKEY et al. 1991). Moreover, a number of other features about T cell traffic into the CNS were defined as well. An activated T cell not only entered the CNS with alacrity, but following stimulation it was embued with the power to enter any organ, including unlikely ones such as the eye and the thymus (HICKEY 1991; HICKEY et al. 1991). This entry was not dependent on the cell's phenotype, antigen specificity,

MHC restriction element compatibility or organ location of the antigen for which the T cell was specific. Only activation state was important. Thus, it seems that activated T lymphoblasts can enter any tissue. Entry seems to be random and in no way antigen specific; in fact, the molecules engaged in the complex process of antigen recognition appear to be irrelevant to T cell migration. The kinetics of the passage of T lymphoblasts into (and most often out of) the nervous system is rapid. The cells appear in the CNS within a few hours after gaining access to the circulation; they reach peak concentration in the CNS between 9 and 12 hours following injection; and by 24 h they have, for the most part, disappeared. If they find the antigen for which they are specific in the CNS, then their migration ends and they begin the process of producing inflammation. This latter process is never initiated unless the lymphocytes recognize their antigen in the appropriate MHC restricted manner, completely compatible with the familiar parameters of antigen presentation to T cells (GERMAIN 1981; SCHWARTZ 1985; ALLEN 1987).

Normally, T cells not in blast phase flow by the CNS endothelium with few if any interactions (Fig. 2A). This situation changes when the T cells become activated (Fig. 2B), when the endothelium is activated (Fig. 2C) or when a lymphoblast encounters an activated endothelial area (Fig. 2D). In the normal situation each of these states subserve different immunological purposes. While the lymphocyte state required for T cell entry seems clear, the specific adhesion molecules involved in this traffic remain to be elucidated. When resting or naive T cells are stimulated to the blast phase, there is a dramatic alteration in the cell surface molecules they express. Some molecules decrease in density, such as the T cell receptor (TcR) and certain "classical" homing receptors, some exhibit elevated levels of expression, and some are expressed de novo. Table 1 gives a list of molecules whose expression changes as the T cell enters blast phase. Some of these molecules may not be crucial to cell trafficking, certainly, MHC and TcR molecules seem not to be, but it is very difficult to define which ones play a critical role in the lymphoblast's acquired ability to penetrate tissue barriers. One would assume that the critical molecules would meet these criteria: new or significantly enhanced expression on lymphoblasts able to enter the CNS, the existence of a ligand for the molecule on nonactivated normal CNS endothelial cells, and independence from antigen recognition mechanisms. To date, certain molecules meeting these criteria are under consideration for key roles in trans-endothelial migration.

Activated T cells that express a very late antigen (VLA-4) are able to bind to VCAM-1 and then enter into the CNS (BARON et al. 1993). VLA-4 expression is upregulated on activated T cells, perhaps explaining in part the differential ability of activated T cells to enter into the CNS parenchyma. Yednock and coworkers have been able to inhibit the clinical and histologic signs of EAE by administering an antibody against the $\alpha4\beta1$ (VLA-4) integrin given 2 days after rats were injected with CNS specific T cells (YEDNOCK et al. 1992) These data were confirmed in a mouse model of EAE (BARON et al. 1993). There is a similar report of the inhibition of the induction of EAE using anti-ICAM-1 mAb (ARCHELOS et al. 1993). However,

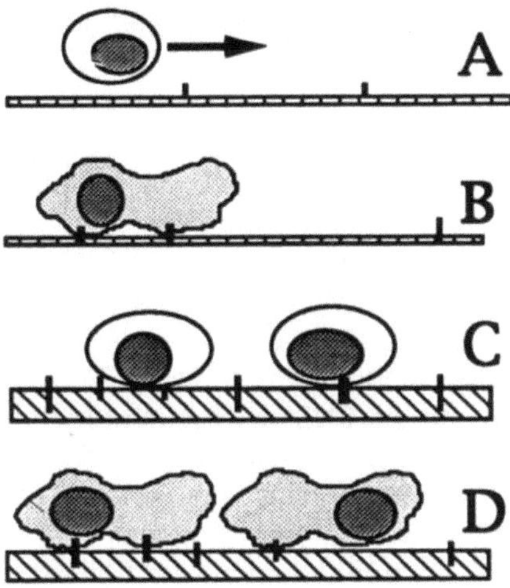

Fig. 2A–D. Interactions of T cells and CNS endothelium during normal and inflammatory conditions. **A** Normal lymphocyte intravascular flow; nonactivated T cells do not significantly interact with CNS endothelium. **B** Initiation of inflammation and immunological surveillance demonstrated by the interaction of an activated T cell blast with nonactivated CNS endothelium. **C** Homing of T cells and recruitment of memory T cells to the activated CNS endothelium at the inflammatory site. **D** Augmentation of inflammation by the interaction of the CNS endothelium with additional T cell blasts resulting in the potentiation of inflammation

Table 1. Cell surface molecules that exhibit alterations in level of expression when a T cell enters blast phase (partial list)

T cell receptor
CD40 ligand (p39)
VLA-4
LFA-1, LFA-2
CD2, CD4, CD5, CD6, CD31, CD44, CD45R0, CD54 (ICAM-1)
L-selectin
ICAM-3
MEL-14·
MHC class II

in situations in which migration blocking antibodies are administered days after the encephalitogenic cells are given, the disease inhibiting effect is most probably not due to the inhibition of initial entry by encephalitogenic T-cells. Their entry occurs within hours of injection. Thus, the disease preventing effect of such adhesion molecule blocking is probably via a mechanism that inhibits the recruitment of additional cells to an inflammatory site, thereby preventing the full development of a pathogenic inflammatory focus. Moreover, it is uncertain that the normal CNS endothelial cells through which activated T cells pass express the molecules VCAM-1 or ICAM-1. It remains an open question whether other known (or unknown) adhesion molecules are the active ligands for that critical initial T lymphoblast entry. This is currently the subject of intense study.

While it is safe to conclude that T cell entry into the CNS and antigen recognition by those T cells are independent processes, it is totally uncertain whether one or both processes are critical in the dissemination of HIV to the nervous system. A subset of the T lymphocytes is the CD4$^+$ T cell, the so-called T helper cell. The CD4 molecule, which is a high affinity receptor for the HIV protein gp 120, has been identified as an essential molecule for HIV infection of cells of the immune system (ROSENBERG and FAUCI 1989). Certainly, activated T cells possess many aspects that would make them prime suspects for the dissemination of HIV into the CNS! CD4⁺ T lymphocytes are the principle cell in the immune system to suffer the ravages of HIV infection; they pass through the CNS (and all other organs) when activated; in an activated state they are secreting cytokines and potentially shedding virus, which in concert may serve to both stimulate and infect microglial cells that are the principle endogenous CNS cell suffering sustained HIV infection (KOENIG et al. 1986; WILEY et al. 1986; VAZEUX et al. 1987; MICHAELS et al. 1988a,b; KURE et al. 1990a,b; GOSZTONYI et al. 1994). Is the infected, activated lymphocyte the principle mode of HIV transport into the nervous system? Can it spread the infection into the CNS as a result of merely "passing through"? Or are the phenomena associated only with antigen recognition and cessation of lymphocyte wandering needed for this virus laden cell to pass the infection to endogenous CNS cells? Answers to these queries are not available, but obviously the entire issue of T cell traffic through the CNS is of central importance.

4 Endothelial Cells and the Blood-Brain Barrier

The elements present at the BBB are the endothelial cells, pericytes, astrocyte foot processes and perivascular "microglial" cells. While the BBB under "basal" conditions is relatively impermeable to the cells, proteins and other macromolecules of the blood, perhaps even more so that its counterparts in other organs (MALE et al. 1990), the BBB changes rapidly in response to various pathological stimuli including acute inflammatory responses and viral infection (LASSMANN et al. 1991a; SOBEL et al. 1991; WEKERLE et al., 1991; SASSEVILLE et al. 1992; SASSEVILLE et al. 1994). The changes of adhesion molecules expressed on endothelial cells as they pass from the basal to the activated state are incompletely defined. Figure 1 depicts some of the changes which occur on CNS endothelia with activation, yet the list is rapidly expanding as new molecules are identified and their regulation elucidated. Unquestionably, the secretion of interleukin (IL)-1β, tumor necrosis factor (TNF)-α and/or interferon-α (IFN-α) in the vicinity of a vessel drastically alters the endothelial adhesion profile (WEKERLE et al. 1991). T lymphocytes and members of the monocyte/macrophage family produce such substances when activated. It has been shown that the BBB can become activated in a variety of CNS diseases where these cytokines (TYOR et al. 1992a) and HIV proteins, including *tat* proteins, are present. IL-1, TNF-α, and *tat*, all of which are elevated in the CNS and serum of HIV infected individuals (WAHL et al. 1989; GRIMALDI et al. 1991; GENIS et al. 1992; TYOR et al. 1992a; NUOVO et al. 1994), can stimulate de novo expression of molecules necessary for trafficking of leukocytes into the CNS (POBER et al. 1987; LASSMANN et al. 1991a; WEKERLE et al. 1991).

In the murine CNS during EAE, adhesion molecules characteristic of the high endothelial venules of lymphoid tissues are expressed de novo (CANNELLA et al. 1990). This coincides with inflammation and illness. Their expression then decreases concomitant with remission (CANNELLA et al. 1990). The up-regulation of BBB adhesion/accessory molecules and MHC class II expression on perivascular and parenchymal microglia has likewise been described in a series of human CNS viral infections. These molecules include ICAM-1, up-regulated on endothelial vessels and parenchymal microglia (SOBEL et al. 1990), and soluble ICAM-1 found in serum and CSF (HARTUNG et al. 1993). It has been recently suggested that soluble ICAM-1 may serve as an activation marker in cases of viral encephalitis (HARTUNG et al. 1993).

In AIDS associated infections of the brain similar endothelial changes are noted. Additionally, the up-regulation of MHC antigens is reported (ACHIM et al. 1991a; ACHIM and WILEY 1992; DECKERT-SCHLUTER et al. 1994; RINALDO et al. 1994). Increased ICAM-1 and MHC class II expression has been documented in cases of PML (progressive multifocal leukoencephalopathy) (ACHIM and WILEY 1992). VCAM-1 also is elevated in the CNS during SIV infection (SASSEVILLE et al. 1992). An in depth examination of adhesion/accessory molecules on the endothelium of murine *Toxoplasmosis* infection has demonstrated a similar up-regulation of the adhesion molecules ICAM-1 and VCAM-1, as well as ligands for the macrophage/monocyte adhesion molecules LFA-1, MAC-1, CD43, VLA-4 (DECKERT-SCHULTER et al. 1994). The difficulties with the interpretation of such molecular expression in human pathological studies of CNS infections are: (a) the inability to discern how long these individuals have been infected and (b) whether the elevated adhesion molecule expression preceded, was coincident with, or occurred as a result of the establishment of the disease focus. Finally, overt or subclinical inflammatory/infectious events may also play a facilitating role in viral spread to the brain. In an animal model of neurotropic viruses it has been demonstrated that bacterial lipopolysaccharide (LPS) in the CNS results in a massive influx of free virus secondary to activation and leakiness of the BBB (LUSTIG et al. 1992). In vitro stimulation of human endothelial cells by recombinant lymphotoxin also directly results in endothelial cell activation (POBER et al., 1987). Additionally, BBB endothelial cells express FcRIII (ULVESTAD et al. 1994b), and these are up-regulated during inflammation. FcRIII can bind immune complexes that might include virus coated with IgG, thereby providing another mechanism of viral entry into the CNS. It must be noted, however, that neither the infection of brain endothelium with HIV nor transendothelial HIV transport has been convincingly demonstrated (VAZEUX et al. 1987; KURE et al. 1990a,b).

In a study of primate SIV infection, probably the best available model of HIV infection (LACKNER et al. 1991), it was demonstrated by immunocytochemistry and competitive binding assays that human monocytes and B cell lines bound preferentially to CNS vessels of infected monkeys and did not bind to the endothelium of noninfected animals (SASSEVILLE et al. 1994). The binding of the B cells and monocytes correlated with the expression of VCAM-1 on the CNS vessels and could be blocked by pretreating the vessels with an antibody against VCAM-1 or by treating the binding cells with an antibody against VLA-4, a VCAM-1 ligand. Preincubating the cells with either antibody resulted in greater than 70% inhibition of binding (SASSEVILLE et al. 1994). These data complement the previous discussion of the parameters of T cell entry into the CNS suggesting that the baseline of endothelial adhesion molecule expression, and presumably the traffic of hematogenous cells through the CNS, also is dramatically altered when SIV infection occurs.

In addition to the established evidence of an activated BBB during encephalitis, there also is evidence, similar to that seen in MS, of BBB breakdown. The observation of BBB defects in HIV infection has been postulated to partially explain the diffuse "demyelination," or myelin pallor, observed in HIV CNS infection (RHODES 1987). Microvascular changes noted in HIV pathology include the observation of mural thickenings, increased vessel wall cellularity, and enlargement and pleomorphism of endothelial cells (SMITH et al. 1990). Further evidence of a compromised BBB has been suggested by the observation of deposited serum proteins including IgG and fixed complement in glial cells and neurons (RHODES 1991). A recent study of the CNS of patients with HIV infection showed complement deposition in the choroid plexus, suggesting a role for complement mediating BBB damage (FALANGOLA et al. 1994). Others have made a similar observation relative to IgG and serum deposition, thus offering the deposition of serum protein across a compromised BBB as a pathological correlate of dementia in ADC (POWERS et al. 1993).

Overall, it is difficult to minimize the importance of endothelial and BBB changes in HIV spread to the CNS and in the associated CNS problems of AIDS.

5 The Macrophage/Monocyte Family

The circulating members of this family are not homogeneous. Likewise in the CNS, based on immunohistochemical staining patterns, morphology and location, there are a minimum of five cell types which qualify for membership in this group (Fig. 3). They are: the meningeal macrophage, the macrophage of the Virchow-Robin space, the perivascular microglial cell, the parenchymal microglial cell, and the choroid plexus macrophage (HICKEY et al. 1992b). All of them are bone marrow derived and are dynamic populations in the CNS (PERRY et al. 1985; OEMICHEN et al. 1978; JORDAN and THOMAS 1988; HICKEY and KIMURA 1988; GRAEBER et al. 1989; LASSMANN et al. 1991b; MATYSZAK et al. 1992; BRAUN et al. 1993; LASSMANN et al. 1993; PERRY 1994). Their kinetics of appearance in the nervous system appears to be relatively distinct for each cell type (HICKEY et al. 1992b; LASSMANN et al. 1993; PERRY 1994). While these could conceivably be various phases in the life of a single cell type, this hypothesis has not been supported. Available evidence from in vitro and in vivo studies seems to indicate that they are distinct subtypes

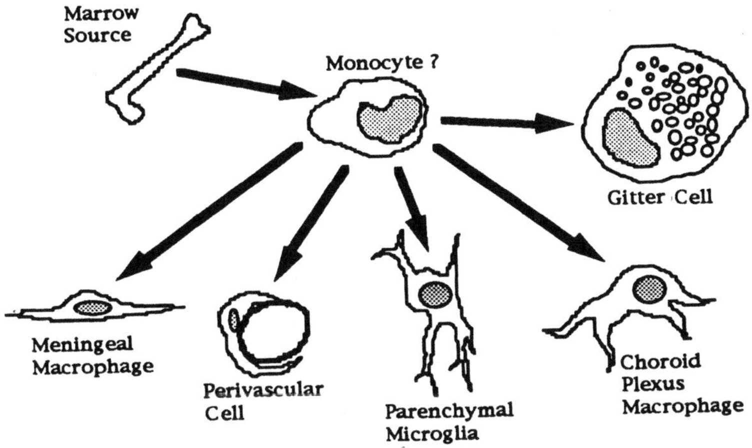

Fig. 3. Monocyte and macrophage populations of the CNS: (1) macrophages of the meninges; (2) perivascular cells or perivascular "microglia" that are continuously repopulated in the CNS and are a candidate cell for carrying HIV into the CNS; (3) resident parenchymal microglia; (4) choroid plexus macrophages that are demonstrated to be an early CNS cell infected with HIV and might represent a reservoir of the virus; (5) gitter cells that are "activated macrophage" with demonstrated phagocytized material in their cytoplasm

(Oemichen et al. 1978; Perry et al. 1985; Dickson et al. 1991; Matyszak et al. 1992; Braun et al. 1993; Perry 1994).

In a number of species including humans, the monocyte/macrophage can express the CD4 molecule (Crocker et al. 1987; Peudenier et al. 1991b). In many organs such cells express this molecule constitutively, and it is inducible on the monocytic elements in the CNS (Crocker et al. 1987; Perry and Gordon 1987). As a cell group that can become infected with HIV in the periphery, the status of this family's members in the nervous system is of great interest (Koenig et al. 1986; Dickson et al. 1991). In contrast to infection of the $CD4^+$ cell by HIV, infection of monocytes/macrophages results in a nonlytic infection enabling the cell to harbor the virus during the latent period of infection, sequestered from normal immune surveillance. Indeed, the macrophages' unique mode of viral infection and growth (Meltzer et al. 1990) have lead them to be considered the major reservoir of HIV (Fauci 1988; Meltzer et al. 1990). Studies of diseased CNS tissue from AIDS patients, as well as elegant in vitro studies, clearly demonstrate that the microglial cells and the perivascular cells require close scrutiny with regard to HIV infection (Watkins et al. 1990; Achim et al. 1991b; Dickson et al. 1991; Jordan et al. 1991). While CNS macrophages in the choroid plexus and the meninges are demonstrated to be HIV positive and possibly represent the earliest cells infected in the CNS (Harouse et al. 1989), cells with the morphology of parenchymal microglia are HIV positive by in situ hybridization for HIV DNA and by immunocytochemistry for HIV proteins (Wiley et al. 1986; Koenig et al. 1986; Vazeux et al. 1987; Michaels et al. 1988a; Price et al 1988; Kure et al. 1990a,b; Dickson et al. 1991). Whether the meningeal or choroid plexus macrophages can function as a reservoir for virus that subsequently infects the CNS is not known (Kure et al. 1990a,b).

5.1 Microglial Cells

Microglia cells represent a population of brain macrophages that are bone marrow derived and establish residence early in development. Their turnover after birth is extremely low in experimental systems (HICKEY et al. 1992b). Also, the factors that initially attract and subsequently retain them in the CNS are undefined. Parenchyma microglia in the CNS have many markers of cells of myeloid lineage, albeit often at low and normally undetectable levels, possibly the result of their being sequestered from serum cytokines and proteins (PERRY and GORDON 1989). Interestingly, there are some notable differences between rodent and human microglia that are important with respect to assumptions made about the role of these cells contributing to CNS HIV pathology. While rodent microglia express variable levels of CD4 molecules and proliferate in vitro and in situ in response to injury, fetal and adult human microglia have low to undetectable levels of CD4 expression (CROCKER et al. 1987; PEUDENIER et al. 1991a; WILLIAMS et al. 1992) and do not undergo significant proliferation in vitro (PEUDENIER et al., 1991a; WILLIAMS et al. 1992). Infection of fetal and adult human derived microglia by HIV has been demonstrated in vitro (WATKINS et al. 1990; JORDAN et al. 1991; PEUDENIER et al., 1991a). Evidence that other CNS cells including oligodendrocytes, endothelial cells, neurons, and astrocytes are susceptible to HIV infection in vitro is less convincing, although it is believed that astrocytes are infected in infants who are born with viral infection (TORNATURE et al. 1991; TORNATURE et al. 1994). The infection of adult human derived microglia in vitro has been demonstrated to be specific to monocyte-tropic HIV isolates possibly determined by a region of the *env* glycoprotein that also controls macrophage tropism (SHARPLESS et al., 1992a, b). The infection of microglia in vitro has been demonstrated to be blocked with soluble anti-CD4 antibodies (WATKINS et al. 1990; JORDAN et al. 1991) although the expression of CD4 on adult human microglia in vitro and in situ has been difficult to demonstrate (WATKINS et al. 1990; DICKSON et al. 1991). Interestingly, these microglia upon infection in vitro can fuse similar to multinucleated giant cells. Also, the infection of adult human derived microglia results in lysis of the cells in vitro (WATKINS et al. 1990), contrary to monocyte/macrophage infection by HIV that is nonlytic (MELTZER et al. 1990). Evidence supporting the infectability of microglia having low to undetectable levels of CD4 is found in studies of the adult thymus where triple negative ($CD3^-$, $CD4^-$, $CD8^-$) cells have been infected via a CD4-dependent mechanism which can be blocked with anti-CD4 antibody (SCHNITTMAN et al. 1990). These studies demonstrate that even an extremely low number of CD4 molecules are sufficient for HIV infection. Indeed, the study that demonstrated in vitro infection of adult human microglia was only able to show CD4 expression by the cells using a highly sensitive PCR protocol with nested primers and 40 cycles of amplification (WATKINS et al. 1990)!

In addition to the CD4 molecule as a route of viral infection, microglia also express Fc receptors for the constant region of IgG and receptors for complement (C) (ULVESTAD et al. 1994a,b; WILLIAMS et al. 1994). Studies transfecting cDNA for FcR into fibroblasts have demonstrated this confers on them the ability to be infected with HIV via an FcR-dependent mechanism (FAUCI et al. 1991). Indeed FcR mediated infection of cells has been previously postulated as an alternative to CD4-dependent infection (TAKEDA et al. 1988; HOMSY et al. 1989). Complement receptors have also been demonstrated to function by enhancing HIV infection of Epstein-Barr Virus (EBV) carrying cells (TREMBLAY et al. 1990). Microglia in vitro can bind and phagocytize IgG or complement opsonized substances (ULVESTAD et al. 1994a, b; WILLIAMS et al. 1994;) and thus could theoretically bind opsonized virus or phagocytize dying cells that are HIV infected. Whether microglial cells in situ become infected by these mechanisms, by contact with infected T cells that pass in the vicinity, by contact with infected perivascular cells, or by free virus crossing in the area of a BBB leak remains to be demonstrated.

5.2 Perivascular "Microglial" Cells

Perivascular cells merit special attention. They are unique hematogenous cells that reside on the parenchymal side of the BBB, juxtaposed to both the CNS vasculature and the foot processes of astrocytes and microglia. These cells have been demonstrated to express MHC class II antigens in both rodents (HICKEY and KIMURA 1988) and humans (GRAEBER et al. 1992), and one of their main functions appears to be antigen presentation to T cells crossing the BBB (HICKEY and KIMURA, 1988). Perivascular cells are a component of the glia limitans (LASSMANN et al. 1991b) and are likely to harbor the HIV virus in the CNS, due to the fact that they are continually repopulated from the infected bone marrow (HICKEY and KIMURA 1988; UNGER et al. 1993), a process that is accelerated during CNS inflammation (LASSMANN et al. 1993). Moreover, they are the cells which accumulate and fuse into virus laden giant cells, a hallmark of HIVE. It is noteworthy that in HIV infected individuals, bone marrow progenitor cells have been demonstrated to be heavily infected with HIV (Connor and Ho 1992). It is probable that perivascular cells provide one major route by which viruses can continuously and diffusely enter the intact CNS. Perivascular cells become activated during a number of pathological states. It has been demonstrated in vitro that coculture of T cells with macrophages that are latently infected with HIV can result in the activation of macrophages and then viral expression (MELTZER et al. 1990; SCHRIER et al. 1990). Whether this mechanism is operative in the CNS where T cells may interact first with perivascular "microglial" cells has yet to be proven experimentally. Nevertheless, the great importance of this cell type should be underscored. Understanding the specific adhesion molecules which govern their physiological turnover in the nervous system may provide insight into the mechanism of HIV infection of the brain, the pathology of HIVE, and potential therapeutic modalities.

6 B Cells

Although B cells are not considered to be a normal CNS constituent in the adult human brain, it is possible that, similar to T cells, they have the ability to pass through a normal nervous system. Consistent but low number of B cells have been demonstrated in normal noninfected mouse brains and the number increases with the age of the mice (GRIFFEN et al. 1992). Low numbers of B cells have been demonstrated in HIV infected human and SIV infected nonhuman primate brain. Similarly, B cells have been demonstrated in the human and murine CNS in a wide variety of other pathological conditions ranging from viral infection to autoimmune disease (CSERR and KNOPF 1992; TYOR and GRIFFIN 1993). It remains enigmatic how these cells arrive on the scene. The precise roles of cell activation, adhesion molecules and antigen specific Ig secreted by these cells needs to be defined. Several questions with regard to the physiology of B cells in the CNS must be answered. They include: Do B cells disseminate in the CNS in a random fashion? Are B cells in the CNS representative of the peripheral immune repertoire or are they unique to the CNS?

Numerous studies have demonstrated that B cells can enter and accumulate in target organs in which their antigen is located. Rigorous investigations examining the kinetics of inflammatory cells entering into the CNS as a result of sindbis virus infections indicate that B cells enter into the CNS after (and perhaps dependent upon) the entry of other cells including CD4 and CD8 positive T cells and NK cells (GRIFFIN et al. 1992). The data in the sindbis encephalitis model, in which the B cell is a crucial component in neutralization of the viral load, suggest that B cell entry depends upon the activation state of the B cells. These studies have also demonstrated a continued secretion of Ig by B cells long after active viral encephalitis has resolved (TYOR et al. 1992b). It has been demonstrated that activated, but not resting, B cells can bind to the extracellular matrix of endothelium via E- and P- selectins paralleling the induction of Lewis^x, sialyl-Lewis^x, CD57, and CDx65 on the activated B cells (POSTIGO et al. 1994). Activation of human B cells in vitro results in the induction, de novo, of ligands for E- and P-selectins (POSTIGO et al. 1991; POSTIGO et al. 1993).

One important question with regard to B cell traffic into the CNS in viral encephalitis models and in autoimmune disease is whether B cell entry is antigen specific or random. In vivo studies in rats immunized in the periphery with a nominal antigen demonstrate that B cells and plasma cells specific for the antigen can be detected in the CNS when the same antigen is subsequently introduced into the CNS (HARLING-BERG et al. 1989; CSERR and KNOPF 1992; KNOPF et al. 1994). Therefore, the existence of an antigen in the CNS in some way serves as a sufficient stimulus to induce B cells to locate the CNS antigen depot. In the absence of antigen in the CNS, although the animal is immunized, no antigen specific B cells are detected in the brain.

Few studies have addressed the issue of the number of B cells in the CNS that secrete viral specific Ig and the life span of these cells. Griffin and colleagues studying the activation state of B cells entering the CNS and the differentiation state of these cells, have shown that B cells recovered from the CNS in early to mid viral infection, and late infections, produce virus specific antibody of the IgG or IgM isotype. This is in contrast to B cells obtained from spleen and peripheral blood of non-infected litter mates which are of the IgM and IgD producing types (TYOR et al. 1989; TYOR et al. 1992b; GRIFFIN et al. 1992). In HIV infection, similar to the phenomenon of elevated IgGs seen in multiple sclerosis, it is uncertain what the role of these IgGs are and whether an increased synthesis of IgG against viral proteins in HIV necessarily correlates with demonstrated pathology. A recent report has suggested that in HIV intrathecal IgG production does not correlate with CNS disease and pathology (GOSWAMI et al. 1991a,b). While in rodent encephalitis models, the IgG response seems to function to clear virus from the CNS (GRIFFIN et al. 1992; NAGRA et al. 1993), in HIV it is possible that the local presence of virus specific antibodies behind the BBB might also increase viral infection of neighboring cells, thereby enhancing HIV infectivity (HOMSY et al. 1989, 1990; TAKEDA et al. 1988; FAUCI et al. 1991; ROBINSON et al. 1990) The formation of opsonized viral complexes suggests an alternative route for viral infection of microglia and endothelial cells via FcRIII. These receptors are up-regulated during CNS inflammation, function to bind immune complexes (ULVESTAD et al. 1994a,b; WILLIAMS et al. 1994) and may allow interaction of virus with the CNS.

7 Natural Killer Cells and Cytotoxic Lymphocytes

Natural killer (NK) cells and cytotoxic lymphocytes (CTLs) are implicated in clearing virus from organs and therefore might serve a protective functional component of the antiviral immune response. NK cells and CTLs can lyse virally infected cells which ultimately may result in the clearance of virus from specific organs. However, the killing of a CNS cell expressing viral peptides may prove to be detrimental when several cells express such antigens.

NK cells are defined as being CD3-, large granular lymphocytes (LGLs) that are CD8 and asilo-GM1 positive (REYNOLDS et al. 1981; CHAMBERS et al. 1989). NK cells function to destroy virally infected and tumor cells in a non-MHC class restricted manner (HERBERMAN et al. 1986; TRINCHIERI et al., 1989). One study has implicated NK cells in the lysis of specific neurons suggesting a possible role for these cells in immune mediated damage (HICKEY et al. 1992a). Cytotoxic lymphocytes are $CD8^+$, $CD3^+$ cells that lyse target cells via MHC class I restricted mechanisms (ZINKERNAGEL and DOHERTY 1974). Both NK cells and CTLs have been demonstrated in the peripheral blood and CSF of HIV infected individuals, but their role in HIV infection and their ability to penetrate the normal BBB need investigation.

The role of NK cells in CNS viral infection has been demonstrated in several animal models including mouse hepatitis virus, sindbis virus encephalitis, Theiler's virus, herpes simplex encephalitis; and lymphocytic choriomenigitis virus (LCMV) (BAEZINGER et al. 1986; GRIFFIN and HESS 1986; PAYA et al. 1989; LINDSLEY et al. 1991; RODRIGUEZ et al. 1991; WILLIAMSON et al. 1991; HUDSON and STREILEIN 1994). Early observations in the sindbis viral encephalitis model noted numerous lymphocytes in the CSF of which a small percent were NK cells. In contrast to an antigen specific immune response that can take between 7 and 10 days to occur, the NK cells were demonstrated in the CSF of these animals within 24 h of viral infections (GRIFFIN and HESS 1986) similar to the quick response of NK cells seen in other organs (GRIFFIN and HESS 1986). The role of NK cells in clearance of virus has been shown in the Theiler's model by comparing the NK cell defective SJL mouse that is susceptible to viral infection with the resistant C57B7/10 mouse (LINDSLEY and RODRIGUEZ 1989; PAYA et al. 1989; RODRIGUEZ et al. 1991). In these studies, Paya and colleagues found that the viral titers of Theiler's in the CNS of SJL mice was higher than that found in the C57B7/10 mice (PAYA et al. 1989). Furthermore, these data correlated with a 50% lower NK cell number in the SJL animals in comparison with the C57BL/10 mice. If the C57BL/10 mice were depleted of NK cells using monoclonal antibodies (mAbs) against NK cells surface antigens, these animals came down with a diffuse encephalitis and meningitis shortly following infection with the virus (PAYA et al. 1989). In the rat EAE, NK cells are a very early participant. They appear to arrive 12–24 h after the disease inducing T cells, before significant endothelial cell activation occurs, and prior to the arrival of any other inflammatory cells type (Zhao, Sedgwick and Hickey, unpublished observations). This indicates that NK cells may utilize a mechanism of selective entry which is operational very early in inflammation and is dependent on T cells. Although there is an extensive literature covering NK cell phenotype and function in the periphery of HIV infected individuals, there are few reports of NK cells in the CNS of these patients. Nevertheless, there is reason to suspect they also come and go in the nervous system.

CTLs are readily demonstrated in the peripheral blood (CLERICI et al. 1991; KOENIG et al. 1988; WALKER et al., 1987) and the CSF (Sethi et al. 1988; JASSOY et al. 1992) of HIV infected individuals and have also been demonstrated, with less frequency, in the brain parenchyma (KURE et al. 1990a; WEIDENHIM et al. 1993) most usually within microglial and glial cell nodules (VAZEUX et al. 1987; ACHIM et al. 1991a; PATSOURIS et al. 1993). In fact, it has been suggested that there is an increased frequency of HIV specific CTLs in the CSF over that found in the peripheral blood (SETHI et al. 1988) Whether the increased frequency of HIV specific CTLs in the CSF represents recruitment of cells to the CNS or in situ expansion is not resolved.

Early studies reported the observation of HLA restricted, HIV specific CTLs recognizing target cells expressing either gp120 or inactivated HIV (SETHI et al. 1988). Interestingly, this report and a subsequent one have suggested that the clones can recognize the same HIV peptides both by MHC class I and II restricted mechanisms (SETHI et al. 1988; CLERICI et al. 1991). These observations support the notion that CTLs can recognize purified gp120 determinants with HLA molecules and that they can generate an immune response against gp120 determinants as a function of the hosts normal immune response (SETHI et al. 1988). A similar observation of MHC class I and II restricted response against gp160 peptides has been reported using clones derived from the peripheral blood of HIV seropositive patients (CLERICI et al. 1991). Subsequent studies have confirmed this observation showing CTL in the CSF of HIV infected individuals that recognized epitopes of *gag*, reverse transcriptase, *env*, and *nef* proteins. Also one finds a higher frequency of such clones in the CSF than in the peripheral blood (JASSOY et al. 1992).

One obvious difficulty with human studies of CNS viral infection is the inability to follow the early events of virus entering the CNS and the specific traffic of NK cells and CTL into the CNS as a result of viral infections. Similarly in animal models, due to the replication of viral gene products in numerous CNS sites, it is also difficult to study discrete patterns of traffic of these cells into the CNS. Oldstone and colleagues (1993) have made use of an elegant transgenic mouse system to address specific questions of CTL traffic to discrete brain areas. Using a SV40 T promoter, these researchers engineered animals that express LCMV proteins containing sequences with defined CTLs epitopes. By the adoptive transfer of MHC class I restricted, LCMV specific CTLs, this group was able to demonstrate the traffic of CTLs that were virus specific into the choroid plexus where the virus resides (OLDSTONE and SOUTHERN 1993). These data extend the observations of HICKEY et al. (1991), that activated T blasts, irrespective of MHC class I or II restriction and antigen specificity, can enter into the CNS parenchyma, demonstrating that if CTLs find their antigen in the context of the correct MHC complex, they will remain in the CNS (OLDSTONE et al. 1986; HICKEY et al. 1991).

Studying the traffic of CTLs in a herpes simplex encephalitis model in SJL mice, it has been demonstrated that they infiltrate the brain, can be recovered, and are cytotoxic to target cells immediately upon ex vivo isolation without the requirement of prior stimulation (HUDSON and STREILEIN 1994). In this model and in the Theiler's model mentioned earlier, CTLs were not recovered from the spleen or the lymph nodes of infected animals, but were readily recovered from the target organ (LINDSLEY et al. 1991; HUDSON and STREILEIN 1994). These observations offer the intriguing possibility that the CTLs might enter into the CNS as precursor cells that subsequently develop into effector cells once migration to the site of viral infection occurs (HUDSON AND STREILEIN 1994). Whether a similar phenomenon occurs in HIV infected brains and the role of NK cells and CTLs in mediating viral clearance or neural cell damage are undefined.

While $CD8^+$ cells are not implicated directly in carrying HIV into the CNS, it has been demonstrated that coculture of T cells with infected macrophages in vitro results in the production of virus by the macrophages (SCHRIER et al. 1990). This report suggests that these macrophages can subsequently infect adjacent cells possibly through cell-cell contact (SCHRIER et al. 1990).

8 Discussion

There are several problems inherent in the assumptions made concerning the function and contribution of hematogenous cells entering the CNS during HIV infection. The least of these is the inability in most cases to accurately assess how long an individual has been infected with virus and what the pathological state of the CNS was at the time the HIV spread occurred. In addition, there are the further confounding factors of genetic differences between individuals with regard to viral susceptibility and secondary infections of the CNS.

Currently, there are three defined stages of pre-AIDS-HIV infection: (1) an early, asymptomatic stage, characterized by B and T cell activation; (2) a stage at which virus proliferates beyond the control of the immune system, characterized by viremia; and (3) a stage of latent infection in which the viremia subsides (PANTALEO and FAUCI 1994). Neurologically, the signs and symptoms of HIVE or ADC do not usually present until later stages of viral infection (JOHNSON et al. 1988; PRICE et al. 1988, 1991). Thus, the neurologic signs of HIVE or ADC, other than initial meningitis, are not present until the late chronic stage, long after the viral infection in the CNS is established (JOHNSON et al. 1988; PRICE et al. 1988, 1991). For this reason it is difficult if not impossible to dissect the role of the $CD4^{+}$T cell and the monocyte in the initial CNS infection in humans. While it is generally assumed that infected monocyte/macrophages are the cells that enter into the CNS, the so called "trojan horse" hypothesis, the possibility of infected T cells effecting viral entry cannot be ruled out. One might surmise, based upon the postulated monocytic tropism of HIV found in the CNS (WATKINS et al. 1990; SHARPLESS 1992a,b), that indeed the monocyte/macrophage was the initial cell bringing virus into the CNS; however, there exists a high rate of viral mutation between and within individuals, making it difficult to assess cell infectivity based upon such CNS isolates (MELTZER et al. 1990). The possibility cannot be excluded that HIV infection of the nervous system is due to HIV shedding T cells that are highly stimulated and therefore able to enter the CNS, doing so and then dying perhaps by apoptosis, as occurs in most T cells entering the CNS during autoimmune models in rodents (PENDER et al. 1992; SCHMIED et al. 1993) and as is seen in HIV infected T cells in vitro (TERAI et al. 1991; GROUX et al. 1992). It has been demonstrated that virus released from $CD4^{+}$T cells in vivo can preferentially infect cells of the monocytic lineage in vitro (MASSARI et al. 1990). The question of whether virus release from T cells entering the CNS infects monocyte/macrophage/ microglia or whether apoptotic, HIV infected T cells are phagocytized by a monocytes/macrophages resulting in the infection of these scavenger cells is unanswered. It is noted that perivascular cells, microglia and microglia in microglial nodules, in addition to other brain macrophages, are the CNS cells most often demonstrated to be infected with HIV (NAVIA et al. 1986; KOENING et al. 1986; VAZEUX et al.1987; MICHAELS et al. 1988a,b; PRICE et al. 1988; KURE et al. 1990a,b). The mechanism of the spread of virus within the brain is unknown. Certainly, in the noninflamed adult CNS, microglia make up a reticular array or network that might suggest the possibility of passage of the virus from cell to cell (MELTZER et al. 1990; MIKOVITS et al. 1990; SCHRIER et al. 1990; PEUDENIER et al. 1991b). A similar mode of transfer of HIV has been demonstrated with T cells and between monocyte/macrophages and T cells in vitro during antigen presentation assays (MELTZER et al. 1990; SALAHUDDIN et al. 1986; MIKOVITS et al. 1990). Likewise, the role of scavenger, FcR and complement receptors on microglia and macrophages in the spread of virus seems important.

What can be said of a cell belonging to the immune system which is found in a tissue like the CNS? As a chance observation, very little can be derived from it. Important questions relative to its presence are: Was its entry selective or random? Does it represent a cell whose program is to search for antigen, or to initiate inflammation, or to respond to some soluble attractant for cells of that type? Was the primary stimulus for its entry some alteration in the cell's state, or a change in the CNS endothelia, or an interplay between both? Is the cell "passing through", or has it reached its destination? While these questions lie beyond the power of any current review to resolve, they must be borne in mind if one is to devise systems and/or collect data that will elucidate the forces driving, and the participants active in, hematogenous cell traffic through the nervous system. The overarching importance of these questions to the pathogenesis of HIV associated CNS disease is obvious.

References

Achim CL, Wiley CA (1992) Expression of major histocompatibility complex antigens in the brains of patients with progressive multifocal leukoencephalopathy. J Neuropath Exp Neurol 51: 257–263
Achim C, Morey MK, Wiley CA (1991a) Expression of major histocompatibility complex and HIV antigens within the brains of AIDS patients. AIDS 5: 535–541
Achim CL, Schrier RD, Wiley CA (1991b) Immunopathogenesis of HIV encephalitis. Brain Pathol 1: 177–184
Allen PM (1987) Antigen processing at the molecular level. Immunol Today 8: 270–273
Archelos JJ, Jung S, Maurer M, Schmied M, Lassman H, Tamatani T, Miyasaka M, Toyka KV, Hartung HP (1993) Inhibition of experimental autoimmune encephalomyelitis by an antibody to the intercellular adhesion molecule ICAM-1. Ann Neurol 34: 145–154
Barker CF, Billingham RE (1977) Immunologically privileged sites. Adv Immunol 25: 1–6
Baron JL, Madri JA, Ruddle NH, Hashim G, Janeway CA (1993) Surface expression of alpha4 integrin by CD4 T cells is required for their entry into brain parenchyma. J Exp Med 177: 57–68
Baezinger J, Hengartner H, Zinkernagel RM, Cole GA (1986) Induction or prevention of immunopathological disease by cloned cytotoxic T cell lines specific for lymphocytic choriomeningitis virus. Eur J Immunol 16: 387–391
Bevilacqua MP, Nelson RM (1993) Selectins. J Clin Invest 91: 379–387
Boyd AW, Wawrk SO, Burns GF, Fecondo JV (1988) Intercellular adhesion molecule 1 has a central rôle in cell-cell contact-mediated mechanisms. Proc Natl Acad Sci USA 95: 3095–3099
Braun JS, Kaissling B, Le Hir M (1993) Cellular components of the immune barrier in the spinal meninges and dorsal root ganglia of the normal rat: immunohistochemical (MHC class II) and electron-microscopic observations. Cell Tissue Res 273: 209–217
Butcher EC (1991) Leukocyte-endothelial cell recognition: three (or more) steps to specificity and diversity. Cell 67: 1033–1036
Cannella B, Cross AH, Raine CD (1990) Upregulation and coexpression of adhesion molecules correlates with relapsing autoimmune demyelination in the central nervous system. J Exp Med 172: 1521–1524
Chambers WH, Vujanovic NL, Deleo AB, Olszowym W, Herbermann RB, Hiserodt JC (1989) Monoclonal antibody to a triggering structure expressed on rat natural killers cells and adherent lymphokine activated killer cells. J Exp Med 169: 1373–1376
Clerici M, Lucey DR, Zajac RA, Boswell RN, Gebel HM, Takahashi H, Berzofsky JA, Shearer GM (1991) Detection of cytotoxic T lymphocytes specific for synthetic peptides of gp120 in HIV-seropositive individuals. J Immunol 146: 2214–2219
Connor RI, HO DD (1992) Etiology of AIDS: biology of human retroviruses In: ReVita VT Jr, Hellmann S, Rosenburg SA (eds) AIDS: etiology, diagnosis, treatment, and prevention, 3rd edn. JB Lippincott, Philadelphia, pp 13–38
Crocker PR, Jeffreries WA, Clark SJ, Chung LP, Gordon S (1987) Species heterogeneity in macrophage expression of the CD4 antigen. J Exp Med 166: 613–618

Cserr HF, Knopf PM (1992) Cervical lymphatics, the blood-brain barrier and the immunoractivity of the brain: a new view. Immunol Today 13: 507–512

Deckert-Schuler M, Schluter D, Hof H, Wiestler OD, Lassmann H (1994) Differential expression of ICAM-1, VCAM-1 and their ligands LFA-1, MAC-1, CD43, VLA-4 and MHC class II antigens in murine toxoplasma encephalitis: a light microscopic and ultrastructural immunohistochemical study. J Neuropath Exp Neurol 53: 457–468

Dickson DW, Mattiace LA, Kure K, Hutchins K, Lyman WD, Brosnan CF (1991) Biology of disease: microglia in human disease, with an emphasis on acquired immune deficiency syndrome. Lab Invest 64: 135–156

Falangola MF, Castro-Filho BG, Petito CK (1994) Immune complex deposition in the choroid plexus of patients with acquired immunodeficiency syndrome. Ann Neurol 36: 437–440

Fauci AS (1988) The human immunodeficiency virus: infectivity and mechanisms of pathogenesis. Science 239: 617–622

Fauci AS, Schnittman SM, Poli G, Koenig S, Panthelo G (1991) NIH Conference. Immunopathogenic mechanism in human immunodeficiency virus (HIV) infection. Ann Inter Med 114: 678–693

Genis P, Jett M, Bernton EW, Boyle T, Gelbard HA, Dzenko K, Keane RW, Resnick L, Mizrachi Y, Volsky DJ, Epstein LG, Gendelman HE (1992) Cytokines and arachidonic metabolites produced during human immunodeficiency virus (HIV)–infected macrophage-astroglia interactions: implications for the neuropathogenisis of HIV disease. J Exp Med 176: 1703–1718

Germain RN (1981) Accessory cell stimulation of T cell proliferation requires active antigen processing- Ia-restricted antigen presentation, and a separate nonspecific second signal. J Immuol 127: 1964–1989

Goswami KK, Kaye S, Miller R, McAllister R, Tedder R (1991a) Intrathecal IgG synthesis and specificity of oligoclonal IgG in patients infected with HIV-1 do not correlate with CNS disease. J Med Virol 33: 106–113

Goswami K, Miller RF, Harrison MJ, Hamel DJ, Daniels RS, Tedder RS (1991b) Expression of HIV-1 in the cerebrospinal fluid detected by the polymerase chain reaction and its correlation with central nervous system disease. AIDS 5: 797–803

Gosztonyi G, Artigas J, Lamperth L, Webster HD (1994) Human immunodeficiency virus (HIV) distribution in HIV encephalitis: study of 19 cases with combined use of in situ hybridization and immunocytochemistry. J Neuropath Exp Neurol 53: 521–534

Graeber MB, Streit WJ, Buringer D, Sparks DL, Kreutzberg GW et al (1992) Ultrastructural location of major histocompatibility complex (MHC) class II positive perivascular cells in histologically normal human brain. J Neuropath Exp Neurol 51: 303–311

Graeber MB, Streit WJ, Kreutzberg GW (1989) Identity of ED2-positive perivascular cells in rat brain. J Neurosci Res 28: 236–243

Griffin DE, Levine B, Tyor WR, Irani DN (1992) The immune response in viral encephalitis. Sem Immunol 4: 111–119

Griffin DE, Hess JL (1986) Cells with natural killer cell activity in the cerebrospinal fluid of normal mice and athymic mice with acute sindbis virus encephalitis. J Immunol 136: 1841–1845

Grimaldi LME, Martino GV, Franciotta DM, Brustia R, Castagna A, Pristera R, Lazzarin A (1991) Elevated alpha-tumor necrosis factor levels in spinal fluid from HIV-1-infected patients with central nervous system involvement. Ann Neurol 29: 21–25

Groux H, Torpier G, Monte D, Mouton Y, Capron A, Ameisen JC (1992) Activation induced death by apoptosis in CD4⁺T cells from human immunodeficiency virus infected asymptomatic individuals. J Exp Med 175: 331–340

Harlan JM, Liu DY (1992) Adhesion: its role in inflammatory disease. WH Freeman, New York

Harling-Berg C, Cserr HF, Knopf PM (1989) The role of cervical lymph nodes in the immune response to human serum albumin microinfused into rat cerebrospinal fluid. J Neuroimmunol 25: 93–105

Harouse JM, Wroblewsha Z, Laughlin MA, Hickey WF, Schonwetter BS, Gonzalez-Scarano F (1989) Human choroid plexus cells can be latently infected with human immunodeficiency virus. Ann Neurol 4: 406–411

Hartung H-P, Michels M, Reiners K, Seeldrayers P, Archelos JJ, Toyka KV (1993) Soluble ICAM-1 serum levels in multiple sclerosis and viral encephalitis. Neurol 43: 2331–2335

Herberman RB, Reynolds CW, Ortaldo J (1986) Mechanisms of cytotoxicity by natural killer cells. Annu Rev Immunol 4: 651–661

Hickey WF (1991) Migration of hematogenous cells through the blood brain barrier and the initiation of CNS inflammation. Brian Pathol 1: 97–105

Hickey WF, Kimura H (1988) Perivascular microglia are bone marrow derived and present antigen in vivo. Science 239: 290–292

Hickey WF, Hsu BL, Kimura H (1991) T-lymphocyte entry into the central nervous system. J Neurosci Res 28: 254–260

Hickey WF, Ueno K, Hiserodt JC, Schmidt RE (1992a) Exogenously-induced natural killer cell-mediated neuronal killing: a novel pathogenetic mechanism. J Exp Med 176: 811–817

Hickey WF, Vass K, Lassmann H (1992b) Bone marrow derived elements in the central nervous system: an immunohistochemical and ultrastructural study of rat chimeras. J Neuropathol Exp Neurol 51: 246–256

Hinrichs DJ, Roberts CM, Waxman FJ (1981) Regulation of paralytic experimental allergic encephalomyelitis in rats: susceptibility to active and passive disease reinduction. J Immunol 128: 1857–1862

Ho DD, Rota TR, Schooley RT, Kaplan JC, Allan JD, Groopman JE, Resnick L, Felsenstein D, Andrews CA, Hirsch MS (1985) Isolation of HTLV-III from cerebrospinal fluid and neural tissues of patients with neurologic syndromes related to the acquired immunodeficiency syndrome. New Engl J Med 313: 1493–1497

Holda JH, Welch AM, Swanborg RH (1980) Autoimmune effector cells I. Transfer of EAE with lymphoid cells cultured with antigen. Eur J Immunol 10: 657–659

Hollander H, Levy JA (1987) Neurological abnormalities and recovery with human immunodeficiency virus from cerebral spinal fluid. Ann Internal Med. 106: 692–695

Homsy J, Meyer M, Levy JA (1990) Serum enhancement of human immunodeficiency virus (HIV) infection correlates with disease in HIV-infected individuals. J Virol 64: 1437–1440

Homsy J, Meyer M, Tateno M, Clarkson S, Levy JA (1989) The Fc and not CD4 receptor mediates the antibody enhancement of HIV infection in human cells. Science 244: 1357–1360

Hudson SJ, Streilein JW (1994) Functional cytotoxic T cells are associated with focal lesions in the brains of SJL mice with experimental herpes simplex encephalitis. J Immunol 152: 5540–5547

Hynes RO (1992) Integrins: versatility, modulation, and signaling in cell adhesion. Cell 69: 11–25

Ihrcke NS, Wrenshall LE, Lindman BJ, Platt JL (1993) Role of heparin sulfate in immune system-blood vessel interactions. Immunol Today 14: 500–505

Jassoy C, Johnson RP, Navia BA, Worth J, Walker BD (1992) Detection of a vigorous HIV-1 specific cytotoxic T lymphocyte response in cerebrospinal fluid from infected person with AIDS dementia complex. 149: 3113–3119

Johnson RT, McArthur JC, Narayan O (1988) The neurobiology of human immunodeficiency virus infections. FASEB J 2: 2970–2981

Jordan CA, Watkins BA, Kufta C, Dubois-Dalcq (1991) Infection of brain microglial cells by human immunodeficiency virus type 1 is CD4 dependent. J Virol 65: 736–742

Jordan FL, Thomas WE (1988) Brain macrophages: question of origin and interrelationship. Brain Res Rev 13: 165–178

Knopf PM, Basu D, Sirulnick E, Nolan S, Cserr HF, Kier G, Thompson EJ (1994) B-cell traffic and intrathecal antibody synthesis in normal brain. FASEB: Abstract 1434

Koenig S, Earl P, Powell D, Pantaleo G, Merli S, Moss Fauci AS (1988) Group specific, major histocompatibility complex class I-restricted cytotoxic responses to human immunodeficiency virus-1 (HIV-1) envelope proteins by cloned peripheral blood T cells from an HIV-1 infected individual. Proc Natl Acad Sci (USA) 85: 8638

Koenig G, Gendelman HE, Orenstein JM, Orenstein JM, Dal Canto MC, Pezeshkpour GH, Yungbluth M, Janotta F, Aksamit A, Martin MA, Fauci AS (1986) Detection of AIDS virus in macrophages in brain tissue from AIDS patients with encephalopathy. Science 233: 1089–1093

Kure K, Lyman WD, Weidenheim K, Dickson DW (1990a) Cellular localization of an HIV-1 antigen in subacute ADIS encephalitis using an improved double-labeling immunohistochemical method. Am J Pathol 136: 1085–1092

Kure K, Weidenheim KM, Lyman WD, Dickson DW (1990b) Morphology and distribution of HIV-1 gp41-positive microglia in subacute AIDS encephalitis. Acta Neuropath 80: 393–400

Lackner AA, Dandekar S, Gardner MB (1991) Neurobiology of simian and feline immunodeficiency virus infections. Brain Pathol 1: 201–212

Lassmann H, Schmied M, Vass K, Hickey WF (1993) Bone marrow derived elements and resident microglia in brain inflammation. Glia 7: 19–24

Lassmann H, Rossler K, Zimprich F, Vass K (1991a) Expression of adhesion molecules and histocompatibility antigens at the blood-brain barrier. Brain Pathol 1: 115–123

Lassmann H, Zimprich F, Vasss K, Hickey WF (1991b) Microglial cells are a component of the perivascular glial limitans. J Neurosci Res 28: 236–243

Lawrence MB, Springer TA (1991) Leukocytes roll on a selectins at physiologic flow rates: distinction from and prerequisite for adhesion through integrins. Cell 65: 859–873

Levy JA, Shimabukuro J, Hollander H, Milles J, Kaminsky L (1985) Isolation of AIDS-associated retroviruses from cerebrospinal fluid and brain of pateints with neurologic symptoms. Lancet ii: 586–588
Lindsley M, Thiemann R, Rodriguez M (1991) Cytotoxic T cells isolated from the central nervous systems of mice infected with Theiler's virus. J Virol 65: 6612–6620
Lindsley M, Rodriguez M (1989) Characterization of the inflammatory response in the central nervous system of mice susceptible or resistant to demyelination by Theiler's virus. J Immunol 142: 2677–2682
Lustig S, Danenberg HD, Kafri Y, Kobiler D, Ben-Nathan D (1992) Viral neuroinvasion and encephalitis induced by lipopolysaccharide and its mediators. J Exp Med 176: 707–712
Mackay CR, Martson WL, Dudler L (1990) Naive and memory T cells show distinct pathways of lymphocyte recirculation. J Exp Med 171: 801–817
Male D, Pryce G, Hughes C, Lantos P (1990) Lymphocyte migration into brain modelled in vitro: control by lymphocyte activation, cytokines and antigen. Cell Immunol 127: 1–11
Massari FE, Poli G, Schnittman SM, Psalliuopoulos MC, Davey V, Fauci AS (1990) T lymphocyte origin of macrophage tropic strains of HIV: role of monocytes during in vitro isolation and in vivo infection. J Immunol 144: 4628–4632
Mathez D, Paul D, De Belilovsky C, Sultan Y, Deleuze J, Gorin I, Saurin W, Decker R, Leibowitch J (1990) Productive human immunodeficiency virus infection levels correlate with AIDS-related manifestations in the patient. Proc Natl Acad Sci USA 87: 7438–7442
Matyszak MK, Lawson LJ, Perry VH, Gordon S (1992) Stromal macrophages of the choroid plexus situated at an interface between the brain and peripheral immune system consititutively express major histocompatibility class II antigens. J Neuroimmunol 40: 173–181
McEver RP (1992) Leukocyte-endothelial cell interactions. Curr Opin Cell Biol 4:840–849
Meltzer MS, Skillman DR, Gomatos PJ, Kalter DC, Gendelman HE (1990) Role of mononuclear phagocytes in the pathogenesis of human immunodeficiency virus infection. Annu Rev Immunol 8: 169–194
Michaels J, Price RW, Rosenblum MK (1988a) Microglia in the giant cell encephalitis of acquired immune deficiency syndrome: proliferation, infection and fusion. Acta Neuropathol 76: 373–379
Michaels J, Sharer LR, Epstein LG (1988b) Human immunodeficiency virus type 1 (HIV-1) infection of the nervous system: a review. Immunodefic Rev 1: 71–104
Mikovits JR, Raziuddin S, Gonda M, Ruth M, Lohery NC, Kung HF, Rusletti FW (1990) Negative regulation of human immunodeficiency virus replication in monocytes. Distinction between restricted and latent expression in cells. J Exp Med 171: 1705–1720
Nagra RM, Wong PKY, Wiley CA (1993) Expression of major histocompatibility complex antigens and serum neutralizing antibody in murine retroviral encephalitis. J Neuropath Exp Neurol 52: 163–173
Naparstek Y, Cohen IR, Fuks Z, Vlodavsky I (1984) Activated T lymphocytes produce a matrix degrading heparin sulphate endoglycosidase. Nature 310: 241–244
Navia BA, Cho E-S, Pettito CK, Price RW (1986) The AIDS dementia complex: II. neuropathology. Ann Neurol 19: 525–535
Nuovo GJ, Gallery F, MacConnel P, Braun A (1994) In situ detection of polymerase chain reaction-amplified HIV-1 nucleic acids and tumor necrosis factor alpha in the central nervous system. Am J Pathol 144: 659–666
Oehmichen M (1978) Mononuclear phagocytes in the central nervous system. Springer, Berlin Heidelberg New York (Schriftenreihe Neurologie, vol 21)
Oldstone MBA, Blount P, Southern PJ, Lampart PW (1986) Cytoimmunotherapy for persistent virus infection: unique clearance pattern from the central nervous system. Nature 321: 239–243
Oldstone MBA, Southern PJ (1993) Trafficking of activated cytotoxic T lymphocytes into the central nervous system: use of a transgenic model. J Neuroimmunol 46: 25–32
Pantaleo G, Fauci AS (1994) Tracking HIV during disease progression. Curr Opinion Immunol: 601–604
Pantaleo G, Graziosi C, Demarest JF, Cohen OJ, Vaccarezza M, Gantt K, Muro-Cacho, Fauci AS (1994) Role of lymphoid organs in the pathogenesis of human immunodeficiency virus (HIV) infection. Immunol Rev 140: 104–129
Pantich HS (1980) Adoptive transfer of EAE with activated spleen cells: comparison of in vitro activation by concavalin A and myelin basic protein. Cell Immunol 56: 163–171
Patsouris E, HA Kretzschmar HA, Stavrou D, Mehraein P (1993) Cellular composition and distribution of gliomesenchymal nodules in the CNS of AIDS patients. Clin Neuropath 12: 130–137
Paya CV, Patick AK, Leibson PJ, Rodriguez M (1989) Role of natural killer cells as immune effectors in encephalitis and demyelination induced by Theiler's virus. J Immunol 143: 95–102

Pender MP, McCombe A, Yoong G, Ngyen KB (1992) Apoptosis of alpha beta T lymphocytes in the nervous system: its possible implications for recovery and acquired tolerance. J Autoimmun 5: 401–410

Perry VH (1994) Macrophages and the nervous system. CRC Press, Austin

Perry VH, Gordon (1987) Modulation of CD4 antigen on macrophages and microglia in rat brain. J Exp Med 166: 1138–1143

Perry VH, Gordon S (1989) Resident macrophages of the central nervous system: modulation of phenotype in relation to a specialized microenvironment. In: Goetzl EJ, Spector NH (eds) Neuroimmune networks: physiology and disease. Liss, New York ,119–125

Perry VH, Hume DA, Gordon S (1985) Immunohistochemical localization of macrophages and microglia in the adult and developing mouse brain. Neurosci 15: 313–326

Peudenier S, Hery C, Montagnier, Tardieu M (1991a) Human microglial cells: characterization in cerebral tissue and in primary culture, and study of their susceptibility to HIV-1 infection. Ann Neurol 29: 152–161

Peudenier S, Hery C, Ng KH, Tardieu M (1991b) HIV receptor within the brain: study of CD4 and MHC II on human neurons, astrocytes and microglial cells. Res Virol 142: 145–149

Pober JS, Lapierre LA, Stolpen AH, Brock TA, Springer TA, Fiers W, Bevilacqu MP, Mendrick DL, Gimbrone MA (1987) Activation of cultured human endothelial cells by recombinant lymphotoxin: comparison with tumor necrosis factor and interleukin-1 species. J Immunol 138: 3319–3224

Postigo AA, Marazuela M, Sanchez-Madri F, de Landazuri MO (1994) B Lymphocyte binding to E- and P-Selectins is mediated through the de novo expression of carbohydrates on in vitro and in vivo activated human B cells. J Clin Invest 94: 1585–1596

Postigo AA, Sanchez-Mateos P, Lazarovits AI, Sanchez-Madrid F, de Landazuri MO (1993) $\alpha 4\beta 1$ integrin mediates B cell binding to fibronectin and vascular cell adhesion molecule-1. Expression and function of $\alpha 4$ integrins on human B lymphocytes. J Immunol 151: 2471–2483

Power C, Kong P-A, Crawford TO, Wesselingh S, Glass JD, McArthur JC, Trapp BD (1993) Cerebral white matter changes in acquired immunodeficiency syndrome dementia: alterations of the blood brain-barrier. Ann Neurol 34: 339–350

Price RW, Brew B, Sidtis J, Rosenblum M, Scheck AC, Cleary P (1988) The brain in AIDS: central nervous system HIV-1 infection and AIDS dementia complex. Science 586: 586–592

Price RW, Sidtis JJ, Brew BJ (1991) AIDS dementia complex and HIV-1 infection: a view from the clinic. Brain Pathol 1: 155–162

Reynolds CW, Sharrow SO, Ortaldo JR, Herberman RB (1981) Natural killer cell activity in the rat. II Analysis of surface antigens on LGL cells by flow cytometry. J Immunol 127: 2251

Rhodes RH (1987) Histopathology of the central nervous system in the acquired immunodeficiency syndrome. Hum Pathol 18: 636–643

Rhodes RH (1991) Evidence of serum-protein leakage across the blood brain barrier in the acquired immunodeficiency syndrome. J Neuropath Exp Neurol 50: 171–183

Rinaldo CR (1994) Modulation of major histocompatibility complex antigen expression by viral infection. Am J Pathol 144: 637–646

Robinson WE, Kawamura I, Lake D, Masuho Y, Mitchell WM, Hersh EM (1990) Antibodies to the primary immunodominant domain of human immunodeficiency virus type 1 (HIV-1) glycoprotein gp41 enhance HIV-1 infection in vitro. J Virol 64: 5301–5305

Rodriguez M, Lindsley M, Pierce ML (1991) Role of T Cells in resistance to Theiler's virus infection. Microb Pathol 11: 269–281

Rosenberg ZF, Fauci AS (1989) The immunopathogenesis of HIV infection. Adv Immunol 47: 377–431

Rossler K, Neuchrist C, Kitz K, Scheiner O, Kraft D, Lassman H (1992) Expression of leucocyte adhesion molecules at the human blood-brain barrier (BBB). J. Neurosci Res 31: 365–374

Salahuddin SZ, Rose RM, Groopman JE, Markham PD, Gallo RC (1986) Human T lymphotropic virus type III infection of human alveolar macrophages. Blood 68: 281–284

Sasseville VG, Newman W, Brodie SJ, Hesterberg P, Pauley D, Ringler DJ (1994) Monocyte adhesion to endothelium in simian immunodeficiency virus-induced AIDS encephalitis is mediated by vascular cell adhesion molecule $1\alpha 4\beta 1$ integrin interactions. Am J Pathol 144: 27–40

Sasseville VG, Newman WA, Lackner AA, Smith MO, Lausen NCG, Beall D, Ringler DJ (1992) Elevated vascular cell adhesion molecule-1 in AIDS encephalitis induced by simian immunodeficiency virus. Am J Pathol 141: 1021–1130

Schnittman SM, Denning SM, Greenhouse JJ, Justement JS, Baseler M, Kurtzberg J, Hayes BF, Fauci AS (1990) Evidence for susceptibility of intrathymic T cell precursors and their progeny carrying

T cell antigen receptor phenotypes TCR alpha beta+ and TCR gamma delta+ to human immunodeficiency virus infection: a mechanism for CD4+(T4) lymphocyte depletion. Proc Natl Acad Sci (USA) 87: 7727–7731

Schmied M, Breitschopf H, Gold R, Zischler H, Roth G, Wekerle H, Lassmann H (1993) Apoptosis of T lymphocytes in experimental autoimmune encephalomyelitis. Am J Pathol 143: 446–452

Schrier R, McCutchan JA, Venable JC, Nelson JA, Wiley CA (1990) T-cell-induced expression of human immunodeficiency virus in macrophages. J Virol 64: 3280–3288

Schwartz RH (1985) T-lymphocyte recognition of antigen in association with gene products of the major histocompatibility complex. Ann Rev Immunol 3:237–261

Sethi KK, Nahler H, Stroehmann I (1988) Phenotypic heterogeneity of cerebrospinal fluid-derived HIV-specific and HLA-restricted cytotoxic T-cell clones. Nature 335: 178–181

Sharpless N, Gilbert D, Vandercam B, Zhou JM, Verdin E, Ronnett G, Friedman E, Dubois-Dalcq M (1992a) The restricted nature of HIV-1 tropism for cultured neural cells. Virol 191: 813–825

Sharpless NE, O'Brien WA, Verdin E, Kufta CV, Chen ISY, Dubois-Dalcq M (1992b) Human immunodeficiency virus type 1 tropism for brain microglial cells is determined by a region of the env glycoprotein that also controls macrophage tropism. J Virol 66: 2588–2593

Shimizu Y, Newman W, Tanhka Y, Shaw S (1992) Lymphocyte interactions with endothelial cells. Immunol Today 13: 106–112

Sloan DJ, Wood MJ, Charlton HM (1992) Leucocyte recruitment and inflammation in the CNS. Trends Neurosci 15: 276–280

Smith TW, DeGirolami U, Henin D, Bolgert F, Hauw J-J (1990) Human immunodeficiency virus (HIV) leukoencephalopathy and the microcirculation. J Neuropath Exp Neurol 49: 357–370

Sobel RA, Mitchell ME, Fondren G (1990) Intercellular adhesion molecule-1 (ICAM-1) in cellular immune reactions in the human central nervous system. Am J Pathol 136: 1309–1316

Springer TA (1990) Adhesion receptors of the immune system. Nature 346: 425-434

Takeda A, Tuazon CU, Ennis FA (1988) Antibody enhanced infection by HIV-1 via Fc receptor mediated entry. Science 242: 580–583

Takenaka A, Minegawa H, Kaneka K, Mori R, Itoyama Y (1986) Adoptive transfer of EAE with lectin activated spleen cells, part 2. Studies on T-cell subsets and interleukin 2 production. J Neurol Sci 72: 337–345

Terai C, Kornbluth RS, Pauza CD, Richman DD, Carson DA (1991) Apoptosis as a mechanism of cell death in cultured T lymphoblasts acutely infected with HIV-1. J Clin Invest 87: 1710–1715

Tornatore C, Chandra R, Berger JR, Major EO (1994) HIV-1 infection of subcortical astrocytes in the pediatric central nervous system. Neurol 44: 481–487

Tornatore C, Nath A, Amemiya K, Major EO (1991) Persistent human immunodeficiency virus type 1 infection in human fetal glial cells reactivated by T cell factors or by the cytokines tumor necrosis factor alpha and interleukin-1 beta. J Virol 65: 6094–6100

Tremblay M, Meloche S, Sekaly R-P, Wainberg MA (1990) Complement receptor 2 mediates enhancement of human immunodeficiency virus 1 infection in Epstein-Barr virus-carrying B cells. J Exp Med 171: 1791–1796

Tjinchieri G (1989) Biology of natural killer cells. Adv Immunol 47: 187

Tyor WR, Griffin DE (1993) Virus specificity and isotype expression of intraparenchymal antibody-screened cells during sindbis virus encephalitis in mice. J Neuroimmunol 48: 37–44

Tyor WR, Glass JD, Griffin JW, Becker PS, McArthur JC, Bezmen L, Griffin DE (1992a) Cytokine expression in the brain during the acquired immunodeficiency syndrome. Ann Neurol 32: 349–360

Tyor WR, Wesselingh S, Levine B, Griffin DE (1992b) Long term intraparenchymal Ig secretion after acute encephalitis in mice. J Immunol 149: 4016–4020

Tyor WR, Moench TR, Griffin DE (1989) Characterization of the local and systemic B cell response of normal and athymic hude mice with sindbis virus encephalitis. J Neuroimmunol 24: 207–215

Ulvestad E, Williams K, Matre R, Nyland H, Oliver A, Antel J (1994a) Fc receptors for IgG on cultured microglia mediate cytotoxicity and phagocytosis of antibody targets. J Neuropath and Exp Neurol 53: 27–37

Ulvestad E, Williams K, Vedler C, Antel J, Nyland H, Morke S, Matre R (1994b) Reactive microglia in multiple sclerosis lesion have an increased expression of receptor for the Fc part of IgG. J Neurol Sci 121: 125–131

Unger ER, Sung JH, Manivel JC, Chenggis ML, Blazer BR (1993) Male donor-derived cells in the brains of female sex-mismatched bone marrow transplant recipients: a Y-chromosome specific in situ hybridization study. J Neuropath Exp Neurol 52: 460–470

Vazeux R, Brousse N, Jarry A, Henin D, Marche C, Vedrenne C, Mikol J, Wolff M, Michon C, Rozenbaum W, Bureau J-F, Montagnier L, Brahic M (1987) Aids subacute encephalitis: identification of HIV-infected cells. Am J Pathol 126: 403–410

Wahl LM, Corcoran ML, Pyle SW, Arthur LO, Harel-Bellan A, Farrar WL (1989) Human immunodeficiency virus glycoprotein (gp120) induction of monocyte arachidonic acid metabolites and interleukin 1. Proc Natl Acad Sci USA 86: 621–625

Walker BD, Chakrabarti S, Moss B, Paradis TJ, Flynn T, Durno AG, Blumberg RS, Kaplan JC, Hirsch MS, Schooley RT (1987) HIV-specific cytotoxic T lymphocytes in seropositive individuals. Nature 328: 345–347

Watkins BA, Dorn HH, Kelly WB, Armstrong RC, Potts BJ, Michaels F, Kufta CV, Dubois-Dalcq M (1990) Specific tropsim of HIV-1 for microglial cells in primary human brain cultures. Science 249: 549–553

Weidenheim KM, Epshteyn I, Lyman WD (1993) Immunocytochemical identification of T cells in HIV-1 encephalitis: implications for pathogenesis of CNS disease. Mod Pathol 6: 167–174

Wekrle H, Engelhardt B, Risau W, Meyermann R (1991) Interaction of T lymphocytes with cerebral endothelial cells in vitro . Brain Pathol 1: 107–114

Wekerele H, Linnington C, Lassmann H, Meyermann R (1986) Cellular immune reactivity with the CNS. Trends Neurosci 9: 271–277

Wiley CA, Schrier RD, Nelson JA, Lampert PW, Oldstone MBA (1986) Cellular localization of human immunodeficiency virus infection within the brains of acquired immune deficiency syndrome patients. Proc Natl Acad Sci USA 83: 7089–7093

Williamson JSP, Sykes K, Stohlman SA (1991) Characterization of brain-infiltrating mononuclear cells during infection with mouse hepatitis virus strain JHM. J of Neuroimmunol 32: 199–207

Yednock TA, Cannon C, Fritz LC, Sanchez-Madrir F, Steinman L, Karin N (1992) Prevention of experimental autoimmune encephalomyelitis by antibodies against $\alpha 4\beta 1$ integrin. Nature 356: 63–66

Williams K, Ulvestad E, Wage A, Antel J, McLaurin J (1994) Activation of adult human derived microglia by myelin phagocytosis in vitro. J Neurosci Res 38: 433–443

Williams K, Bar-Orr A, Ulvestad E, Olivier A, Antel J, Yong VW (1992) Biology of adult human microglia in culture: comparisons with peripheral blood monocytes and astrocytes. J Neuropath Exp Neurol 51: 538–549

Zinkernagel RM, Doherty PC (1974) Restriction of in vitro T cell-mediated cytotoxicity in lymphocyte choriomeningitis within syngeneic or semiallogeneic system. Nature 248: 701–702

Principles of Cytotoxic T Lymphocyte Induction and Recognition

J.L. WHITTON and J. ZHANG

1 Introduction: Why Study Vaccine-Induced Immunity?

In this chapter I will discuss neither HIV nor its associated dementia; instead I will focus on the principles which underlie the importance of cytotoxic T lymphocytes (CTLs) in controlling viral infections and the means by which these critical antiviral effectors can be induced. Viral infections remain major causes of human morbidity and mortality. Vaccination has eradicated smallpox and has diminished the incidence of the more severe viral infections (mumps, rubella, measles and polio), at least in the "developed" countries. However, in "underdeveloped" countries the incidence of many of these diseases remains high, and the morbidity and mortality consequent upon the infections remains significant. For example, measles virus infections in the third world lead directly or indirectly to an estimated 1–2 million deaths per year. Furthermore, since the appearance of HIV the contribution of viruses to human suffering has escalated enormously, and forecasts suggest that this will increase in the years to come. At present of course there is no vaccine for HIV, and this drives much of the current effort to molecularly dissect the antiviral immune response.

In addition to the impetus provided by HIV, several other considerations mandate a full understanding of vaccine induced antiviral immunity. First, new

Department of Neuropharmacology, CVN9, The Scripps Research Institute, 10666 North Torrey Pines Road, La Jolla, CA 92037, USA approaches have been suggested to augment or even to supplant traditional vaccines. The latter have been remarkably successful public health tools and should not be abandoned or radically altered without good reason and an adequate understanding of the probable consequences. Secondly, "inappropriate" vaccination can be harmful. For example, administration of killed vaccines for respiratory syncytial virus and measles virus induced high levels of antiviral antibodies, along with much optimism about the protection which would have been conferred. Sadly, the vaccines were not protected against pathogenic virus challenge, and instead when exposed to the "street" agent exhibited more severe signs and symptoms of disease (FULGINITI et al. 1967; KAPIKIAN et al. 1969). Although the cause of this phenomenon has not been fully dissected, it is likely that an "imbalanced" immune response induced by a killed virus vaccine may have been responsible. The important lesson here is that vaccination is not always beneficial, or even neutral; it can be detrimental.

2 Antiviral Immunity: Experiments of Nature

The two antigen-specific arms of the immune response, antibodies and T cells, complement each other in controlling virus infections. Antibodies reduce the load of infectious virus in the fluid phase, thereby decreasing the number of infected cells with which T cells must interact; and in a complementary fashion T cells eradicate virus infected cells soon after infection, thus diminishing the amount of released virus and easing the load on antibodies. The role of antiviral antibodies in vaccine induced protection has long been unquestioned. In contrast, the role of T cells in vaccine-induced immunity has long been unheralded. One reason for this imbalance is most probably pragmatic; it has been, and remains, much easier to analyze antibody responses than to analyze T cell responses. Furthermore, there has been an excellent correlation between vaccine immunogenicity (as judged by induction of high levels of specific antibody) and vaccine efficacy (as judged by the vaccine's protective effect). This correlation has led some investigators to conclude that antibodies play the major protective role in vaccine-induced antiviral protection. Although the potential for antibodies to protect in primary and secondary infection has been clearly demonstrated in many viral systems (ENRIA et al. 1984; LIN et al. 1988; LAUSCH et al. 1990), data suggest that their functions may in fact be inessential for many viruses. For example, children born with no detectable immunoglobulins show no increased susceptibility to most viral diseases (with the exceptions of a few rare enteroviral meningitides; MCKINNEY et al. 1987; MISBAH et al. 1992). Both the incidence of viral diseases and disease severity in these antibody-deficient children are indistinguishable from those of normal children. Instead these agammaglobulinaemic children show a marked increase in susceptibility to bacterial infection (GOOD and ZAK 1956; GOOD 1991). Thus the absence of antibodies does not appear to result in an increase in primary viral infections and disease. However, the antibodies required for vaccine induced antiviral immunity? Consideration of measles infection in agammaglobulinaemic children suggests not. Agammaglobulinaemic children infected by measles virus have a normal course of disease; they develop the characteristic rash (which is T cell mediated) and clear infection in a manner similar to that of normal children. In contrast, in T cell deficient children the disease is often fatal (NAHMIAS et al.1967; SIEGEL et al. 1977; GRAY et al. 1987). Furthermore agammaglobulinaemic children are subsequently immune to measles; thus in this instance antibodies are required neither for control of the primary infection nor for resistance to secondary disease (GOOD and ZAK 1956). Thus it appears that T cells alone may be able to stimulate recovery from primary virus challenge and to confer resistance to disease following secondary exposure.

3 Why Are T Cells So Important in Controlling Virus Infections?

To understand this, one must first understand how T cells recognize viruses. T cells, unlike antibodies, cannot recognize antigen in its intact "native" form; thus, while antibodies can recognize whole microbes (bacteria or viruses), T cells are unable to do so. Instead, T cells recognize antigen only when it is presented as a short peptide in association with a host glycoprotein molecule encoded in the MHC. This phenomenon, termed MHC restriction, was discovered by analyses of host responses to infection by the mouse lymphocytic choriomeningitis virus (LCMV) (ZINKERNAGEL and DOHERTY 1974a,b).

3.1 MHC Restriction and Antigen Presentation Pathways

There are two major classes of MHC molecule, class I and class II. MHC class I molecules are present on almost all nucleated cells, with the exception of neurons (JOLY et al. 1991), while the expression of class II is much more restricted, being found mainly on immunologically important cells, most often specialized antigen-presenting cells (APCs). There is a close relationship between the type of MHC/peptide complex recognized by a particular type of T cell, the surface marker (CD4 or CD8) carried by that T cell, and the T cell's function. $CD8^+$ T cells recognize complexes of peptide with MHC class I and are generally cytotoxic in function. In contrast, $CD4^+$ T cells recognize peptide in association with MHC class II and are generally helper T cells (T_h). The CD4/II and CD8/I relationships are determined by direct interactions between the respective molecules: CD4 binds to a specific conserved region of MHC class II (KONIG et al. 1992; VIGNALI et al. 1992), and a similar interaction occurs between CD8 and class I (SALTER et al. 1990). The peptide/MHC complexes are, at least superficially, remarkably similar. Both MHC class I and MHC class II present short peptides in a groove at the top of the glycoprotein molecule (BJORKMAN et al. 1987; BROWN et al. 1993). In both cases the MHC molecules are polymorphic around the groove, thus allowing individual alleles to bind different subsets of peptides. Why has evolution produced two different types of MHC/peptide complex which are at least superficially very similar? Conversely, why has evolution produced two classes of T cell which are distinguished mainly by which of the two MHC/peptide complexes each can recognize? Presumably there must be some fundamental difference between the class I and class II complexes, and the ability of the immune response to discriminate between these two complexes must be of great importance. Initially, it was thought that the two classes of MHC glycoprotein might bind very different peptide sequences, thus expanding the peptide universe available for perusal by T cells. There is, however, no indication that certain peptide sequences are preferably presented by class I compared to class II MHC molecules, or vice versa. Indeed certain peptide sequences appear to be presented by both class I and class II molecules (TAKAHASHI et al. 1990). Instead the major difference between class I and class II complexes is the source of antigen which is associated with the particular MHC class. Peptides associated with MHC class I are almost always synthesized within the cell, while peptides associated with MHC class II originate from soluble proteins taken up by an APC. Therefore, in discriminating between class I and class II complexes, T cells are in fact discriminating between different antigen sources; $CD8^+$ CTLs recognize endogenous antigen made within the target cell, while $CD4^+$ T_h cells recognize exogenous antigen which originated outside the presenting cell. Thus, in terms of antimicrobial responsiveness, $CD8^+$ T cells will recognize cells infected by intracellular organisms (usually viruses), while $CD4^+$ T cells will recognize specialized APCs which have taken up material shed by intracellular or extracellular organisms (i.e., most foreign pathogens, including viruses and bacteria). Therefore evolution appears to have honed the interactions between the invading microbe and the host to insure that, upon infection, the immune response which predominates is that which is most appropriate to deal with the specific type of microbe being faced; and the antigen presentation pathway is the fulcrum which determines the nature of this response. The relationship between microbe, antigen presentation pathway, and immune response is represented in Fig. 1. Thus bacterial (extracellular) infections will present antigen through the MHC class II pathway, and as a consequence will induce a predominately $CD4^+$ T cell and antibody response. Antibodies can recognize intact bacteria and can use various effector mechanisms to destroy them. In contrast during bacterial infections $CD8^+$ CTLs, which recognize antigen only when presented through the "endogenous" MHC class I pathway, would be of little value. Conversely, viral (intracellular) infections present antigen via both the class I and class II MHC pathways, and therefore both antibodies and $CD8^+$ CTLs are induced, both of which may play important roles in controlling the viral infection. Thus, the experiments of nature alluded to above can now be understood in a molecular light. Removal of the antibody arm of the immune response (in agamma-

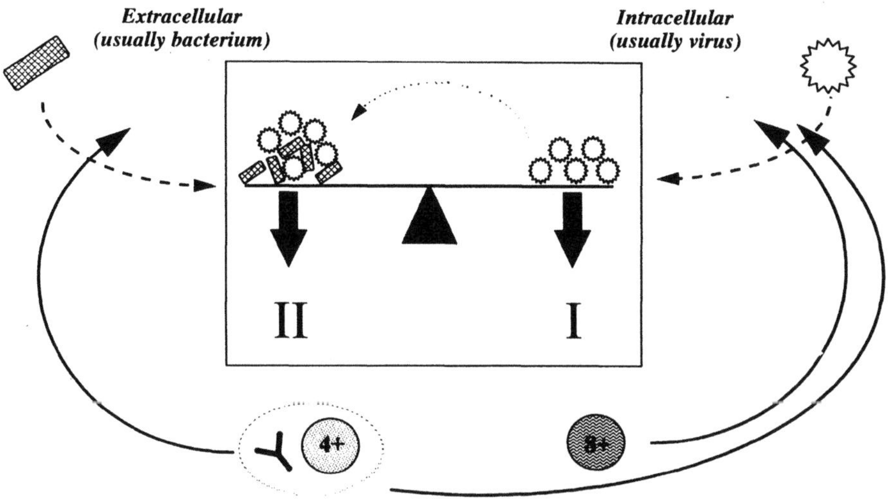

Fig. 1. The antigen presentation pathways act as a fulcrum to ensure the optimal immune response to a given type of microbe. Extracellular microbes (e.g. most bacteria) are shown as *hatched rectangles* and intracellular microbes (e.g., viruses) as *open stars* (not to scale). Viral antigens enter both class I and class II pathways, while bacterial antigens predominantly enter the latter. The type of immune response induced (CD8⁺ CTL, or CD4⁺ Th/Ab) is shown below the appropriate MHC class; the *arrows* emanating from these immune responses represent the effector mechanisms which eradicate the microbes, thus terminating introduction of further antigen into the processing pathways

globulinaemic children) allows bacterial disease to progress almost unhindered. In contrast, removal of the antibody arm of the immune response has minimal effect on the control of virus infection since the $CD8^+$ CTL pathway is present.

3.2 Transgenic Mouse Studies

Transgenic mouse studies have allowed further dissection of the role of T cells in control of virus infection. Stable cell-surface expression of MHC class I requires the protein β2 microglobulin, and mice in which this protein has been deleted by "knockout" techniques express little or no class I MHC on their cell surfaces. These mice do not develop $CD8^+$ T cells and thus cannot mount $CD8^+$ CTL responses but develop readily detectable $CD4^+$ CTLs. This model allowed the first in vivo demonstration of $CD4^+$ antiviral CTLs; such cells had been previously demonstrated, but only following extensive secondary in vitro stimulation. It is tempting to suggest that these cells represent an antiviral function which was less efficient than necessary and which was superceded by the class I/$CD8^+$ system; but when the latter was removed the former system expanded in an attempt to offset the immunological deficit. The outcome of virus infection in these $CD8^-CD4^+$ mice varies with the virus used but in general there is delayed clearance of viruses including LCMV, Sendai, and influenza virus (Muller et al. 1992; Doherty et al. 1993; Hou et al. 1992; Eichelberger et al. 1991). These studies support the contention that CD8⁺ CTLs play an important role in controlling virus infection and show that their ablation reduces but does not altogether eliminate the host's ability to clear virus. They are, therefore, consistent with the scheme outlined in Fig. 1.

3.3 Advantages and Limitations of $CD8^+$ T Cells in Control of Virus Infection

The presence of antiviral $CD8^+$ T cells confers several advantages upon the host when compared to antibodies alone. First, antibodies (for the most part) are biologically effective only if they recognize surface viral proteins (usually glycoproteins). In contrast, for T cells essentially any virus protein is a potential target. Thus the entire virus genome is open to potential immune surveillance, not only those regions encoding proteins exposed to the extracellular milieu. The ability of T cells to recognize all viral proteins confers an additional advantage. Many viruses exhibit a strictly regulated cascade of gene expression, in which the cell membrane glycoproteins are expressed late in infection; if T cells (like antibodies) were restricted to recognition of these proteins then the advantage to the host might be minimal, since CTL killing would hasten release of mature virions without markedly diminishing the quantity. However, T cells can take advantage of the "window of opportunity" provided by the virus and can recognize those proteins made immediately following infection; the infected cells are lysed before virus replication/maturation can occur. The phenomenon is exemplified in herpesviruses, which upon infection synthesize a class of immediate-early (IE) proteins. CTLs directed towards members of this IE family have been demonstrated in animal models and in humans (Borysiewicz et al. 1988), and CTLs specific for a family member can confer protection against infection by murine cytomegalovirus (Jonjic et al. 1988). An additional advantage of CTLs is their ability, often, to cross-react across virus serotypes; that is, they may confer broader immunity than antibodies. Finally, $CD8^+$ CTLs are likely to be better effector cells than their $CD4^+$ counterparts because class I MHC is expressed on almost all cell types, while class II MHC is much more restricted, being present on specialized APCs.

All of the advantages of CTL recognition described above result from the ability of the class I MHC molecules to present peptides originating in the infected cell. However the process also places some severe constraints on T cell recognition, from which antibodies are free. First, the polymorphism at the peptide-binding groove means that each MHC allele will bind and present different peptides. Consequently, different individuals (who differ in their MHC alleles) will respond to different regions of a virus, a situation which greatly complicates subunit vaccination (as discussed below). Second, this same polymorphism ensures that T cells are specific for specific peptide in association with a particular MHC molecule and thus, unlike antibodies, T cells cannot be freely transferred between individuals.

The above hypotheses of the importance of CTLs in vaccine-induced immunity remained largely unproven until the past several years. We have used the LCMV mouse model system to determine whether the induction of CTLs alone (in the absence of antiviral antibodies) can confer sufficient advantage upon the host to allow it to resist subsequent LCMV challenge. We have published several manuscripts on this topic, and the results will be briefly reviewed below.

4 Cytotoxic T Lymphocyte Vaccines Can Be Remarkably Effective

When these studies were begun, the induction of $CD8^+$ CTLs in the absence of antibody was a difficult procedure. The problem resulted from the antigen processing pathways. In order to induce $CD8^+$ CTLs, the desired antigen must be presented by class I MHC; this is readily achieved by using a live virus vector, but antigen thus delivered is then shed, taken up by APCs, and presented via the class II pathway, inducing $CD4^+$ T cells and antibodies. We circumvented this hurdle by mapping precisely the CTL epitopes of our model virus, LCMV, and then expressing these, in the absence of any LCMV antibody epitopes, in recombinant vaccinia viruses (WHITTON 1990). By this means we demonstrated that: (1) fragments of the LCMV nucleoprotein (NP) could induce CTLs which, in the absence of LCMV-specific antibodies, conferred upon the vaccine sufficient advantage that it was fully protected against subsequent LCMV challenge; and (2) such protection mapped with the presence or absence in the vaccine of a CTL epitope (KLAVINSKIS et al. 1990). We demonstrated, too, the risk of subunit vaccination; a truncated NP molecule containing an epitope presented by the L^d class I MHC molecule (carried by BALB/c mice) conferred 100% protection upon these animals, while the same vaccine was entirely ineffective in mice of the H2bb MHC haplotype, which are unable to present any viral sequences from the truncated NP employed in this study (KLAVINSKIS et al. 1989). Furthermore, remarkably short sequences could be expressed from recombinant vaccinia viruses and still could be effective vaccines. "Minigenes" as short as 11 amino acids (essentially, the CTL epitope preceded by an initiator methionine and followed by a stop codon) could be transcribed, translated and appropriately presented by class I MHC to allow recognition of recombinant-infected targets and to induce protective immune responses in vivo (WHITTON et al. 1993; WHITTON 1993, 1994). The role of CTLs in vaccine-induced protection is therefore clear. While the role may vary from virus to virus, the experiments of nature indicate that there are few viral diseases in which T cells are wholly ineffective (if such diseases exist, one would expect them to appear in the agammaglobulinaemic children), and thus we argue that, to confer balanced and "natural" immunity, an antiviral vaccine must induce both T cells and antibodies.

5 DNA Immunization

In designing new vaccines, many criteria are important. Of primary importance, of course, is the induction of long-lasting and effective antiviral immune responses (both antibody and T cell). However other considerations exist. For example, an ideal vaccine would be stable (allowing its worldwide distribution), inexpensive and safe. One approach which has been subjected to much recent study has been administration of plasmid DNA by direct inoculation. The original studies demonstrated that expression occurred when the constructs were directly inoculated into skeletal (WOLFF et al. 1990) or cardiac (LIN et al. 1990) muscle. The reason for the apparent facility of muscle cells to take up and express these materials is uncertain, but they are not unique in this regard. Inoculation of DNA complexed with cationic lipids resulted in protein expression in lung, liver and spleen; indeed in essentially all assayed tissues (ZHU et al. 1993). Such injections do, as might be anticipated, induce immune responses against encoded foreign proteins. Both antibody and T cell responses can be detected, and protection against viral challenge has been demonstrated for influenza virus in mice (ULMER et al. 1993) and in chickens (FYNAN et al. 1993; ROBINSON et al. 1993).

However, the immune responses induced by this new approach are often far more difficult to detect than those induced by more traditional methods (using recombinant vaccinia, for example). For example, anti-influenza antibodies were undetectable in animals receiving a plasmid encoding the viral hemagglutinin, even after booster vaccination; nevertheless, most of the animals survived viral challenge (ROBINSON et al. 1993). Given the apparently low immunogenicity of these vaccines as currently conceived, why are they worthy of further study?

5.1 Advantages of DNA Immunization

There are several advantages of DNA immunization: (1) *Antigen is synthesized within the vaccinee.* As described above, synthesis of antigen within host cells allows presentation by class I MHC and consequent induction of $CD8^{+}$ T cell responses. Plasmid DNA is not, of course, alone in this ability; any live virus delivery system can do the same. However, delivery systems such as vaccinia virus have other disadvantages as described in points (3) and (4) below. (2) *Low/no immune response to vector.* When plasmid DNA is inoculated as naked material, or when complexed with cationic lipid, there is little or no anti-DNA response. (Injection of bacterial DNA can induce antibody responses in normal mice (GILKESON et al. 1989), but the DNA in that study was complexed to methylated BSA and administered in adjuvant.) The low anti-vector response has two great advantages. First, one can give multiple doses of plasmid DNA, either using identical constructs (therefore subsequent doses act as boosters) or using the same plasmid DNA vector but carrying new antigen-coding regions. This contrasts with recombinant viral delivery systems, where there has been a concern that the anti-vector immunity induced by first inoculation will result in a poor response to a second antigen if delivered by the same vector system. Second, the absence of anti-DNA antibodies holds promise for inducing immune responses in infants carrying maternal antiviral antibodies. For example, an infant carrying passively transferred antibody to measles virus cannot be effectively immunized with live measles vaccine, because the vaccine virus is neutralized before infection can occur. However, a DNA vaccine encoding measles immunogens will not, of course, be detected by the anti-measles antibodies and thus may circumvent this problem. (3) *Less antigenic competition.* Antigens administered in other live viral systems must compete with other antigens for MHC presentation, etc. For instance, foreign proteins encoded in recombinant vaccinia viruses have to compete with approximately 200 kb of encoded vaccinia virus proteins for antigen presentation and induced immunity: this is not the case with a small plasmid DNA, in which single proteins may be encoded. (4) *The vector is nonpathogenic.* In contrast to many recombinant delivery systems, plasmid DNA should be nonpathogenic, and studies to date have suggested that this is indeed the case. (5) *Ease of production.* Plasmid DNA can be obtained at high purity, in great quantity, and relatively inexpensively. Unlike the production of attenuated or recombinant viruses eukaryotic cells are not involved in its production, so there is little risk of transfer of adventitious agents.

Collectively the above benefits mandate the careful and thorough evaluation of plasmid DNA in immunization. However, in so doing several concerns and questions must be addressed. For example, what are the levels of gene expression? Low levels may yield inadequate immune responses which may be harmful when the vaccinee is exposed to the pathogenic agent (OEHEN et al. 1991). Secondly, should gene expression be directed to specific tissues? Antigens, particularly if expressed at low levels, may benefit from being expressed in "immunologically appropriate" tissues. Finally, what is the fate of the inoculated DNA? One concern is that the inoculated DNA may integrate into the host genome. This could lead to a potentially harmful up-regulation of endogenous genes; such a mechanism has been shown to explain several retroviral-induced tumors. Secondly the DNA may persist, perhaps leading to long-term antigen expression, the immunological consequence of which is unclear. Presence of the encoded protein has been demonstrated many weeks (at least 4–9) post-inoculation (ZHU et al. 1993; LIN et al. 1990). We have begun to dissect this technique in the LCMV model.

5.2 Experimental Results in the LCMV System

We have begun preliminary studies on the efficacy of DNA immunization in the LCMV system. H2bb (C57BL/6) mice mount CTL responses to both the NP and glycoprotein (GP) of LCMV. Thus, these proteins were cloned into a plasmid vector driven by the IE promoter from human cytomegalovirus. The construct expressing LCMV NP is shown in Fig. 2. Transcription is driven by the IE promoter of human cytomegalovirus, which is known to direct expression of marker

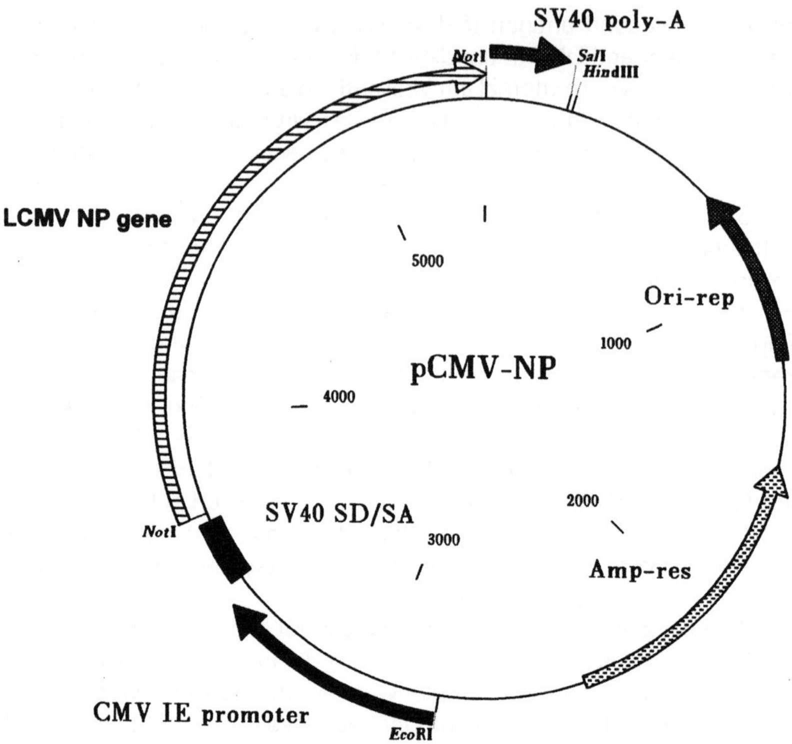

Fig. 2. Plasmid construct expressing lymphocytic choriomeningitis virus (LCMV) protein. A representative plasmid, *pCMV-NP*, is shown. This plasmid encodes full-length LCMV NP, transcription of which is driven by the immediate early (*IE*) promoter of human cytomegalovirus (*CMV*). Upstream of the LCMV sequences is a splice donor/splice acceptor site from simian virus 40(*SV40 SD/SA*); and downstream lies a transcription termination/processing/polyadenylation site from the same virus (*SV 40 poly-A*). The plasmid replication origin (*ori-rep*) and β-lactamase (*Amp-res*) gene are shown

proteins in many tissues in vivo (FURTH et al. 1991); the LCMV sequences are preceded by a splice donor/acceptor sequence from SV40 and are followed by a transcription termination/processing/polyadenylation site from the same source. Expression of the LCMV proteins was confirmed using transient expression assays and immunofluorescent detection methods (data not shown). The plasmids were inoculated intramuscularly into H2bb C57BL/6 mice (100 μg DNA dissolved in 1 *N* saline, per animal). Two negative control groups were established, one immunized with medium alone and the other immunized with the parental plasmid pCMV, which lacks any LCMV sequences. A positive control group comprised mice immunized intraperitoneally with LCMV. Six weeks later all mice were challenged with a normally lethal dose of LCMV administered intracranially. As shown in Fig. 3, almost all mice immunized with medium alone, or with the parental plasmid pCMV, succumbed to infection. In contrast, all mice

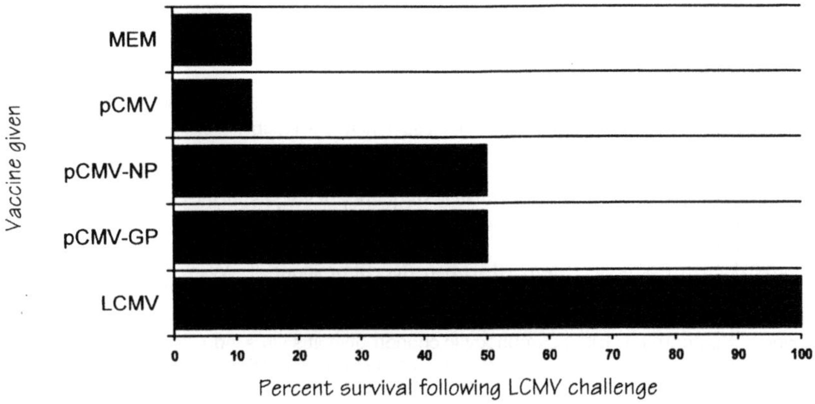

Fig. 3. DNA immunization protects C57BL /6 mice against challenge with a normal lethal dose of lymphocytic choriomeningitis virus (*LCMV*). C57BL/6 (H2bb) mice were immunized as shown (*vaccine given*). Vaccine routes were: medium (*MEM*) and live virus (2×10^5 p.f.u. LCMV) were delivered i.p.; while DNAs (*pCMV, pCMV-NP, pCMV-GP*) were dissolved in 1 *N* saline to a concentration of 1 mg/ml, and 100 µl (100 µg DNA) was injected into the muscle mass of the lower limb. Six weeks later, mice were challenged with a normally lethal intracranial dose of LCMV. Mice were observed daily for 14 days. All deaths occurred on days 7 or 8 postchallenge, and percent surviving mice is shown for each vaccine group

immunized with LCMV survived. Fifty percent of mice which had received either NP or GP under the control of the pCMV promoter survived challenge, indicating the potential of DNA immunization in this system. H2bb mice can present CTL epitopes from both NP and GP, and we have previously shown, using our recombinant vaccinia viruses, that these proteins can protect these mice by induction of CTLs. Thus our present findings are consistent with, though they do not prove, the hypothesis that these DNA vaccines induce protective CTLs; experiments to test the hypothesis are under way. The level of protection induced by a single DNA injection appears lower than that induced with a single inoculation of recombinant vaccinia virus (in which we routinely obtain 85%–100% protection). The reason for this difference is unknown and under investigation.

The potential of DNA immunization in induction of antiviral immunity is clear, but many questions remain. How can we optimize the system? What are the potential problems, and how may they be circumvented? We are currently comparing different routes and vehicles of administration, attempting to target the DNA to specific tissues, and investigating the precise nature and magnitude of the immunity so induced. In so doing we hope to establish principles which may expand and improve the utility of this new technique.

Acknowledgments. This is manuscript #8664 from the Scripps Research Institute. The authors thank Terry Calhoun for excellent secretarial assistance. The work described in this manuscript was supported by NIH grants AI 27028 and AG 04342.

References

Bjorkman PJ, Saper MA, Samraoui B, Bennett WS, Strominger JL, Wiley DC (1987) The foreign antigen binding site and T cell recognition regions of class I histocompatibility antigens. Nature 329: 512–518

Borysiewicz LK, Hickling JK, Graham S, Sinclair J, Cranage MP, Smith GL, Sissons JGP (1988) Human cytomegalovirus-specific cytotoxic T cells. Relative frequency of stage-specific CTL recognizing the 72-kD immediate early protein and glycoprotein B expressed by recombinant vaccinia viruses. J Exp Med 168: 919–931

Brown JH, Jardetzky TS, Gorga JC, Stern LJ, Urban RG, Strominger JL, Wiley DC (1993) Three-dimensional structure of the human class II histocompatibility antigen HLA-DR1. Nature 364: 33–39

Doherty PC, Hou S, Southern PJ (1993) Lymphocytic choriomeningitis virus induces a chronic wasting disease in mice lacking class I MHC glycoproteins. J Neuroimmunol 46: 11–17

Eichelberger M, Allan W, Zijlstra M, Jaenisch R, Doherty PC (1991) Clearance of influenza virus respiratory infection in mice lacking class I major histocompatibility complex-restricted CD8⁺ T cells. J Exp Med 174: 875–880

Enria DA, Fernandez NJ, Briggiler AM, Levis SC, Maiztegui JI (1984) Importance of dose of neutralizing antibodies in treatment of argentine hemorrhagic fever with immune plasma. Lancet 2: 255–256

Fulginiti VA, Eller JJ, Downie AW, Kempe CH (1967) A typical measles in children previously immunized with inactivated measles virus vaccine. JAMA 202: 1075–1080

Furth PA, Hennighausen L, Baker C, Beatty B, Woychick R (1991) The variability in activity of the universally expressed human cytomegalovirus immediate early gene 1 enhancer/promoter in transgenic mice. Nucleic Acids Res 19: 6205–6208

Fynan EF, Robinson HL, Webster RG (1993) Use of DNA encoding influenza hemagglutinin as an avian influenza vaccine. DNA Cell Biol 12: 785–789

Gilkeson GS, Grudier JP, Karounos DG, Pisetsky DS (1989) Induction of anti-double stranded DNA antibodies in normal mice by immunization with bacterial DNA. J Immunol 142: 1482–1486

Good RA (1991) Experiments of nature in the development of modern immunology. Immunol Today 12: 283–286

Good RA, Zak SJ (1956) Disturbance in gamma-globulin synthesis as "experiments of nature". Pediatrics 18: 109–149

Gray MM, Hann IM, Glass S, Eden OB, Jones PM, Stevens RF (1987) Mortality and morbidity caused by measles in children with malignant disease attending four major treatment centres: a retrospective review. Br Med J [Clin Res] 295: 19–22

Hou S, Doherty PC, Zijlstra M, Jaenisch R, Katz JM (1992) Delayed clearance of Sendai virus in mice lacking class I MHC-restricted CD8⁺ T cells. J Immunol 149: 1319–1325

Joly E, Mucke L, Oldstone MBA (1991) Viral persistence in neurons explained by lack of major histocompatibility class I expression. Science 253: 1283–1285

Jonjic S, del Val M, Keil GM, Reddehase MJ, Koszinowski UH (1988) A nonstructural viral protein expressed by a recombinant vaccinia virus against lethal cytomegalovirus infection. J Virol 62: 653–1658

Kapikian AZ, Mitchell RH, Chanock RM, Shvedoff RA, Stewart CE (1969) An epidemiologic study of altered clinical reactivity to respiratory syncitial (RS) virus infection in children previously vaccinated with an inactivated RS vaccine. Am J Epidemiol 89: 404–421

Klavinskis LS, Whitton JL, Oldstone MBA (1989) Molecularly engineered vaccine which expresses an immunodominant T-cell epitope induces cytotoxic T lymphocytes that confer protection from lethal virus infection. J Virol 63: 4311–4316

Klavinskis LS, Whitton JL, Joly E, Oldstone MBA (1990) Vaccination and protection from a lethal viral infection: identification, incorporation, and use a cytotoxic T lymphocyte glycoprotein epitope. Virology 178: 393–400

Konig R, Huang LY, Germain RN (1992) MHC class II interaction with CD4 mediated by a region analogous to the MHC class I binding site for CD8. Nature 356: 796–798

Lausch RN, Staats H, Metcalf JF, Oakes JE (1990) Effective antibody therapy in herpes simplex virus ocular infection. Characterization of recipient immune response. Intervirology 31: 159–165

Lin FT, Chen SB, Wang YZ, Sun CZ, Zeng FZ, Wang GF (1988) Use of serum and vaccine in combination for prophylaxis following exposure to rabies. Rev Infect Dis 10 [Suppl 4]: S766–S770

Lin H, Parmacek MS, Morle G, Bolling S, Leiden JM (1990) Expression of recombinant genes in myocardium in vivo after direct injection of DNA. Circulation 82: 2217–2221
McKinney REJ, Katz SL, Wilfert CM (1987) Chronic enteroviral meningoencephalitis in agammaglobulinemic patients. Rev Infect Dis 9: 334–356
Misbah SA, Spickett GP, Ryba PC, Hockaday JM, Kroll JS, Sherwood C, Kurtz JB, Moxon ER, Chapel HM (1992) Chronic enteroviral meningoencephalitis in agammaglobulinemia: case report and literature review. J Clin Immunol 12: 266–270
Muller D, Koller BH, Whitton JL, La Pan K, Brigman KK, Frelinger JA (1992) LCMV-specific, class II-restricted cytotoxic T cells in b2-microglobulin-deficient mice. Science 255: 1576–1578
Nahmias AJ, Griffith D, Salsbury C, Yoshida K (1967) Thymic aplasia with lymphopenia, plasma cells, and normal immunoglobulins. Relation to measles virus infection. JAMA 201: 729–734
Oehen S, Hengartner H, Zinkernagel RM (1991) Vaccination for disease. Science 251: 195–198
Robinson HL, Hunt LA, Webster RG (1993) Protection against a lethal influenza virus challenge by immunization with a haemagglutinin-expressing plasmid DNA. Vaccine 11: 957–960
Salter RD, Benjamin RJ, Wesley PK, Buxton SE, Garrett TPJ, Clayberger C, Krensky AM, Norment AM, Littman DR, Parham P (1990) A binding site for the T cell co-receptor CD8 on the a3 domain of HLA-A2. Nature 345: 41–46
Siegel MM, Walter TK, Ablin AR (1977) Measles pneumonia in childhood leukemia. Pediatrics 60: 38–40
Takahashi H, Germain RN, Moss B, Berzofsky JA (1990) An immunodominant class I-restricted cytotoxic T lymphocyte determinant of human immunodeficiency virus type 1 induces CD4 class II-restricted help for itself. J Exp Med 171: 571–576
Ulmer JB, Donnelly JJ, Parker SE, Rhodes GH, Felgner PL, Dwarki VJ, Gromkowski SH, Deck RR, DeWitt CM, Friedman A, Hawe LA, Leander KR, Martinez D, Perry HC, Shiver JW, Montgomery DL, Liu MA (1993) Heterologous protection against influenza by injection of DNA encoding a viral protein. Science 259: 1745–1749
Vignali DA, Moreno J, Schiller D, Hammerling GJ (1992) Species-specific binding of CD4 to the beta 2 domain of major histocompatibility complex class II molecules. J Exp Med 175: 925–932
Whitton JL (1990) Lymphocytic choriomeningitis virus CTL. Semin Virol 1: 257–262
Whitton JL (1993) The use of minigenes in antiviral vaccination. In: Talwar GP (ed) Recombinant and synthetic vaccines. Narosa, New Delhi, pp 427–437
Whitton JL (1994) Induction of protective immunity using minigenes. Ann NY Acad Sci 730: 107–117
Whitton JL, Sheng N, Oldstone MBA, McKee TA (1993) A "String-of-beads" vaccine, comprising linked minigenes, confers protection from lethal-dose virus challenge. J Virol 67: 348–352
Wolff JA, Malone RW, Williams P, Chong W, Acsadi G, Jani A, Felgner PL (1990) Direct gene transfer into mouse muscle in vivo. Science 247: 1465–1468
Zhu N, Liggitt D, Liu Y, Debs R (1993) Systemic gene expression after intravenous DNA delivery into adult mice. Science 261: 209–211
Zinkernagel RM, Doherty PC (1974a) Immunological surveillance against altered self components by sensitised T lymphocytes in lymphocytic choriomeningitis. Nature 251: 547–548
Zinkernagel RM, Doherty PC (1974b) Restriction of in vitro T cell-mediated cytotoxicity in lymphocytic choriomeningitis within a syngeneic or semiallogeneic system. Nature 248: 701–702

Virus-Neuron-Cytotoxic T Lymphocyte Interactions

G.F. RALL[1] and M.B.A. OLDSTONE

1 Prologue: Virus-Neuron Interactions

Neurons have evolved unique strategies to evade immune recognition which allow viruses to infect and persist, yet resist attack by antiviral cytotoxic T lymphocytes (CTLs). However, the ability of activated CTLs to percolate through the normal central nervous system (CNS) (HICKEY et al. 1991), coupled with focal up-regulation of immune regulatory proteins on resident CNS cells in response to CNS insults (CANELLA et al. 1990), has suggested that the immune response can serve either in the resolution of viral infections of the CNS or in the precipitation of CNS disease.

CTLs, either alone or through the release of cytokines, can cause immune-mediated disease and are generated in abundance in patients infected with HIV (SCHRIER et al. 1989; JASSOY et al. 1992). Because such anti-HIV CTL have been noted in the cerebrospinal fluid (CSF) of such patients, we have studied the immune response against a neurotropic virus in vivo in order to understand how such viruses, including HIV-1, may cause CNS damage.

Neurotropic viruses can enter the CNS by two main routes, as outlined in Table 1. Most viruses, including HIV, gain access to the CNS by a hematogenous,

The Scripps Research Institute, Department of Neuropharmacology, Division of Virology, 10666 North Torrey Pines Road, Mail Drop IMM-6, La Jolla, CA 92037, USA

[1]Present address: The Fox Chase Cancer Center, 7701 Burholme Avenue, Philadelphia, PA 19111, USA

Table 1. Route of neurotropic viral access to the central nervous system

Hematogenous	Neural Pathway
Herpes simplex	Herpes simplex
Cytomegalovirus	Varicella zoster
Epstein-Barr virus	Rabies
Poliovirus	Poliovirus
Measles	Reovirus
Coxsackieviruses	
Human immunodeficiency virus (HIV)	
Mumps	
Echoviruses	
Lymphocytic choriomeningitis	
Arboviruses	

or blood-borne, route which requires them to cross from the blood vessel into the brain parenchyma via the blood brain barrier. These viruses can travel in plasma, bind to cells (i.e., red blood cells) or be contained inside cells (i.e., lymphocytes and/or macrophages). Other viruses, such as rabies, travel in a retrograde fashion along axon processes, transmitted transsynaptically into the CNS (Weiner and Fleming 1984; Schlitt et al. 1991)

A number of reasons make immune recognition of virally infected neurons interesting. Neurons of the CNS exist in a milieu long considered immune-privileged by virtue of the blood-brain barrier. This barrier is a tightly packed wall of endothelial cells forming the brain capillary lining, restricting access of hydrophilic blood-borne molecules, such as antibodies and cytokines, and nonactivated lymphocytes into the CNS (reviewed in Streilein 1993). However, the barrier, though highly selective, is not impermeable. Studies by Hickey, Wekerle, and others (Wekerle et al. 1986; Hickey and Kimura 1987; Hickey et al. 1991) indicate that the blood-brain barrier is traversed by activated T cells entering the brain parenchyma. Further, in several disease states, destabilization of the blood-brain barrier occurs, usually due to inflammation of the endothelial-meningeal cell layer, allowing increased access of cells and potential neurotoxic molecules into the CNS (reviewed in Greenwood 1991).

Neurons are essential to the host's survival and are irreplaceable once destroyed. Therefore, neurotropic viral infections that directly cause neuronal death, or induce an aggressive antineuronal or antiviral immune response, will deplete an irreplaceable pool of vital cells. Likely through selection, and because of their essential role in host survival, CNS neurons do not express class I major histocompatibility (MHC) molecules (Joly et al. 1991; Keane et al. 1992; Mucke and Oldstone 1992), which are required for recognition by CTL of virally infected target cells. Therefore, neurons cannot be seen by effector CTLs despite being infected and expressing viral proteins on their surfaces.

In this chapter, we discuss the relationship between the inability of neurons to express class I MHC molecules and the ability of viruses to establish persistent infections in these cells. In order for a virus to establish a persistent infection it must be non-cytolytic and must be able to avoid immune surveillance and recognition (reviewed in OLDSTONE 1989). Viruses avoid immune recognition either actively (by synthesizing gene products that paralyze a specific step of antigen presentation; see GOODING 1992 for review) or passively, by residing in cells that fail to serve as targets for CTL-mediated lysis. Thus, many DNA and RNA viruses that establish persistent infections do so in neurons, which shelter them from recognition and attack by CTLs.

2 Virus-Neuron Interactions In Vivo: Cytotoxic T Lymphocytes Fail to Recognize and Lyse Virally Infected Neurons

We have utilized lymphocytic choriomeningitis virus (LCMV), a prototype of the arenavirus family, as a model for understanding the molecular and biological mechanisms of viral persistence in neurons. LCMV consists of a bisegmented ambisense genome encoding four gene products. The L RNA encodes the viral polymerase (L) and the viral Z protein which has a putative zinc binding function. The S RNA encodes the glycoprotein, the molecule which interacts with the cellular receptor, and the viral nucleoprotein. In tissue culture, LCMV is non-cytolytic, yet when injected into animals can result in disease (Fig. 1). Whether an animal develops disease as a consequence of an LCMV infection depends upon whether the animal mounts an anti-LCMV immune response (BUCHMEIER et al. 1980). When LCMV infection occurs prior to the development of the immune system (in utero or within 24 h after birth), or in immunosuppressed animals, injury of infected cells by the virus does not occur and virus persists throughout the host's life. Alternatively, when infection occurs in a host that is immunocompetent and can mount a brisk immune response, a predominately CD8+, anti-LCMV response is raised (BYRNE and OLDSTONE 1984; MOSKOPHIDIS et al. 1987). A balance between clearance of virus and immunopathology occurs, depending on the site of viral infection and the vigor of the antiviral immune response, which either leads to clearance of virus and lifelong immunity or immune-mediated cytopathology and death (BUCHMEIER et al. 1980).

In the adult mouse persistently infected with LCMV, viral nucleic acid can be detected in the majority of tissues examined (Fig. 2). Unmanipulated, persistently infected animals shed virus throughout their lifetime and have overt consequences of virus-induced immune complex deposits in several tissues (OLDSTONE 1975; TISHON et al 1991). Deposits of such immune complexes in the renal glomeruli and arteries occurs uniformly in persistent viral infection, including humans infected with HIV, EBV, HBV and CMV (OLDSTONE 1975; STAGNO et al. 1977; MCDOUGAL et al. 1985). Similar scenarios are likely to occur if attenuated HIV-1 (e.g., nef deletion variants) which establish persistent infection are used.

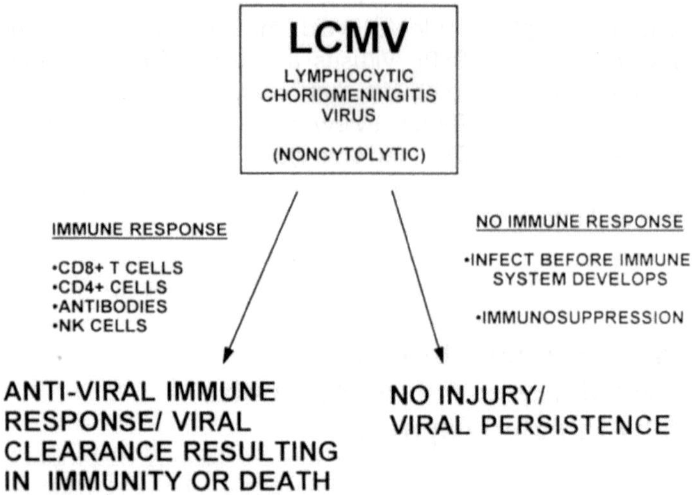

Fig. 1. Mechanisms of lymphocytic choriomeningitis virus (*LCMV*)-induced disease. LCMV can establish a persistent infection when inoculated into neonatal animals or transmitted from infected mother to foetus in utero. The persistent infection is not characterized by an antiviral immune response and, consequently, no disease occurs; the animal remains persistently infected throughout life. Alternatively, infection of immunocompetent mice results in an antiviral response, which is predominately a CD8+ cytotoxic T lymphocyte response, but which encompasses other arms of the immune response as well. The outcome of this clearance, either immunity or death of the host, depends upon the route of viral infection and the location of the antiviral response

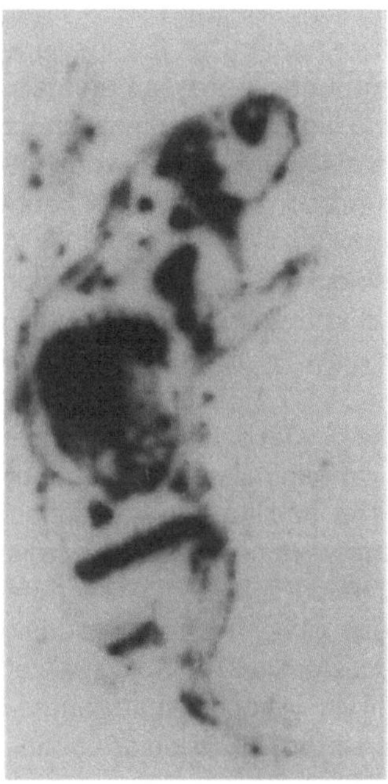

Fig. 2. Viral presence in persistently infected mice. A longitudinal cross-section of an adult, persistently infected mouse given 100 pfu of lymphocytic choriomeningitis virus (LCMV) at birth. In situ hybridization to detect the presence of viral nucleic acid indicates that virus is found throughout the animal, present in most major organs, including brain, thymus, spleen, liver and kidney

Introduction of antiviral CTLs by adoptive transfer into LCMV persistently infected mice clears virus with reproducible kinetics, in which clearance of the virus systemically is rapid but delayed from the CNS (OLDSTONE et al. 1986; AHMED et al. 1987). Hence, clearance of virus from spleen, liver, etc., occurs within 21 days and is profiled by lymphocytic infiltration and tissue destruction. In contrast, clearance from neurons occurs more slowly, requiring 60–120 days post-adoptive transfer before viral nucleic acid is undetectable by in situ hybridization. Further, clearance from neurons is not associated with signs of immune infiltration (MUCKE and OLDSTONE 1992).

Viral sequences are also present in the kidney, but by a different mechanism (OLDSTONE and DIXON 1971; BUCHMEIER and OLDSTONE 1978). Virus-specific CTLs lyse renal tubular and glomerular cells which express virus, but the resultant released virus binds with antiviral antibodies to form immune complexes that deposit in the renal glomeruli. Thus, detection of viral nucleic acid in the kidney at late times postadoptive transfer is due to the presence of virus in the form of antigen-antibody complexes.

3 Virus-Neuron Interactions In Vitro: Molecular Definition of Defects in Neurons That Prevent Antigen Processing and Allow Neurons to Resist Cytotoxic T Lymphocyte-Mediated Injury

The reason for the absence of immune infiltration in the CNS and the absence of neuronal CTL-mediated lysis despite the presence of virus is due the lack of molecules necessary for antigen processing. In Fig. 3A and B, antigen presentation usually seen in most cells is compared to that observed within neurons. As shown in Fig. 3A, when a virus infects a cell, the virus uncoats, viral proteins are synthesized and either are incorporated into progeny virus or degraded into small peptides of nine to 11 amino acids in length, probably mediated by the proteasome complex. These peptides are shuttled into the endoplasmic reticulum by TAP1 and TAP2 proteins where they encounter the host cell-encoded class I MHC: the heavy chain and the light chain (β2 microglobulin). The ternary complex consisting of the class I heavy chain, the β-2-microglobulin, and the viral peptide, is transported to the cell surface, where recognition by the T-cell receptor of a specific cytotoxic T cell occurs, leading to the death of the infected cell by the CTL (reviewed in YEWDELL et al. 1993).

In vitro study of OBL-21 neuronal cells indicates that at least two defects exist within neurons which make antigen presentation of viral peptides to the CTL impossible (JOLY et al. 1991; JOLY and OLDSTONE 1992). As shown in Fig. 3B, these cultured neuronal cells fail to transcribe the mRNA encoding the class I MHC heavy chain and fail to transcribe the mRNA encoding the TAP1 and TAP2

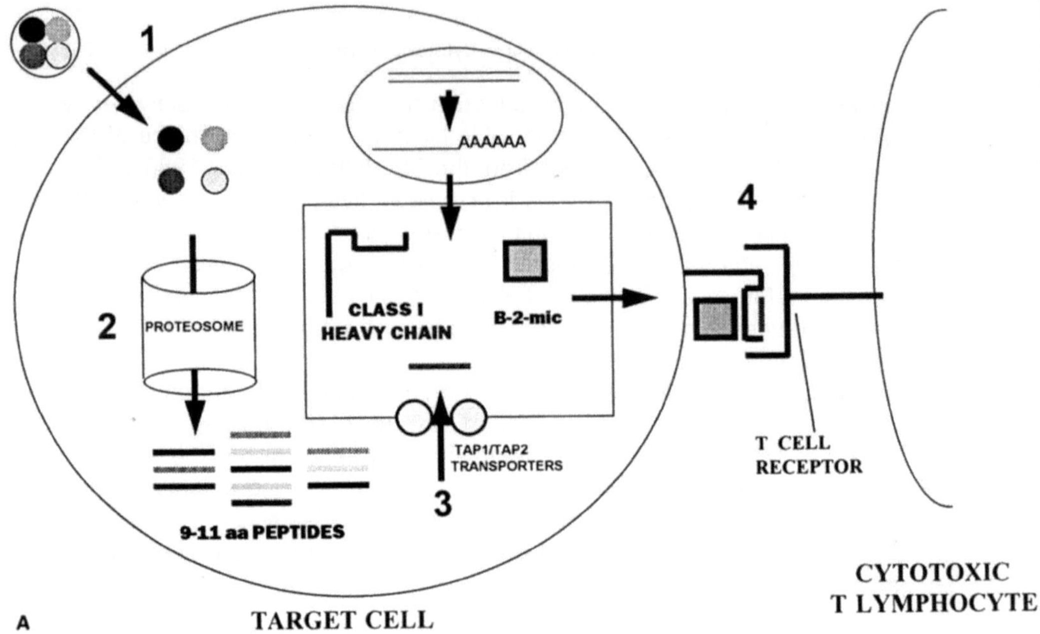

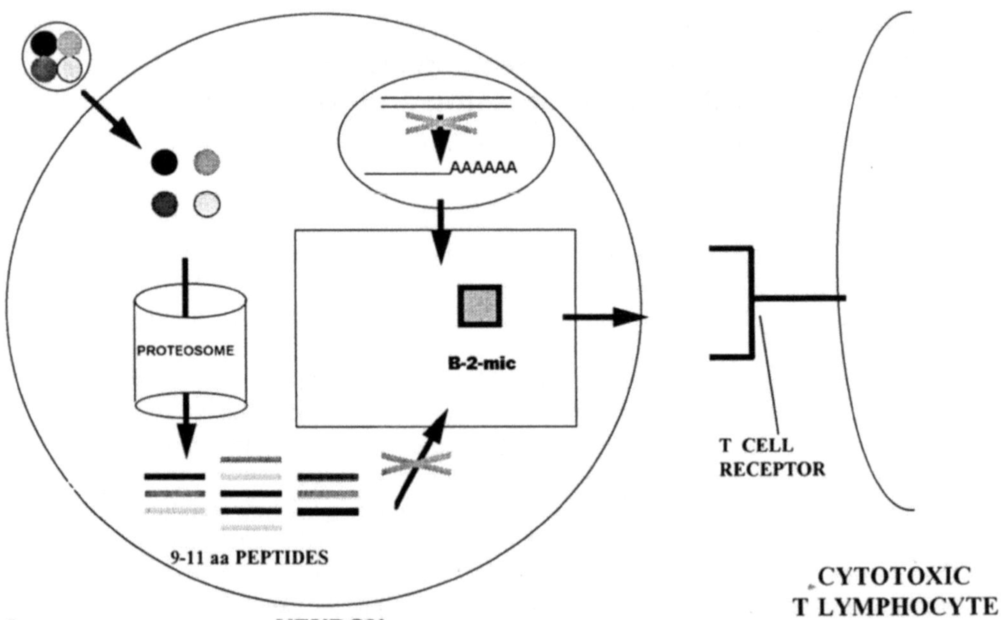

Fig. 3A,B. Antigen presentation in target cells and in neurons. **A** The major steps in antigen presentation in normal target cells and in neurons in vitro are shown. As the virus enters the cell and uncoats (*1*), viral proteins are synthesized and degraded (*2*) into 9–11 residue peptides, which are shuttled, via the TAP transporters (*3*) into the ER, where association with the class I MHC molecule occurs. The ternary complex is then transported to the cell surface, where recognition by the T cell receptor (4) confers the specificity of T cell-mediated lysis of the target. **B** Neurons are unable to present peptide because of a lack of transcription of both the heavy chain and the TAP transporters molecules. The consequence is that these cultured neurons are unable to present viral peptide in the context of class I MHC and therefore do not serve as targets for CTL-mediated lysis (JOLY et al. 1991; Table 2). However, when OBL-21 cells are treated with interferon-γ (IFN-γ), a molecule that up-regulates endogenous MHC, or transfected with a retroviral vector encoding a heterologous MHC molecule (D^b), surface expression of the class I MHC molecule occurs and CTL-mediated lysis is now possible. Thus, reconstitution of class I MHC in vitro restored lysability of neurons by CTLs.

4 Complementation and Consequences of MHC Gene Expression in Neurons In Vivo: Use of the Transgenic Approach

The consequences of class I MHC expression in neurons in vivo were explored (RALL, MUCKE and OLDSTONE, in press). Transgenic mice expressing a class I MHC molecule driven by a neuronal-specific promoter (FORSS-PETTER et al. 1990) were generated (Fig. 4). Our hypothesis was that mouse neurons expressing a functional class I MHC molecule could serve as targets for virus-specific CTL-mediated lysis. Further, such a model would allow one to address questions of trafficking of sensitized CTLs across the blood-brain barrier and the consequences to the host of expressing an MHC molecule in neurons with and without viral infection and in the presence of an antiviral CTL response. Detection of the transgene mRNA was accomplished by reverse transcriptase PCR and noted in brains of transgenic mice but not in other tissues (thymus, spleen, pancreas, lung, liver, testes) or in the peripheral nervous system (sciatic and optic nerves). Protein synthesis of the transgene was detected in primary hippocampal neurons with a monoclonal antibody to the D^b molecule. Using a panning technique (O'MALLEY and MACLEISH 1993), in which neurons were cultured onto cover slips precoated either with an antibody directed against the transgene, (D^b) or against a heterologous class I MHC gene (L^d), only transgenic neurons in contact with the anti-D^b antibody adhered and arborized. Nontransgenic neurons did not stick to either the anti-D^b or the anti-L^d coated slips, and transgenic neurons did not stick to the heterologous (anti-L^d) antibody.

Table 2. Induction of MHC expression in OBL 21 cells. (Modified from JOLY et al. 1991)

Cell	IFN-γ	MHC	Virus	CTL Lysis
OBL (H-2^q)	–	Nil	–	–
OBL	–	Nil	+	–
OBL	+	H-2^q	+	+
OBL-D^b (transfected with a retroviral vector encoding D^b)	–	H-2^b	+	+

IFN, interferon; CTL, cytotoxic T lymphocyte; MHC, major histocompatibility complex

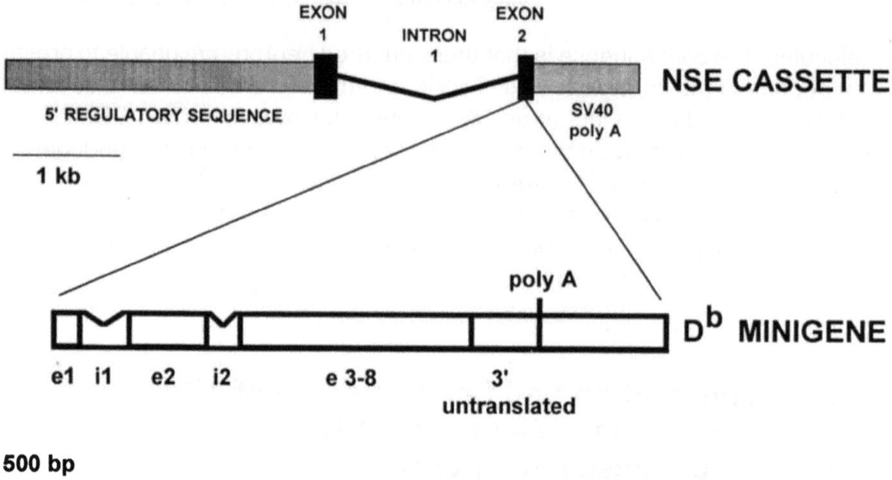

Fig. 4. The neuron-specific enolase-class I major histocompatibility complex molecule (D^b) (NSE-D^b) construct, used to establish transgenic mice which express the class I MHC molecule under the control of regulatory sequences derived from the NSE gene. The D^b minigene (consisting of all eight exons and the first two introns of the D^b gene) was subcloned into exon 2 of the NSE construct and 2.8 kilobases of 5′ NSE regulatory sequence were linked to the first noncoding exon and intron of the NSE sequence. A 960 bp fragment encoding the SV40 polyadenylation signal is at the 3′ end of the construct

Antiviral memory CTLs were adoptively transferred into persistently infected transgenic mice, allowing the following questions to be addressed:

1. Can CTLs that are transferred into transgenic mice cross the blood-brain barrier and enter the parenchyma?
2. Can these CTLs home to infected neurons within the CNS?
3. Can these activated CTLs lyse infected neurons expressing MHC molecules in vivo?
4. What are the consequences of this interaction to the host?

Approaching these questions experimentally, three groups of persistently infected mice were used: (1) transgenic mice expressing the D^b class I MHC gene in CNS neurons; (2) transgenic mice expressing a non-MHC gene, amyloid precursor protein (APP), under the control of the same neuron-specific enolase promoter; and (3) nontransgenic littermates. These groups of $H2^b$ mice were crossed to SJ/L ($H2^s$) mice to create heterozygous (b x s) mice. Newborn litters were infected with LCMV, and persistent infections were confirmed by documenting the presence of infectious virus by plaque assay in sera. Adult mice received an adoptive immunocytotherapy using immune CTLs taken from either $H2^b$ or $H2^s$ inbred mice. Adoptive transfer of either CTL pool should clear virus from the (b x s) animals with equal efficiency. However, if the transgenic neurons expressing the D^b class I MHC molecule are able to present viral peptide in vivo, only the $H2^b$ restricted CTL pool, and not the $H2^s$, would be able to react with the

D^b-expressing neurons. Thus, the expectation is that all animals would display CTL infiltration and immunopathologic-mediated injury in non-CNS tissues, but only the NSE-D^b transgenics given the $H2^b$-restricted anti-LCMV CTLs would develop CNS disease. Indeed, this is what occurred and the results of three such adoptive transfer experiments are summarized in Table 3. As anticipated, all animals clear systemic virus by day 20, but only the class I MHC transgenic mice showed signs of CNS disease. CNS disease was initially measured by clinical symptoms of ataxia, paralysis, ruffling and weight loss. All (100%) of transgenic mice from two different lineages expressing a functional, class I MHC in neurons became ill when given the $H2^b$ restricted anti-viral CTLs with 40% of these mice dying. By contrast, only a small percentage of mice became sick with the anti-$H2^s$ restricted CTLs, with fewer than 10% dying. Likewise, mice expressing the APP transgene or no transgene failed to become ill when given D^b-restricted CTLs. The mild sickness noted in animals of all groups is a consequence of viral clearance associated with immunopathologic injury in liver, spleen, kidney, etc., where virus is found in all persistently infected animals (Fig. 2). In the NSE-APP and non-transgenic groups, health is restored following systemic viral clearance within 20 days postadoptive transfer. In contrast, transgenic mice expressing the neuronal D^b MHC molecule remain sick long after the initial clearance, with illness observed as long as mice were monitored, i.e.,120 days post-CTL transfer.

Immunohistochemically, infiltrating mononuclear cells were detected at several times postadoptive transfer although no signs of CTL-mediated neuronal dropout were detected. Thus, though transgenic mice express a class I MHC molecule in CNS neurons, and become clinically sick as a consequence of CTL interaction with neurons expressing the D^b molecule, signs of CTL-mediated neuronal lysis did not occur.

To understand the CNS disease in these transgenic mice, we evaluated the permeability of the blood-brain barrier. This was accomplished by the use of Evans Blue, which binds to serum proteins and is excluded from the CNS parenchyma normally, but, crosses the barrier and stains the brain parenchyma blue when the blood-brain barrier is disturbed. Transgenic mice expressing the class I MHC molecule in virally infected neurons and given the $H2^b$-restricted anti-LCMV CTLs showed a marked increase in the permeability of the blood-brain barrier. This increased permeability was found only in persistently infected, transgenic mice expressing the D^b molecule and challenged with the $H2^b$ CTLs (Table 4).

Table 3. Mice persistently infected with LCMV

CTL	Assay	$H2^{bxs}D^b$	$H\text{-}2^{bxs}APP$	$H\text{-}2^{bxs}$
$H\text{-}2^b$	Viral clearance	+++	+++	+++
	CNS disease	+++	No	No
$H\text{-}2^s$	Viral clearance	+++	+++	+++
	CNS disease	No	No	No

LCMV, lymphocytic choriomeningitis virus.

Table 4. Increased blood brain barrier permeability in brains of transgenic mice given H-2 matched adoptive immunocytotherapy

Experimental Model			Result
D^b		Persistent infection	Increased BBB permeability
Neuron	CTL		
+	+	+	+
+	–	–	–
+	+	–	–
+	–	+	–
–	+	+	–
–	–	+	–

BBB, blood brain barrier.

5 Conclusions

Activated T cells are able to cross the blood-brain barrier and percolate through the brain parenchyma of normal mice (Fig. 5A). However, since neurons do not normally express class I MHC molecules, these CTLs are unable to recognize virally infected neurons. When infected neurons express a functional class I molecule (Figure 5B), the CTLs may be able to bind to the virally infected neuron, but presumably because of a lack of ancillary molecules on the surface of the neuron (selectins, integrins) or due to immunosuppressive factors within the CNS milieu or the stability and repair of the neuronal plasma membrane, the lysis of neurons by CTLs does not occur. Because neurons are essential for host homeostasis and survival, these and other events likely have evolved over time to prevent neuronal destruction. However, the T cell receptor-MHC interaction likely stimulates the CTL to focally produce cytokines, such as IFN-γ or tumor necrosis factor (TNF-α). Intrathecal production of these and other cytokines may then lead to a variety of changes within the CNS, including permeabilization of the blood-brain barrier. It remains to be seen whether these observations can be extended to other CNS cells and other viral infections.

Although it has been suggested that neurons may be infected with HIV-1, the overwhelming evidence implicates infection of microglia and endothelial cells (reviewed in PRICE 1994). Anti-HIV-1 CTLs are frequently generated in HIV-1 infected patients (SCHRIER et al. 1989; JASSOY et al. 1992) and have been located in the CSF. In the SIV model, similarly, anti-SIV CTLs can be located in blood, CSF, and brain parenchyma (Reumann et al. 1994; von Herrath and Oldstone, unpublished results). The AIDS dementia complex is modulated by anti-viral therapy, such as AZT, suggesting that the pathogenetic mechanism is pharmacologic. Having now demonstrated that CTLs can cause injury in the CNS by permeabilizing the blood-brain barrier, the question of whether a similar phenomena occurs in AIDS dementia and other dementias is raised. Such an increase in permeability could result in several of the symptoms in such patients

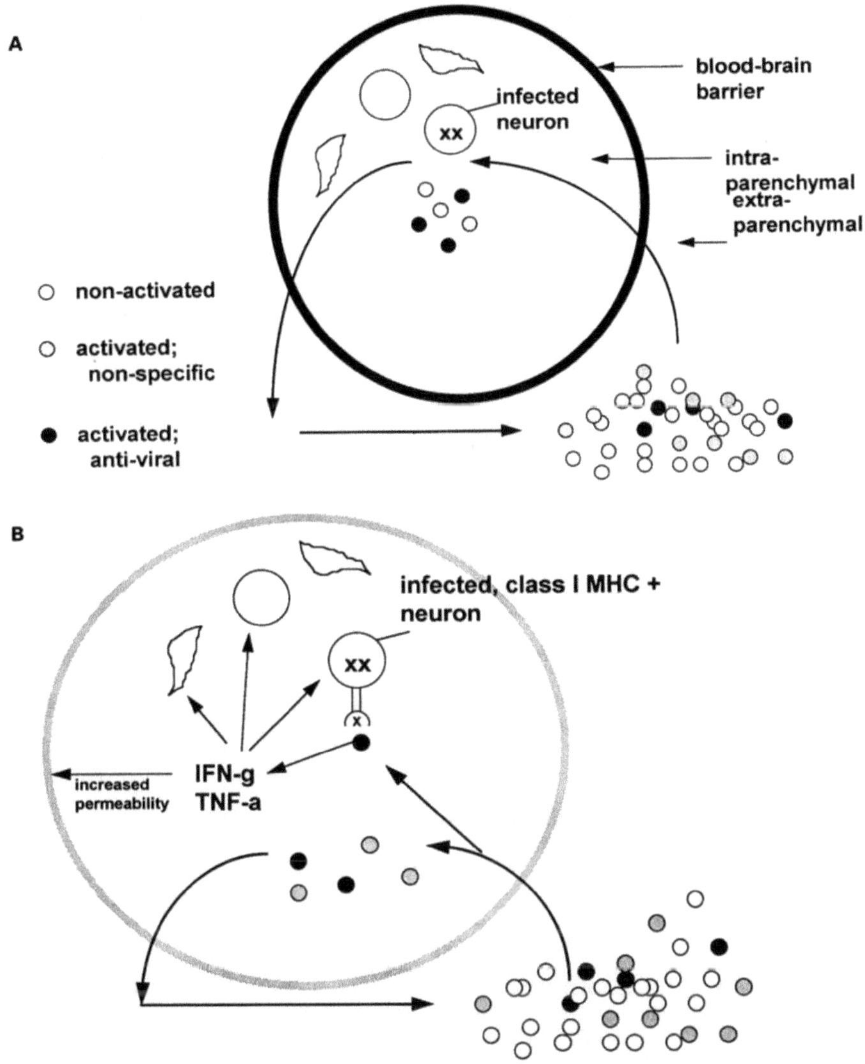

Fig. 5A,B. Model for CNS disease in the absence of overt neuronal damage, mediated by increased permeability of the blood-brain barrier. **A** In the brain of a persistently infected mouse, activated T cells percolate through the CNS parenchyma, but, since neurons cannot present viral peptide in the context of MHC, no cell lysis occurs. **B** When neurons are complemented with the class I MHC molecule, cytotoxic T lymphocyte (CTL) may be able to bind and express some functions (e.g., release of cytokines), but because of the absence of all the cell surface molecules required for CTL-mediated target killing, no lysis occurs. Rather, intrathecal production of cytokines may impact on the overall metabolism of the brain and may have pleiotropic effects, such as the increased permeability of the blood-brain barrier, as seen in the transgenic mice reminiscent of the pathophysiology and accompanying psychoses observed in some patients with lupus erythematosis (PETZ et al. 1973).

Acknowledgments. G.F.R. supported by N.I.H. postdoctoral training grant MH19185 and the Leiper Trust. M.B.A.O. supported by N.I.H. grants NS12428, MH47680, and AI09484.

References

Ahmed R, Jamieson BD, Porter DD (1987) Immune therapy of a persistent and disseminated viral infection. J Virol 61: 3920–3929

Buchmeier MJ, Oldstone MBA (1978) Virus-induced immune-complex disease: identification of specific viral antigens and antibodies depostied in complexes during chronic lymphocytic choriomeningitis virus infection. J Immunol 120: 1297–1304

Buchmeier MJ, Welsh RM, Dutko FJ, Oldstone MBA (1980) The virology and immunobiology of lymphocytic choriomeningitis virus infection. Adv Immunol 30: 275

Byrne JA, Oldstone MBA (1984) Biology of cloned cytotoxic T lymphocytes specific for lymphocytic choriomeningitis virus: clearance of virus in vivo. J Virol 51: 682–691

Canella B, Cross AH, Raine CS (1990) Upregulation and coexpression of adhesion molecules correlate with relapsing autoimmune demyelination in the central nervous system. J Exp Med 172: 1521–1524

Forss-Petter SJ, Danielson PE, Catsicas S, Battenberg E, Price J, Nerenberg M, Sutcliffe JG (1990) Transgenic mice expressing beta-galactosidase in mature neurons under neuron-specific enolase promoter control. Neuron 5: 187–197

Gooding LR (1992) Virus proteins that counteract host immune defenses. Cell 71: 5–7

Greenwood J (1991) Mechanisms of blood-brain barrier breakdown. Neuroradiology 33: 95–100

HIckey WF, Kimura H (1987) Graft versus host disease elicits expression of class I and class II MHC antigens and presence of scattered T lymphocytes in rat central nervous system. Proc Natl Acad Sci USA 84: 2082–2086

Hickey WF, Hsu BL, Kimura H (1991) T lymphocyte entry into the central nervous system. J Neurosci Res 28: 254–260

Jassoy C, Johnson RP, Navia BA, Worth J, Walker BD (1992) Detection of a vigorous HIV-1 specific cytotoxic T lymphocyte response in cerebrospinal fluid from infected persons with AIDS dementia complex. J Immunol 149: 3113–3119

Joly E, Oldstone MBA (1992) Neuronal cells are deficient in loading peptides onto MHC class I molecules. Neuron 8: 1185–1190

Joly E, Mucke L, Oldstone MBA (1991) Viral persistence in neurons explained by lack of major histocompatibility class I expression. Science 253: 1283–1285

Keane RW, Tallent MW, Podack ER (1992) Resistance and susceptibility of neural cells to lysis by cytotoxic lymphocytes and by cytolytic granules. Transplantation 54: 520–526

McDougal JS, Hubbard M, Nicholson JK, Jones BM, Holman RC, Roberts J (1985) Immune complexes in the acquired immunodeficiency syndrome (AIDS): relationship to disease manifestation, risk group and immunologic defect. J Clin Immunol 5: 130–138

Moskophidis D, Cobbald SP, Waldmann H, Lehmann-Grube F (1987) Mechanism of recovery from acute virus infection: treatment of lymphocytic choriomeningitis virus-infected mice with monoclonal antibodies reveals that Lyt2+ T lymphocytes mediate clearance of virus and regulate the antiviral antibody response. J Virol 61: 1867–1874

Mucke L, Oldstone MBA, (1992) The expression of MHC class I antigens in the brain differs markedly in acute and persistent infections with lymphocytic choriomeningitis virus. J Neuroimmunol 36: 193–198

Oldstone MBA (1975) Virus neutralization and virus-induced immune complex disease. Prog Med Virol 19: 84–119

Oldstone, MBA (1989) Viral persistence. Cell 56: 517–520

Oldstone MBA, Dixon FJ (1971) Virus-anti-viral antibody complexes. In: Amos B (ed) Progress in immunology. Academic, New York, pp 763–777

Oldstone MBA, Blount P, Southern PJ, Lampert PW (1986) Cytoimmunotherapy for persistent virus infection reveals a unique clearance pattern from the central nervous system. Nature 321: 239–243
O'Malley MB, MacLeish PR (1993) Induction of class I MHC antigens on adult primate retinal neurons. J Neuroimmunol 43: 45–48
Petz LD, Sharp GC, Cooper NR, Irvin WS (1973) Serum and cerebrospinal fluid complement and serum autoantibodies in systemic lupus erythematosus. Medicine 50: 259–275
Price RW (1994) Understanding the AIDS dementia complex: the challenge of HIV and its effects on the central nervous system. In: Price RW, Perry SW (eds) HIV, AIDS and the brain. Raven, New York, pp 1–46
Reumann KA, Racz KT, Racz P, Montefiori DC, Yasutoni Y, Lin W, Ransil BJ, Letvin NL (1994) Immunopathogenic events in acute infection of rhesus monkeys with simian immunodeficiency virus of macaques. J Virol 68: 2362–2370
Schlitt M, Chronister RB, Whitley RJ (1991) Pathogenesis and pathophysiology of viral infections of the central nervous system. In: Scheld WM (eds) Infections of the central nervous system. Raven, New York, pp 7–18
Schrier, RD, Gnann JW, Landes R, Locksin C, Richman D, McCutcheon A, Kennedy C, Oldstone MBA, Nelson JA (1989) T cell recognition of HIV synthetic peptides in a natural infection. J Immunol 142: 1166–1176
Stagno S, Volanakis JE, Reynolds DW, Stroud R, Alford CA (1977) Immune complexes in congenital and natal cytomegalovirus infections of man. J Clin Invest 60: 838–845
Streilein JW (1993) Immune privilege as the restriction of local tissue barriers and immunosuppressive microenvironments. Curr Opin Immunol 5: 428–432
Tishon A, Salmi A, Ahmed R, Oldstone MBA (1991) Role of viral strains and host genes in determining levels of immune complexes in a model system: implication for HIV infection. AIDS Res Hum Retroviruses 7: 963–969
Weiner LP, Fleming JO (1984) Viral infections of the nervous system. J Neurosurg 61: 207–224
Wekerle H, Linington C, Lassmann H, Meyermann R (1986) Cellular immune reactivity in the central nervous system. TINS June: 271–277
Yewdell JW, Anderson R, Cox JH, Eisenlohr LC, Esquivel F, Lapham C, Restifo NP, Bennick JR (1993) The multiple uses of viruses for studying antigen processing. Semin virol 4: 109–116

Subject Index

Current Topics in Microbiology and Immunology

Volumes published since 1989 (and still available)

Vol. 159: **Gray, David; Sprent, Jonathan (Ed.):** Immunological Memory. 1990. 38 figs. XII, 156 pp. ISBN 3-540-51921-1

Vol. 160: **Oldstone, Michael B. A.; Koprowski, Hilary (Eds.):** Retrovirus Infections of the Nervous System. 1990. 16 figs. XII, 176 pp. ISBN 3-540-51939-4

Vol. 161: **Racaniello, Vincent R. (Ed.):** Picornaviruses. 1990. 12 figs. X, 194 pp. ISBN 3-540-52429-0

Vol. 162: **Roy, Polly; Gorman, Barry M. (Eds.):** Bluetongue Viruses. 1990. 37 figs. X, 200 pp. ISBN 3-540-51922-X

Vol. 163: **Turner, Peter C.; Moyer, Richard W. (Eds.):** Poxviruses. 1990. 23 figs. X, 210 pp. ISBN 3-540-52430-4

Vol. 164: **Bækkeskov, Steinnun; Hansen, Bruno (Eds.):** Human Diabetes. 1990. 9 figs. X, 198 pp. ISBN 3-540-52652-8

Vol. 165: **Bothwell, Mark (Ed.):** Neuronal Growth Factors. 1991. 14 figs. IX, 173 pp. ISBN 3-540-52654-4

Vol. 166: **Potter, Michael; Melchers, Fritz (Eds.):** Mechanisms in B-Cell Neoplasia. 1990. 143 figs. XIX, 380 pp. ISBN 3-540-52886-5

Vol. 167: **Kaufmann, Stefan H. E. (Ed.):** Heat Shock Proteins and Immune Response. 1991. 18 figs. IX, 214 pp. ISBN 3-540-52857-1

Vol. 168: **Mason, William S.; Seeger, Christoph (Eds.):** Hepadnaviruses. Molecular Biology and Pathogenesis. 1991. 21 figs. X, 206 pp. ISBN 3-540-53060-6

Vol. 169: **Kolakofsky, Daniel (Ed.):** Bunyaviridae. 1991. 34 figs. X, 256 pp. ISBN 3-540-53061-4

Vol. 170: **Compans, Richard W. (Ed.):** Protein Traffic in Eukaryotic Cells. Selected Reviews. 1991. 14 figs. X, 186 pp. ISBN 3-540-53631-0

Vol. 171: **Kung, Hsing-Jien; Vogt, Peter K. (Eds.):** Retroviral Insertion and Oncogene Activation. 1991. 18 figs. X, 179 pp. ISBN 3-540-53857-7

Vol. 172: **Chesebro, Bruce W. (Ed.):** Transmissible Spongiform Encephalopathies. 1991. 48 figs. X, 288 pp. ISBN 3-540-53883-6

Vol. 173: **Pfeffer, Klaus; Heeg, Klaus; Wagner, Hermann; Riethmüller, Gert (Eds.):** Function and Specificity of γ/δ TCells. 1991. 41 figs. XII, 296 pp. ISBN 3-540-53781-3

Vol. 174: **Fleischer, Bernhard; Sjögren, Hans Olov (Eds.):** Superantigens. 1991. 13 figs. IX, 137 pp. ISBN 3-540-54205-1

Vol. 175: **Aktories, Klaus (Ed.):** ADP-Ribosylating Toxins. 1992. 23 figs. IX, 148 pp. ISBN 3-540-54598-0

Vol. 176: **Holland, John J. (Ed.):** Genetic Diversity of RNA Viruses. 1992. 34 figs. IX, 226 pp. ISBN 3-540-54652-9

Vol. 177: **Müller-Sieburg, Christa; Torok-Storb, Beverly; Visser, Jan; Storb, Rainer (Eds.):** Hematopoietic Stem Cells. 1992. 18 figs. XIII, 143 pp. ISBN 3-540-54531-X

Vol. 178: **Parker, Charles J. (Ed.):** Membrane Defenses Against Attack by Complement and Perforins. 1992. 26 figs. VIII, 188 pp. ISBN 3-540-54653-7

Vol. 179: **Rouse, Barry T. (Ed.):** Herpes Simplex Virus. 1992. 9 figs. X, 180 pp. ISBN 3-540-55066-6

Vol. 180: **Sansonetti, P. J. (Ed.):** Pathogenesis of Shigellosis. 1992. 15 figs. X, 143 pp. ISBN 3-540-55058-5

Vol. 181: **Russell, Stephen W.; Gordon, Siamon (Eds.):** Macrophage Biology and Activation. 1992. 42 figs. IX, 299 pp. ISBN 3-540-55293-6

Vol. 182: **Potter, Michael; Melchers, Fritz (Eds.):** Mechanisms in B-Cell Neoplasia. 1992. 188 figs. XX, 499 pp. ISBN 3-540-55658-3

Vol. 183: **Dimmock, Nigel J.:** Neutralization of Animal Viruses. 1993. 10 figs. VII, 149 pp. ISBN 3-540-56030-0

Vol. 184: **Dunon, Dominique; Mackay, Charles R.; Imhof, Beat A. (Eds.):** Adhesion in Leukocyte Homing and Differentiation. 1993. 37 figs. IX, 260 pp. ISBN 3-540-56756-9

Vol. 185: **Ramig, Robert F. (Ed.):** Rotaviruses. 1994. 37 figs. X, 380 pp. ISBN 3-540-56761-5

Vol. 186: **zur Hausen, Harald (Ed.):** Human Pathogenic Papillomaviruses. 1994. 37 figs. XIII, 274 pp. ISBN 3-540-57193-0

Vol. 187: **Rupprecht, Charles E.; Dietzschold, Bernhard; Koprowski, Hilary (Eds.):** Lyssaviruses. 1994. 50 figs. IX, 352 pp. ISBN 3-540-57194-9

Vol. 188: **Letvin, Norman L.; Desrosiers, Ronald C. (Eds.):** Simian Immunodeficiency Virus. 1994. 37 figs. X, 240 pp. ISBN 3-540-57274-0

Vol. 189: **Oldstone, Michael B. A. (Ed.):** Cytotoxic T-Lymphocytes in Human Viral and Malaria Infections. 1994. 37 figs. IX, 210 pp. ISBN 3-540-57259-7

Vol. 190: **Koprowski, Hilary; Lipkin, W. Ian (Eds.):** Borna Disease. 1995. 33 figs. IX, 134 pp. ISBN 3-540-57388-7

Vol. 191: **ter Meulen, Volker; Billeter, Martin A. (Eds.):** Measles Virus. 1995. 23 figs. IX, 196 pp. ISBN 3-540-57389-5

Vol. 192: **Dangl, Jeffrey L. (Ed.):** Bacterial Pathogenesis of Plants and Animals. 1994. 41 figs. IX, 343 pp. ISBN 3-540-57391-7

Vol. 193: **Chen, Irvin S. Y.; Koprowski, Hilary; Srinivasan, Alagarsamy; Vogt, Peter K. (Eds.):** Transacting Functions of Human Retroviruses. 1995. 49 figs. IX, 240 pp. ISBN 3-540-57901-X

Vol. 194: **Potter, Michael; Melchers, Fritz (Eds.):** Mechanisms in B-cell Neoplasia. 1995. 152 figs. XXV, 458 pp. ISBN 3-540-58447-1

Vol. 195: **Montecucco, Cesare (Ed.):** Clostridial Neurotoxins. 1995. 28 figs. XI, 278 pp. ISBN 3-540-58452-8

Vol. 196: **Koprowski, Hilary; Maeda, Hiroshi (Eds.):** The Role of Nitric Oxide in Physiology and Pathophysiology. 1995. 21 figs. IX, 90 pp. ISBN 3-540-58214-2

Vol. 197: **Meyer, Peter (Ed.):** Gene Silencing in Higher Plants and Related Phenomena in Other Eukaryotes. 1995. 17 figs. IX, 232 pp. ISBN 3-540-58236-3

Vol. 198: **Griffiths, Gillian M.; Tschopp, Jürg (Eds.):** Pathways for Cytolysis. 1995. 45 figs. IX, 224 pp. ISBN 3-540-58725-X

Vol. 199/I: **Doerfler, Walter; Böhm, Petra (Eds.):** The Molecular Repertoire of Adenoviruses I. 1995. 51 figs. XIII, 280 pp. ISBN 3-540-58828-0

Vol. 199/II: **Doerfler, Walter; Böhm, Petra (Eds.):** The Molecular Repertoire of Adenoviruses II. 1995. 36 figs. XIII, 278 pp. ISBN 3-540-58829-9

Vol. 199/III: **Doerfler, Walter; Böhm, Petra (Eds.):** The Molecular Repertoire of Adenoviruses III. 1995. 50 figs. Approx. XIII, 324 pp. ISBN 3-540-58987-2

Vol. 200: **Kroemer, Guido; Martinez-A., Carlos (Eds.):** Apoptosis in Immunology. 1995. 14 figs. XI, 242 pp. ISBN 3-540-58756-X

Vol. 201: **Kosco-Vilbois, Marie H. (Ed.):** An Antigen Depository of the Immune System: Follicular Dendritic Cells. 1995. 39 figs. IX, 209 pp. ISBN 3-540-59013-7

MIX
Papier aus verantwortungsvollen Quellen
Paper from responsible sources
FSC® C105338

If you have any concerns about our products,
you can contact us on
ProductSafety@springernature.com

In case Publisher is established outside the EU,
the EU authorized representative is:
Springer Nature Customer Service Center GmbH
Europaplatz 3, 69115 Heidelberg, Germany

Printed by Libri Plureos GmbH
in Hamburg, Germany